CONSERVATION BIOLOGY AND BIODIVERSITY

GEOMATICS AND CONSERVATION BIOLOGY

Conservation Biology and Biodiversity

Additional books and e-books in this series can be found
on Nova's website under the Series tab.

CONSERVATION BIOLOGY AND BIODIVERSITY

GEOMATICS AND CONSERVATION BIOLOGY

MICHAEL O'NEAL CAMPBELL
EDITOR

nova science publishers
New York

Additional color graphics may be available in the e-book version of this book.

Library of Congress Cataloging-in-Publication Data

ISBN: 978-1-53614-468-0

Published by Nova Science Publishers, Inc. † New York

CONTENTS

PREFACE

This book covers selected topics from the interlinkages between the two multi-disciplinary fields of geomatics and conservation biology. This field is vast and too large for one volume. However, one common theme is considered: the contribution of geomatics to new developments in conservation biology, and the ways in which these applications have enhanced the understanding of the issues of conservation biology. The field of geomatics in this case is considered in a broad sense, to include the fields of remote sensing, mapping, global positioning systems (GPS) and geographical information systems (GIS). Conservation biology is also considered in its broadest sense, to include all possible ecological relations for animals and plants, including the spatial distribution and movement patterns of organisms that may be documented with geomatics techniques. Case studies are cited to support the theoretical constructions of both conservation biology and geomatics.

Conservation biology is assessed as the leader of the interactions between the two fields, with geomatics seeking to adapt to the changing themes and priorities of conservation biology. As such, an examination of the main parameters of conservation biology that are measurable by geomatics techniques is important. These may include animal movements and behavior, locations and attributes of habitats. As conservation biology concerns both animals and plants, at individual and population scales, various geomatics-based techniques are applicable, including satellite imaging and aerial photography for map-making, georeferencing and tabulation of locations and other data for statistical analysis, and global positioning systems for tracking and recording animal behavior and movements. These are described in all the chapters, using both theoretical and technical descriptions and case studies to assess the status of the various applications of geomatics to conservation biology.

The first chapter, *The Fields of Geomatics and Conservation Biology* by Michael O'Neal Campbell, introduces the fields of geomatics and conservation biology. These are argued to be two new scientific fields that emerged from earlier knowledge groups, such as ecology, applied biology, conservation studies and mapping. Geomatics, defined as largely a North American term, fairly corresponds to the mostly European term of geoinformatics and includes a broad range of techniques, including the fields and subfields of image capture, associated spatial data acquisition, spatial assessment and analysis, locational definition monitoring, the technologies used for these processes and the associated knowledge systems on geographical (natural and social) issues. Linked fields include remote sensing, geographical information science (GIS), geographical information systems (GIS) and global positioning systems (GPS). These are supported by the theoretical and applied disciplines of geodesy, cartography and mapping, and by applications to the natural and social sciences. Conservation biology, similarly broad and multidisciplinary, also emerged from several older, developing disciplines, including the biological sciences (especially ecology and biogeography, and later natural resource management and wildlife biology), the conservation movement and from this environmental economics and sociology, political ecology and even history, anthropology and planning. The spatial issues of species, population and individual organism location, behavior and movement, and habitats, landcover change and the regionalization of ecological change are crucial. Hence, conservation biology shares much with geomatics, which must evolve techniques to identify, measure and analyze those parameters that define the dynamics of life systems. Case studies from Canada and Ghana are included to illustrate these themes. The chapter concludes that more research is required to further develop links between these fields.

The second chapter, offers a case study titled *Mapping and Management of Vultures in an Indian Stronghold* by Professor Kaushalendra Kumar Jha, a world and erudite authority on the ecology of vultures. This chapter looks at the possible applications of geomatics to some of the more important issues of conservation biology, taking an example from the application of a GIS and mapping methodology to the study of endangered vulture species in India. Vulture populations in India have declined markedly, due to the incidence of diclofenac poisoning, habitat degradation and climatic variation. An accurate assessment of the population and ecological status of Indian vultures needs a methodology that includes literature study, field research and the mapping of the ranges and habitat distributions of the vulture species. This chapter investigates the evidence for the decline of vulture populations in a case study of Central India, an important location for vulture populations. The study provides a synthesis of relevant literature sources on vulture population counts, ranging and habitat mapping and predictive models of vulture presence. It investigates the status of endangered vulture species, their conservation requirements, the data collection methodologies and technology required for such assessments and the importance of the results of such a study. Maps of vulture distribution and ecological relations are presented, analyzed and discussed. Vulture conservation policies and recommendations for future action are also explored. The chapter therefore illustrates the application of geomatics-based tools to a relevant, current issue within the themes of conservation biology.

The third chapter, titled *Geospatial Modelling of Human-Elephant Conflict in Dalma Wildlife Sanctuary and its Surroundings in India*, by four top wildlife scientists Somak Das, Jeganathan, C., Saptarshi Mondal and Richa Sharma, applies geomatics techniques to study conflicts between people and elephants in India, taking the case of Dalma wildlife sanctuary, located near Jamshedpur city in Jharkhand State of India, a state well known for its elephant (*Elephas maximus*, Linnaeus, 1758) habitat and large elephant population. The study examines and maps the problems associated with the migration and habitat use of the Dalma elephants, which conflict with increased human landuse, mostly agriculture and settlement expansion. Based on a background of field surveys, published records and image-based analyses, the study uses GIS methods to create models that produce habitat suitability maps for elephant habitats and possibilities of conflict with people. The objective of the study is to identify potential conflict spots which may assist in the avoidance of human-elephant conflict in study area. This makes a contribution to the conservation biology of elephants in India, the tools of geomatics for such studies and the general issues of human-large wildlife conflict.

The fourth chapter, *Remote Sensing and Conservation Biology* by Michael O'Neal Campbell, introduces remote sensing applications to conservation biology. Remote sensing is one of the most important aspects of geomatics, as images and recorded visual data are at least as important as numerical counted data or tabular based statistical analysis for the evaluation of environmental features and dynamics. The chapter examines the origins of remote sensing as a form of visual recording of environmental features, based on artistic skills, and documents its evolution through primitive photography to more advanced space-based 'spy' technology, which further developed through recent software and computerized technologies. Recent technological developments include LIDAR and many competing computer software systems, all of which are relevant to the main topics of conservation biology. This chapter also examines the key problems and prospects of the application of remote sensing techniques to conservation biology, examining the history and current strands of development. The adaptation of remote sensing techniques to changes in the research foci of conservation biology is also examined. The development of remote sensing techniques relevant to conservation biology is mostly on the theme of increased flexibility, to enable the measurement of more complex patterns of distribution, smaller scales and longer range observation possibilities; for example, observing smaller species at greater distances, more accurate recording of animal behaviors such as migratory and foraging movements, breeding, mating and adult-juvenile contrasts, landcover change (including temperature, moisture, vegetation pattern and species change), and the integration and syntheses of such variables for the documentation of ecosystem variation. The chapter concludes that the future of conservation biology may be determined by the flexible application of remote sensing techniques to current and topical issues.

The fifth chapter, *GIS and Conservation Biology* by Michael O'Neal Campbell, looks at the discipline of geographical information systems or geographical information science (GIS). The chapter defines GIS as a system for the processing and production of information, about earth surfaces features, organisms and processes, to objective being to increase knowledge of the variables in the context under study. The chapter starts with an examination of the definitions of GIS and the applications of this science to conservation biology. It presents a case study of the application of GIS methods to the analysis of multidirectional environmental change. Multi-directional change in this sense refers to landcover changes that are too complex to be described with the common, standardized assumption of linear progression (for example, deforestation, from a densely forested cover type to open, mixed grassland). With multidirectional change, there may what can be termed patch dynamic change, as multiple trajectories of change occur simultaneously in different parts of the studied context, such changes including for example deforestation, reforestation, species increases and decreases etc. GIS techniques, including statistical and matrix-based analyses are useful for the documentation of such change, as they offer more accurate measurements of the trajectories of landcover change.

The sixth chapter, *GPS, Animal Tracking and Conservation Biology* by Michael O'Neal Campbell, looks applications of GPS and associated technologies to selected conservation biology topics. Global Positioning Systems (GPS) are the most important tools available for advanced animal tracking. They record the movement of animals, including individual movements and migratory patterns, to assess and evaluate their ecological relations with their life spaces and the spatial dynamics that may influence their presence and movements. This chapter examines the development of Global Positioning Systems (GPS) from more primitive beginnings to more advanced technologies for animal tracking and their applications to the main themes of conservation biology. Case studies of wildlife conservation are presented, that illustrate the parameters that would be the focus of advanced GPS and associated techniques. These are hypothetical cases, which represent the main issues that would be applicable to other studies relevant to conservation biology. The object of GPS and associated technologies would be to record the movement, location and mobile behavior of animals, as individuals or as part of migratory patterns, to determine their ecological relations with their environments and the dynamics that may affect their behavior. This would require the examination of the parameters that would be measured by the applications of geomatics techniques, especially the recent, advanced techniques for tracking and locational analyses of target animals. It is concluded that GPS systems are vital for current studies of conservation biology, especially if supported by related geomatics methods, parameter identification and social studies.

The seventh chapter, *Geomatics for Micro-Spaces of Avian Foraging, Subjective Behavior and Human Presence*, by Michael O'Neal Campbell, looks at the measurements required for the study of avian foraging in micro-spaces, which include limited areas such as gardens, shrubs, and parks. This chapter argues that such small-scale ecology, an important aspect of conservation biology, currently uses mostly human observation or 'eye balling' methods to record bird behavior. Recent developments in geomatics technology are discussed, to identify the applications that would enable the more precise and reliable study of avian ecology. The most relevant developments concern techniques for the more detailed measurements, using sharper resolutions and flexibility in the movements of animals. These include satellite images and aerial photographs with much sharper resolutions, and global positioning systems (GPS) and motion activated cameras which record bird movements. It also argues that the methodologies of the sciences that examine individual or group-based bird behavior may guide the application of geomatics techniques. Animal geography methodologies that examine the subjective and unpredictable behavior of animals are discussed as complementary to such detailed studies. Alert and flight distances, which record avian reactions to people are also examined as measurable parameters. It is concluded that current advancements in geomatics techniques are suitable for the measurement of bird behavior at micro-levels, with a supportive background comprised of older methodologies.

The eighth chapter, *Application of Geographical Technology in Modelling of Migratory Bird Habitat in Mangrove Forest: A Review*, by the leading geomatics and ecology scholars Azimah Abd Rahman and Tarmiji Masron, examines the modelling of the habitat of migratory birds to support the preservation and conservation of migratory birds' habitat. The habitat model proposed helps in the identification of differentials in migratory population densities, thereby mapping distribution pattern, habitat requirements, and potential locations for future bird habitats of migratory birds. The chapter also examines the process of analysis for habitat modelling and contrasts the contribution of GIS applications for mapping out the field data, and that of remote sensing for modelling the parameters of the study that affect the migratory birds' habitat preferences. The result of this complex and combined modelling technique is the construction of habitat models that not only determine bird habitat choices and current population distributions, but also enables the prediction of the trends in future bird habitat choices through visual mapping. This may create simple models that enable easier understanding of the parameters and issues that underlie preservation and conservation programs in the relevant environments, which in this case are the sea mangrove zones, the main habitat area.

The ninth chapter, *Problems of Vegetation Identification Using Historical Aerial Photographs in West Africa*, by Michael O'Neal Campbell, looks at historic vegetation change, using older forms of data such as black and white, coarse-grained aerial photographs, which are usually the only data available for such historic time periods.

Feature identification, and the production of maps illustrating landcover change must therefore emphasize the visual analysis of such photographs, in some cases with stereoscopic equipment. Elements that may be used in such a situation are the visual elements of tone, shape, size, pattern, texture, shadow and association. In studies of problematic landcover, such analytical parameters may be difficult to utilize. An example is used of the analysis of landcover change in a tropical savanna, which is a complex case for several reasons. There is no clear definition of what constitutes savanna vegetation and terms such as 'tree' or 'shrub' may be subjective and may even overlap. Some shrubs may be trees that have lost branches to firewood cutters or been burnt by bushfires. There are also disputes as to what is natural savanna (if it exists), and whether savanna is a transitional landcover between grassland and forest, or a derived landcover from human modified forestland. The coastal savanna of Ghana, the case study area, is a complex, perhaps unique environment which has disputed origins. Some writers claim that it has a forested origin and others attribute its open landcover to a naturally dry climate. The evidence of landcover change, as recorded in black and white aerial photographs and the methodology and results of this study are presented. It is concluded that these older methods offer a good support for the more technically advanced methods used for studies of more recent landcover change.

The tenth chapter, *Unravelling the Complexity of Protected Area Management in Two Developing Countries: Issues of Human Displacement and Wildlife Conflict*, by Kaushalendra Kumar Jha and Pradeep Chaudhry, looks at the complexities of protected area management, taking the example of forest dwelling communities and biodiversity conservation in PAs and the use of modern technologies like geoinformatics for land tenure related and human wildlife conflict issues. The case studies are of India and Thailand, which have similar problems in this regard. The protected areas in these countries are experiencing major problems with human landuse and conflicts with large mammals such as elephants, tigers and leopards. Human landuse is important in these areas, as people depend on the land for economic activities. Geomatics-based technologies may be required for the analysis of landuse problems. Participatory methods may also be needed, as there is evidence that the involvement of the local people creates greater likelihood of success for conservation planning and management. This constitutes a contribution to conservation biology and management.

Michael O'Neal Campbell

In: Geomatics and Conservation Biology
Editor: Michael O'Neal Campbell

ISBN: 978-1-53614-468-0

Chapter 1

THE FIELDS OF GEOMATICS AND CONSERVATION BIOLOGY

Michael O'Neal Campbell
Camosun College, Victoria, Canada

ABSTRACT

Geomatics and conservation biology are two new scientific fields that emerged from earlier knowledge groups. Geomatics (mostly a North American term), sometimes also termed geoinformatics (especially in Europe), is defined variously, usually including the fields of spatial data acquisition, assessment and analysis, and technologies used for these processes and the associated knowledge systems on geographical (natural and social) issues. Linked fields include remote sensing, geographical information science (GIS), geographical information systems (GIS), global positioning systems (GPS), geodesy, cartography and mapping, and applications to the natural and social sciences. Conservation biology, also variously defined, likewise emerged from several root disciplines, such as the biological sciences (especially ecology and biogeography), the conservation movement, environmental economics and sociology, political ecology and even history and anthropology. The spatial issues of landcover change and the regionalization of ecological change are crucial, hence conservation biology shares much with the geographical bases of geomatics. This chapter introduces these links between the two stellar, disciplinary conglomerates and serves as the background to the following chapters. Case studies are provided on geomatics and conservation in Canada and Ghana. It concludes that more research is required on this topic within both fields.

INTRODUCTION

The application of geomatics to conservation biology is a potentially strong, yet recently developing area of study (Decker et al., 2012; Goodfellow, 2014; Morzillo et al.,

2014; Jackson & Fahrig, 2015; Campbell, 2017). Geomatics is one of the most important, emerging tools in the study of conservation biology and there is much room for expansion. These two fields are closely linked, largely because of the large scale of the problems involved, see Figure 1. Implicit in these two fields are certain shared concepts and components. Firstly, there is the concept of space (Campbell, 2017). Space concerns the area within which the entities of interest are located. For both conservation biology and geomatics, this space covers the whole planet earth, including the icecaps, oceans and land surface, as all are either directly or indirectly linked to living beings, human action on the Earth and/or environmental management, and may require measurement and analysis for deeper understanding or preservation.

Secondly, there is pattern (Jensen, 2007; Campbell, 2017). This describes how the components or issues arrange themselves in space. Pattern, like space, is a key component of the geographical sciences. For geomatics, patterns are recognizable entities that may be identified, recorded, analyzed and predicted, the better to understand the dynamics of the identified environment. Geographical patterns are also at the root of conservation biology, as the main concerns are with the areal differentiation of human impacts on the ecosystems, and the spatially variable solutions required. Ecosystems vary across space and so do peoples and their cultures, and human-environment relations, hence different policies are applicable to different problems in different regions.

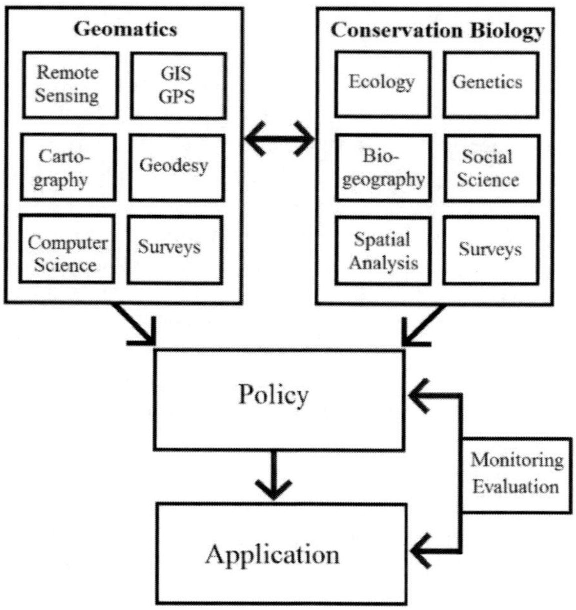

Figure 1. A modelled relationship between geomatics and conservation biology.

Thirdly, implicit in the concepts of space and pattern is the concept of place (Heywood et al., 2006; Campbell, 2017). Place refers to a localized point, where certain possibly unique or uniquely variable entities combine to create a recognizably different component of a larger pattern. Single patches of forest, or unique groups of such patches may create places, and may be measured and analyzed by the tools of geomatics such as aerial photographs, satellite images, or point count surveys and statistical tables. In harboring endangered species, they may also comprise important components for conservation and ecology (Campbell, 2015, 2016).

A fourth concept is that of integrated analysis. Analytical tools from several disciplines or theoretical/methodological complexes must be integrated to assess such complicated entities. Image-based methods may capture visible and non-visible (to the naked human eye) light bands, and record and analyze these to derive descriptive, analytical and predictive models. Additionally, point counts of animals, measurable parameters of vegetation, soils and climate may be taken, and analyzed with tables within GIS software, creating maps that may be integrated with the derived images. There may also be the georeferencing of points using GPS, the points representing important issues such as locations of habitats such as nests, dens or even carcass locations, human actions or trees. These may be located with the images created by the methods above.

The last concept, emerging from the rest, is therefore that of analytical mapping. Analytical maps created through the methods of geomatics form the backbone of conservation biology, acting as reference points, authoritative policy supports and derivable databases. These maps, formed through plural methodologies and hybrid research, may also form the basis of predictive modelling in conservation biology. The idea that authoritative spatial databases and information conglomerates are crucial supports for environmental policy has been current since the end of the 20th century. This is especially relevant in cases of multi-directional change (environmental, especially vegetation type change that is not unilinear but may include increases and reductions in species or habitat forms) (Elliot & Campbell, 2002). The modelled relations between geomatics and conservation biology are illustrated in Figure 1.

This integrated relationship acknowledges the recent paradigmatic developments in ecology that emphasize non-equilibrium dynamics over the older ‘balance of nature’ perspectives. The ‘new ecology’ recast older concepts such as mechanistic succession, vegetation climaxes and human/nature separation, and developed a new paradigm of non-linear and multi-directional ecological change (Pickett et al., 1992), including “multiple interpretations” (Scoones, 1997) and the natural centrality of chaotic change within ecosystems (Campbell 1998). As argued by Scoones, 1997:162) there is a need for further studies that emphasize “non-linear, multi-directional, punctuated... dynamic and non-equilibrium possibilities for processes of trans-formation.” Such developments are currently possible, with the increased sophistication of geomatics technology and the associated increased complexity of the requirements of conservation biology (Campbell, 2017a).

Conservation biology, and its main predecessor protected area and wildlife management and assessment was traditionally based on field surveys and measurement and in some cases, social impact assessments (Campbell, 1998, Elliott & Campbell, 2002). Current studies of landscape change frequently use remotely sensed time series images (Campbell, 2005). These, however are arguably at inappropriate scales and may be too irregular in time coverage for application to the parameters of intensive micromanagement of local contexts required by many conservation biology developments. Solutions are required for poor conservation management, and the links with other landcover activities such as forestry, urbanization and tourism (Solecki, 1994; Page et al., 1996; Machlis & Force, 1997; Aikman et al., 1998; Diduck, 1999; Campbell 1998, 2002a, b). Regular, integrated, multidisciplinary research focused on the field ecology and the cultural and legal issues relevant to the area are necessary for effective protected area management (Wells et al., 1992; Leong, et al., 2012; Campbell, 2017a).

DEFINITIONS AND PERSPECTIVES ON GEOMATICS

One illustration of the links between geomatics and conservation biology may be derived from the definitions of the two fields and the related theoretical links. Geomatica (2002) defines geomatics as a "field of activities which, using a systematic approach, integrates all the means used to acquire and manage spatial data required as part of scientific, administrative, legal and technical operations involved in the process of the production and management of spatial information." Natural Resources Canada (2018) defines geomatics as "the modern discipline which integrates the tasks of gathering, storing, processing, modeling, analyzing, and delivering spatially referenced or location information… surveying, hydrography, mapping, remote sensing (often called earth observation) and geographic information processing, often called geographic information systems (or GIS)." This breadth of focus and multiplicity of tools gives geomatics, including geographical information systems, remote sensing, cartography and related statistical analyses important roles supporting the interdisciplinary fields of conservation biology and protected area management (Dempsey, 2011; Wedding et al., 2011; Goodfellow, 2014).

The Journal of Remote Sensing and GIS (2018) defines geomatics, which is also known as "geospatial technology or geomatics engineering, or geomatic engineering" as …the discipline of gathering, storing, processing, and delivering geographic information, or spatially referenced information. It consists of products, services and tools involved in the collection, integration and management of geographic data. The rapid progress and increased visibility of geomatics since the 1990s has been made possible by advances in computer hardware, computer science, and software engineering, as well as by airborne and space observation remote-sensing technologies.

The British Columbia Institute of Technology (2018) gives a similar definition, defining geomatics as: "the art and science of measuring, storing, managing and presenting of geospatial data. It has a rich history and an exciting future, where rapidly changing technology is applied directly to the sustainable management of the world's resources. Geomatics consists of a wide range of topics, including GPS technology, field surveying, cartography, geodesy, land use planning, photogrammetry, hydrography and cadastral studies."

These two definitions point to some important issues. Firstly, geomatics is both an art and science. This means that it may be based on mathematical, scientific principles, and the more human issues such as subjective classification, visual presentation and strategic application. Secondly, geomatics is comprised of many disciplines or sub-disciplines. Fields of knowledge, such as computer hardware, computer science, software engineering, space observation, GPS (global positioning systems) technology, field surveying, cartography, geodesy, land use planning, photogrammetry, hydrography and cadastral studies are all vast fields of knowledge in their own right. Thirdly, the field of geomatics is advancing in terms of technology, both in terms of its own technology and the applications in the wider fields of the natural and social sciences.

An important consideration in the definition of geomatics is the relationship of this field to the disciplines or fields of geoinformatics, remote sensing, geographical information systems, geographical information science, geodesy, spatial statistics, geography and the wider fields of the geosciences. Geoinformatics has been defined as virtually the same as geomatics. Ehlers (2008, 17) defines geoinformatics as "as a new scientific discipline designed for handling of geospatial information" ... not a part of Geography, Surveying, or Computer Science, but a new self-contained scientific discipline…and including "the acquisition and storing of geospatial data, the modelling and presentation of spatial information, geoscientific analyses and spatial planning, and the development of algorithms and geospatial database systems."

Ehlers (2008, 20) notes that Geoinformatics is synonymous with geomatics (which is named from the French word Géomatique, coined by the French photogrammetrist Dubuisson) (Dubuisson, 1975, Gagnon & Coleman, 1990, Campbell, 2002a,b). The distinction is largely stylistic, as geoinformatics the term usually used in Germany and the rest of Europe, while geomatics is generally used in North America. Nevertheless, Ehlers (2008, 20) argues that "Geoinformatics… emphasizes a formal scientific approach to handle geoinformation which Geomatics does not imply."

Geographic (or Geospatial) Information Sciences may also be synonymous with geomatics or geoinformatics, but usually indicates a connection with geography (Goodchild 1992; Ehlers 1993). Gomarasca (2010, 137) notes that the term geomatics was crafted at Laval University in Canada during the early 1980s, in acknowledgment of the increasing role and potential of electronic computing and software in the geographical sciences, and the importance of geographical location in such studies. According to

Gomarasca (2010, 137) geomatics was "erected on the scientific framework of geodesy" …using "terrestrial, marine, airborne, and satellite-based sensors to acquire spatial and other data." Furthermore, as an activity it is defined as "a systemic, multidisciplinary, integrated approach to selecting the instruments and the appropriate techniques for collecting, storing, integrating, modeling, analyzing, retrieving at will, transforming, displaying, and distributing spatially georeferenced data from different sources with well-defined accuracy characteristics and continuity in a digital format" (Gomarasca, 2010, 137).

From this initial genesis, geomatics has spread along several lines (Campbell, 2017). There are regulations based on GeoSpatial Information or (also termed GeoInformation) and Earth Observation data. There are also internationally developed Spatial Data Infrastructures (SDIs), with the ultimate objective being a Global Spatial Data Infrastructure (GSDI), which would be totally inter-operable for all data forms and sets, based on the harmonizing of the different formats of SDIs at the local, national, and continental levels. Examples are the Global Earth Observation System of Systems (GEOSS) (February 16, 2005), and the Infrastructure for Spatial Information in the European Community (May 15, 2007) (Gomarasca, 2010).

Samuelsen (2008) lists 10 disciplines as components within geomatics: surveying, cartography, geodesy, hydrography and ocean mapping, photogrammetry, remote sensing, global positioning systems (GPS), geographic information systems (GIS) automated mapping/facilities management (AM/FM) and computer and communications science. Gomarasca (2010), supported by other writers (Heywood et al., 2006; Jensen, 2007), concurs and lists 17 disciplines within geomatics, each of which may be regarded as important subjects. Computer science is vital, as the current information processing needs would be impossible without the massive calculating capabilities of computers and associated software. Most current geomatics models would be impossible without the software packages. Geodesy is also important, as its basic research components (such as the geoid or the surface of reference of the complete Earth, and the simplified ellipsoid, and the external gravitational field) are a framework for spatial studies (Gomarasca, 2010). The related field of topography is included, as is planimetry, which determines the relative positions of points on the Earth's surface within the same reference surface. Altimetry measures the height of the points on the Earth's surface in relation to the geoid surface and tachymetry determines the distances, bearings, and elevations of distant objects in surveying. Land surveying measures the three-dimensional positions of points and the angles and distances between them on the Earth's surface (Jensen, 2007).

Other disciplines within geomatics include cartography, which is the science and practice of map-making, and photogrammetry which uses photographs to take measurements mostly to determine the precise positions of surface points. Remote sensing acquires environmental data without making physical contact with the object, through methods such as photography; hence it is the basis of some of the other disciplines such as photogrammetry and some aspects of land surveying. There is also global positioning systems (GPS) which provides the three-dimensional position of fixed or moving objects, in space and time, over the entire surface of the Earth. Laser-scanning systems (LS) mostly use LIDAR (named from the combination of light and radar, and light detection and ranging) to map topography, vegetation, urban areas and other landcover features. It measures the distance to a target by illuminating a target with rapid pulses of laser light and measuring the reflected pulses with a sensor (Jensen, 2007; Gomarasca, 2010; Campbell, 2017a).

There is also geographical information systems (GIS), which records, stores, transforms, represents and processes spatial data, for value added information analyses. When using internet and web-based methods, it is termed WebGIS. GIS may support the development of decision support systems (DSS), which create models supportive of organizational decision-making. They may also support expert systems, which are computer systems that approach the duplication of a human expert's abilities. Finally, as in any knowledge system, there is ontology, which in geomatics and related information sciences is defined as the representation and definition of the categories of concepts and data behind the working of the system (Gomarasca, 2010).

All these disciplines are relevant to conservation biology, in that they identify objects (which could be plants, animals, or associations of these in landcover, or landscapes, or environmental parameters such as water, moisture, temperature or light) and represent such entities in point, line, polygon, georeferenced and/or image format, for the processing of such data for information support (Jensen, 2007). As conservation biology covers virtually the whole biosphere, and associated components of the hydrosphere, lithosphere and atmosphere, the potential applications are exponentially limitless, especially considering the relations among these environmental variables (Campbell, 2017).

DEFINITIONS AND PERSPECTIVES ON CONSERVATION BIOLOGY

Conservation biology has two main goals: the study of the anthropogenic impacts on biodiversity, and the design of methods for species protection (Campbell 2017a). Conservation biology is an extremely broad subject, mostly because of its complex, multidisciplinary and interdisciplinary subject matter and because of the polarized, political debates that underpin much of conservation science (Soulé & Wilcox, 1980; Soulé 1986a,b). Conservation biology is not merely a branch of biology as its name suggests; it links with disciplines across the natural and social sciences. Conservation biology plays an important role in the theoretical and scientific background for conservation, and "is rooted in older scientific, professional, and philosophical traditions" (Meine et al., 2006, 631). It may be defined as the integrated discipline that applies the methodologies of the biological, environmental and social science to the solution of conservation issues, especially biodiversity (Campbell, 2017a).

As such, conservation biology is a new focus, and from such broad roots it is much broader than biology; every aspect of biology may have a related conservation aspect, as all life may require some degree of protection or at least acknowledgment of such a possibility (Campbell, 2015, 2017). The inclusion of the term 'biology' is however important, as it reminds the reader or protagonist of the focus on living organisms, which remains regardless of other disciplinary involvement and application (Campbell, 2017a). Meine (2010, 7) argues that the "main thread" of conservation biology is "the description, explanation, appreciation, protection, and perpetuation of biological diversity." This issue is also examined in other, older disciplines, especially in those linked to ecology.

Nevertheless, a detailed history of conservation may explain the unique approach of conservation biology, especially the science, practice, policy and philosophy positions that combined with the biological approach (Meine, 2004; Campbell, 2015). The main shift concerns the transition from the dominance of genetics, species demographics, population and habitat viability and landscape fragmentation, to include issues such as species interactions, global warming and geomatics techniques, enabling new, wider applications (see also Odenbaugh, 2003; Meine et al., 2006; Proulx et al., 2014). Much of the recent and current debate on the form, topics, applications and future of conservation biology concerns its recent origin (Sodhi & Ehrlich, 2010; Campbell, 2017a). Meine et al., (2006, 631), argue that conservation biology "has gained its contemporary definition only in the last three decades" and that it has not yet matured enough to have a clear definition. Three agreed components underpin the philosophy of conservation biology; the important role of biology, the methodology of conservation applications and the justification of conservation. Debates cluster around these components.

Concerning the prevailing character of the discipline, some point to its political strengths. Soulé and Wilcox (1980) define conservation biology as "a mission driven discipline comprising both pure and applied science." This focus allows biologists to participate in the solution of problems that may have wider impacts. This approach largely emerged from the initial foundation of conservation biology. Part of the focus is due to the development of the discipline, which started largely within academic structures. Work (2015, 368) notes that "the field of conservation biology has grown substantially from its inception in 1985, through a proliferation of journals, research articles, societies, and institutions of higher education that offer conservation biology courses or programs."

Aspects of conservation biology developed within geography, biology and legal structures during the 19th century, and in disciplines such as sociology, political ecology and planning during the early 20th century (Campbell, 2015). The core philosophy of conservation biology, which was a unique mix of biological science and social disciplinary methods, specifically focused on conservation topics, developed more slowly

(Campbell, 2015). One early contribution was the work "Man and Nature" by George Perkins Marsh, the second chapter of which was "Transfer, Modification, and Extirpation of Vegetable and of Animal Species," examining people as a new geographic force. Meine et al., (2006, 633) point to this as one of the first explorations of nature-society relations, with socio-economic development examined as an agent for environmental change. One factor for this perspective was the rapid industrial development of the post-Civil War United States, which was followed by the Progressive Era, in which conservation issues emerged (Hays, 1959; Kolko, 1963; Leonard, 2009).

Even after the emergence of conservation issues, a division in conservation policy emerged between the earlier, more anthropocentric, socially oriented approaches and the later, more biological applications. The former was dominated by the utilitarian Resource Conservation Ethic, based on scientific natural resource management, and the Romantic-Transcendental Preservation Ethic, focused more on the spiritual and aesthetic benefits of nature-society relations (Hays, 1959; Callicott, 1990). In the latter half of the 20th century a more nature-focused philosophy emerged within the rudiments of conservation biology (Campbell, 2015).

As argued by Whittaker & Richard (2011, 5) "conservation biology gained huge momentum during the 1970s and early 1980s, when it was formally identified as a sub – discipline" with relevant journals and textbooks. Other recent developments were new texts, such as Biological Conservation by Ehrenfeld (1970), MacArthur and Wilson (1967) and the journals of Conservation Biology and Biological Conservation (Meine et al., 2006; Campbell, 2015). Supportive conferences included the First International Conference on Conservation Biology in September 1978, the papers there published as Conservation Biology: An Evolutionary- Ecological Perspective (Soulé & Wilcox 1980); and a Second International Conference on Conservation Biology at the University of Michigan in May 1985 (Soulé, 1986a). The Society for Conservation Biology (SCB) was approved at the end of this meeting (Soulé, 1986a). The National Forum on BioDiversity, September 21–24, 1986 in Washington, DC., also took place, and the papers were published as Biodiversity (Wilson & Peter, 1988). In June 1987, the first annual meeting of the Society for Conservation Biology was held in Bozeman, Montana (Meine, 2010).

These supportive structures influenced the academic development of conservation biology and associated fields. These developments were associated with the breaking down of academic disciplinary boundaries and the rise of transdisciplinary, interdisciplinary, cross-disciplinary and multidisciplinary fields of study. Within geography, these included island and historical biogeography, which networked within the social and natural sciences. Within biology, these included population biology and genetics, which linked with the medical and chemical sciences. Applied and interdisciplinary subjects also emerged, such as landscape ecology, forestry, wildlife, range and fisheries management, environmental ethics, economics and history, and technological disciplines such as geomatics, remote sensing and geographical information systems (Crisci et al., 2003; Meine, 2004; Posadasa et al., 2006). Pure and applied research slowly integrated (Meffe & Carroll 1994; Meffe & Viederman 1995), supporting practitioners in wildlife, fish and forestry management (Temple et al., 1988), planning processes (Primack, 1993; Craighead 1998; Orr, 2004) and biological conservation (Soulé 1985, 1986; Burley, 2002).

Despite the emergence of conservation issues across several disciplines, Meine et al., (2006, p. 639) noted that ...conservation biology reflected essential qualities that set it apart from predecessor and affiliated fields." These qualities include its broad scientific base in the biological sciences and the wide range of studies, with a "primary focus on the conservation of genetic, species, and ecosystem diversity (rather than those ecosystem components with obvious or direct economic value)" (ibid.). Conservation biology initiated a paradigm shift to a more inclusive science, which networked more with greater inclusivity than before (Roebuck & Phifer, 1999; Campbell, 2015).

At the practitioner level, these developments initiated or increased the engagement between academics (biological and physical scientists, economists, political scientists, sociologists) and practitioners (policy makers, organizers and social workers, managers, businesspeople and members of non-governmental organizations) (Barry & Oelschlaeger, 1996). These links created common causes and policies that challenged existing practices, and sought to make a difference in a networked, socio-environmentally changing world (Soulé & Orians, 2001). As argued by Ehrenfeld (1992, 1625) "conservation biology is not defined by a discipline but by its goal — to halt or repair the undeniable, massive damage that is being done to ecosystems, species, and the relationships of humans to the environment...many specialists in a host of fields find it difficult, even hypocritical, to continue business as usual, blinders firmly in place, in a world that is falling apart."

Efforts within conservation biology towards environmental management nevertheless must encounter and ameliorate arenas "of multiple, conflicting societal agendas" (Gibeau, 2000, 106; Brunner & Clark, 1997; Orr, 2004). Brunner and Clark (1997, 49) note that in such scenarios of cooperation and conflict the role of environmental management "is not to evaluate action alternatives as if the decision problems were merely technical, but to guide a collective and continuing evolution process that can reduce ambiguities and uncertainties and improve alternatives through the comparative evaluation of practical experience." For solutions, further advancements in the methodologies of conservation biology may be required, especially based on knowledges derived from the monitoring and evaluation of policies and projects, and new methods of data acquisition and information analysis (Brunner & Clark, 1997; Gibeau, 2000; Mazzocchi, 2006). Some social and scientific developments have supported these possibilities (Gerlach & Bengston, 1994; Jensen & Bourgeron, 1994; Stanley, 1995; Bengston et al., 1999; Mazzocchi, 2006; Meine et al., 2006; Campbell, 2015, 2017).

LINKS BETWEEN GEOMATICS AND CONSERVATION BIOLOGY

The spatial nature of many conservation issues provides the justification for most of the geomatics-based applications to conservation biology (Campbell, 2017a). This interdisciplinary, research, teaching and policy interface is dominated by topics covered by the science of biogeography, which is a broad discipline with an interdisciplinary, spatial emphasis. Conservation biogeography is an important, emerging sub-field; basically, it is conservation biology with a spatial focus (Campbell, 1998). This follows the trend of the increased integrative and spatial character of the environmental sciences through recent history. Whittaker et al. (2005, 4) cite a "chronology of emergence" for the related sciences, starting with biogeography (1800 AD), then wildlife conservation (1900 AD), nature conservation (1950 AD) and conservation biology (2000 AD). The concept of biodiversity is complementary, emerging during the 1980s.

This gives an "approximate chronology of the emergence of conservation biogeography, positioning the subject as a subset of conservation biology focused on pattern and process at coarser scales of analysis" (Whittaker et al., 2005, 4). Whittaker et al., (2005, 3) describe the core of conservation biogeography as the "application of biogeographical principles, theories, and analyses, being those concerned with the distributional dynamics of taxa individually and collectively, to problems concerning the conservation of biodiversity." They further add that "as an applied and interdisciplinary science concerned with the conservation of nature, conservation biogeography can be seen as a product both of biogeography and of conservation biology" (Whittaker & Richard, 2011, 4).

Implicit in the justification on biogeographical studies is the recognition of the importance of scale in all ecological studies. As argued by Jackson and Fahrig (2015, 52). The spatial extent (scale) at which landscape attributes are measured has a strong impact on inferred species–landscape relationships. Consequently, researchers commonly measure landscape variables at multiple scales to select one scale (the 'scale of effect') that yields the strongest species–landscape relationship. Biogeography considers all scales of analysis and is deeply rooted in many sciences.

Whittaker & Richard (2011) point to the origins of many core biogeographical components from the 19th century under the twin fields of zoogeography and phyto-geography. Scholars such as Alfred Russel Wallace (possibly the 'father' of zoogeography), Charles Darwin and Alexander von Humboldt played important roles in the development of biogeographical methodologies and philosophies (Lomolino et al., 2004, 2006). The resultant discipline has a wide focus, including long term temporal change (historical biogeography), more recent changes (palaeoecology) and contemporary patterns and processed (ecological biogeography). The key issue is the description, explanation and prediction of distribution and diversity, from the level of orders to sub-species.

Conservation biogeography largely developed after a conference of the International Biogeography Society in Washington in January 2005 (Lomolino & Heaney, 2004), and through the publications of top scholars such as Whittaker (Whittaker et al. 2005; Whittaker & Richard, 2011). These authors argue that the concepts of biogeography were implicit in conservation biology from the beginning, even if the formal recognition of the term conservation biogeography was later in coming (e.g., Primack, 2002, Campbell, 2017a). The role of biogeography in conservation biology may have been underestimated but may now attain full recognition due to recent technological advances within the geographical sciences, especially geomatics. Whittaker & Richard (2011, 7) point to "the recent and huge technological advances in biogeographical data collection, storage and analysis, which have enabled rapid progress in many areas of the field, both pure and applied"; additionally, there have been related theoretical and conceptual advances in biogeography (Williams et al., 2000; Lomolino & Heaney, 2004). Therefore, "there is a need for more biogeographers to engage with the problems of conservation science, and for the injection of more biogeography into training for conservation scientists and practitioners" (Whittaker & Richard, 2011, 7).

Geographical scale is one of the most important links between biogeography and conservation biology (Whittaker & Richard, 2011). The role of biogeography is particularly important at the coarser scales, such as the mapping and modelling of biogeographical patterns, the distribution and explanation of geographical patterns in diversity, which is " unambiguously within the heartland of biogeography....it is here where there is greatest need for critical attention to our science and for greater interaction between those involved in theory and application" linking to other aspects of conservation biology such as community, population and behavioral ecology, macroecology, and genetics (ibid. 7, see also Lourie & Vincent, 2004). The work of biogeography illustrates the importance of coarser scales of classifications, such as zoogeographic regions, areas of endemism, patterns in species richness, or phylogeographic structures related to conservation foci, and finer scales such as habitat corridors or metapopulation dynamics.

Data analysis, using geomatics and/or statistical software may also be used "on 'ecological' traits (e.g., growth rates, propagule size, breeding system, body size) and 'biogeographical' traits (e.g., range size, region of origin) (Whittaker & Ladle, 2011, 7).

Another important link between geomatics and conservation biology is within the fields of integrated research (IR) and adaptive management (AM). Both focus on the spatial issues of conservation, in which geomatics analysis is particularly suited, and examine many of the spatial issues of conservation biology (Campbell, 2016, 2017a). Integrated research methods (IR) (the integration of field, social, geomatics methods) is important for the monitoring and management of conservation areas. The application of IR methods, including geomatics to intensive protected area management is a developing area of study (Decker et al., 2012; Goodfellow, 2014; Morzillo et al., 2014; Jackson &

Fahrig 2015). Protected area management is traditionally based on field surveys, measurements and sometimes, social impact assessments (Campbell, 1998, Elliott & Campbell, 2002). Some studies of landscape change use remotely sensed, time series images which are at scales too coarse for detailed work on conservation. Detailed images would be required for problems such as poor conservation management, and forestry, agriculture, urbanization and tourism impacts (Campbell 1998, 2002a; Leong et al., 2012).

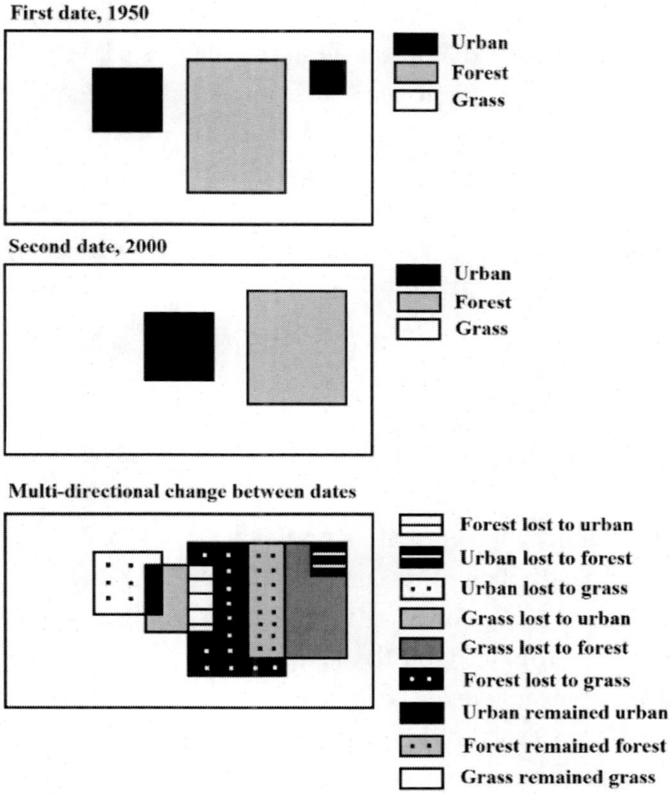

Figure 2. Theoretical example of the mapping of multidirectional change. Each landcover type gained and lost cover against the others, making an analysis based on linear change inaccurate.

Adaptive management is a system which developed during the 1970s (Holling, 1978; Hales, 1991; Johnson, 1999), defined as "an innovation that implements policies as experiments and mobilizes available information for objectives that are less susceptible to unexpected outcomes" (Agrawal, 2000, 326), and "probes the responses of ecosystems as people's behavior changes (Lee, 1999, 3). Several researchers have studied the prospects for the application of adaptive management within the larger system of conservation biology and ecological management (see Callicott et al., 1999; Morzillo et al., 2014). Gunderson (1999) and Agrawal (2000) contrasted conventional management and adaptive management, defining the former as focused on prediction and short-term equilibrium, with linear trajectories of ecological change; and the latter as integrating knowledge from several sources (social, technical, natural), with an experimental approach for future improvement. For both IR and AM, more information sources are needed, including geomatics techniques and products. For example, in the relationship between wildlife management and landscape patch change, Brennan et al. (2001, 68) argue that "with ever-increasing loss and degradation of wildlife habitat, wildlife management decisions depend on a solid understanding of the influence of both patch characteristics and landscape structure on populations." Field research to support such work requires appropriate scales, examination of both landscape and patch scale factors, multiple landscapes examined at multiple scales and balances between sampling intensity and suitable sample sizes. Required attributes include short term survey regularity, comprehensiveness, social awareness, and avoidance of overgeneralization (Campbell, 2002, 2015, 2016). For such applications, geomatics offers increased quantification and evidence of time series images, deriving more reliable documentation of trajectories of environmental change and impacts. This also enables the assessment of multidirectional change (Campbell 1998, 2005, 2013). Figure 2 shows a theoretical example of multi-directional change.

CASE STUDY OF CANADA AND GHANA

This case study examines the current and possible role of geomatics in conservation biology applications (mostly natural resources policy and management) in Canada and Ghana. Both countries have large areas of undeveloped land, substantial wildlife/human contact, and strong human impacts on the land. In both cases, the geomatics tools are sufficiently robust for comparative analysis of such widely spaced and variable contexts. There is often a gap between developed and developing countries in access to such technology. For example, in Canada, geomatics is making an increasing contribution to protected area management, forestry, wildlife studies and nature/society relations. In Ghana, geomatics applications have proceeded more slowly with applications to land management. Can lessons be learned from cross-contextual studies? Technology (in this case geomatics) must be related to the changing dynamics in its environment (both social and natural). As argued by Norgaard (1994) technology coevolves with the environment, socio-cultural values, organization, knowledge and the natural environment (see Figure 3 and Table 1 above). The adapted model described in Tables 1 to 4 seek to integrate the issues in Ghana and Canada within this framework.

The social variables also coevolve with the natural environment. Social issues influence the way technology is perceived, the method of environmental modification and the priorities for environmental assessment using the relevant technology. For integrative studies, links between technological, environmental and social dynamics must be constructed. Tables 2, 3 and 4 illustrate technology/environment/social relations in more detail. Environmental systems may be measured using parameters (Table 2) and the available technology varies in Canada and Ghana (Table 3). For applications to policy actions, there have been contributions, but (especially in Ghana, but also in Canada) more advances are needed (Table 4).

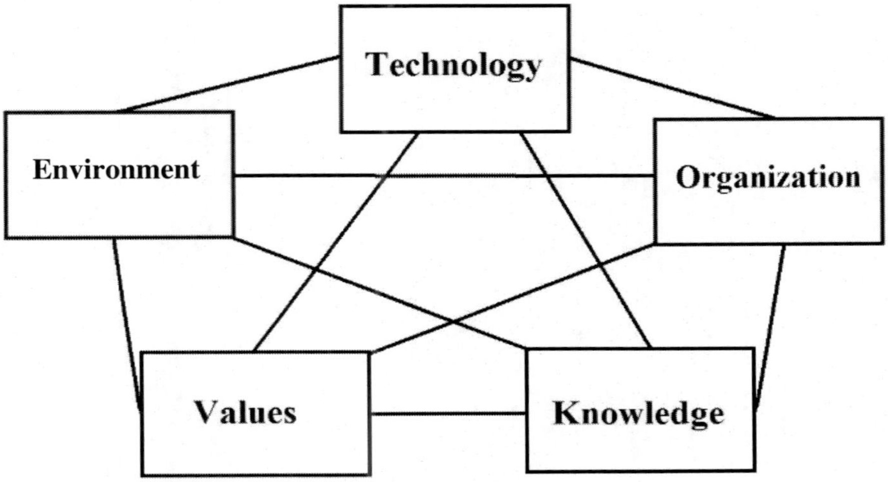

Figure 3. The coevolutionary relationship of technology and its context Derived from Norgaard (1994). Here technology centers around geomatics and environment is seen through the lens of conservation biology.

In such cases, landcover change moves in several directions, sometimes in opposite trajectories in different locations, rendering a linear change analysis inadequate. Assessments of such change is usually based on the tools of geomatics (GIS methods, time series images, field data, and related social surveys) to describe more dynamic environmental changes (Campbell & Palmer-Jones, 1999; Campbell, 2002, 2005, 2013). For example, Campbell (2005) used two sets of aerial photographs (1960 and 1986) and a SPOT satellite image (1998) to document multidirectional forest/savanna border change in coastal Ghana, West Africa.

Canada

Canada is a global leader in both environmental conservation and the development of geomatics technology and applications. Environmental awareness, top rate scientific education and computer literacy are strong; these are combined with vast land areas with interesting ecosystems (prairies, forests, mountains, lakes and rivers) and a low human population density.

Table 1. Geomatics and the Coevolutionary Framework in Canada and Ghana. This illustrates the application of the components of the Coevolutionary Framework on geomatics issues in the two countries. It shows how the variable in the first column engages with that in the top row (i.e., how the former influences the latter)

Components	Technology	Environment	Values	Organization	Knowledge
Technology	Geomatics etc.	Degradation/change	Change in outlook	Interest groups	Specialization applied
Environment	Influences applications	Ecology	Outlook change	Spatial reorganization	Adapts to changes
Values	Changed focus (e.g., valued species)	Prioritization, contacts	Old values change, new ideas	New group identities	Prioritization of knowledge to social interests)
Organization	Area of application (e.g., urban, forest)	Area of effect (e.g., proximity of forest)	Groups, views stakeholder classes (hiking groups)	Reorganization of groups and class relations	Systematization of knowledge, disciplines
Knowledge	Sophistication, relevance, viability (e.g., change in detail)	Environmental management (e.g., new applications)	Support for outlooks, interests	Support for organized groups, classes	Evolution of knowledge interfaces

Source: Literature cited in References List.

Table 2. The choice of environmental indicators for geomatics analysis

Variables	Parameters/indicators	Time frame of study
Forestry	Vegetation classes, species distribution/density, growth rate, multidirectional change	Mainly long term
Wildlife	Distribution, migration, landscape relations (specialist/generalist) human proximity impact, numbers, density	Long term/short term
Landuse (agriculture, hunting, tourism)	Frequency, impact, numbers, patterns, distribution, trails, vegetation patch change	Long term/short term

Source: Literature cited in References List.

Table 3. Canada and Ghana: Summary of contrasts and similarities in geomatics applications

Methods/Comments	Geomatics	Canada	Ghana
Time series Images	Reliance on time series images	More images and maps	Fewer images. Use of field and social data.
Telemetry	Integrated telemetry, image analysis	Common use of telemetry	Less or no use of telemetry
Social analysis and use of maps and images	Integration of social analysis/images, maps	Common product of land use studies	Rarely a product of land use studies
Problems	Too general	More needed for regular monitoring	Not enough for local planning/ monitoring
Recommended geomatics	Regular field studies, data field	Conservation monitoring advanced	Conservation monitoring rudimentary
Social issues	Dissemination and integration	Social participation advanced	Social participation needed

Source: Literature cited in References List.

Table 4. Geomatics and conservation policy Issues in Canada and Ghana

Country	Main Issues	Available resources	Needed Information
Ghana	Decentralization/participation	Little data, few surveys	Social data, regular monitoring
	Environmental conservation	Broad, long term, scanty data	Detailed images, regular monitoring, telemetry
	Urbanization	Very little, some survey information	Social data, regular monitoring
Canada	Decentralization/participation	Adequate structures	More social surveys, consultations
	Environmental Conservation	Long term data, some micro data (e.g., telemetry, images)	More micro data (regular, dense dates, monitoring).
	Urbanization	Long term data, some micro data (social, environmental data)	More micro data (regular, dense dates, monitoring)

Source: Literature cited in References List.

National institutions concerned with landuse documentation and geomatics applications include Environment Canada, Natural Resources Canada and Geomatics Canada. Inter-organizational networks include the Ecological Monitoring and Assessment Network (EMAN), coordinated by Environment Canada, based on about 100 sites, which conducts multidisciplinary environmental studies on many protected areas. Priorities are the development of information bases for planning and management applications (Craig et al., 2000: Conservation Halton, 2017). Environment Canada's Lands Directorate's Canadian Geographic Information System (CGIS) initiated land surveys of the Canadian National parks, including data from Agriculture Canada's Canadian Soil Information System (CANSIS).

Currently, scores of geomatics sites undertake such work (Woodley & Carruthers, 2000; Campbell, 2017a). Applications include wild population census, ecological mapping, forest fire research and protection, wildlife migration modeling, habitat management, eco-tourism development, biodiversity analysis, land-use planning, temporal changes analysis, habitat fragmentation analysis and environment impact assessment. The next three sections explore three important environmental issues in Canada: forestry and environmental change, wildlife ecology and conservation and rural nature/society relations. These are the primary environmental parameters that must be measured for data provision, to which geomatics methods may be applied for more accurate documentation and reliable information synthesis.

Forestry

In Canada (Figure 4a), the vegetation is dominated by forest (48 percent) (Maini & Carlisle, 2000), with about 40 percent of the total national area covered by forest (Statistics Canada, 2011, Figure 4b). Deforestation is one of the key forms of environmental change, especially in Ontario during the last two centuries (Larson et al., 1999). About 90 percent of older wooded areas changed to open landscape by the 1920s. Factors for this change included woodcutting, intensive grazing, and watershed losses. The wooded areas in the peri-urban and rural contexts of Ontario (Figure 4a,b) have expanded in recent times. Reasons for this include plantation development, public education, stream protection, fencing and changing agricultural practices (Larson et al., 1999).

Parameters for measurement may include vegetation type classification and biodiversity change, at specific scales and time sequences, sometimes supported by time series images. For example, Rhemtulla and Higgs (2000) conducted a study of Jasper National Park (Figure 4a), based on vegetation maps derived from aerial photographs (1949 and 1991) and analyzed with geomatics software. This formed the basis for ecological restoration work. Some woodlands have acquired increased areas of younger stands and new species, while fragmentation and reduction in species numbers still occurs (Riley & Mohr, 1994, Campbell 2017a). Such changes must be monitored through dense, spatial and temporal data sources, to support conservation policies.

Figure 4a. Map of Canada and selected parks.

Other methodologies may focus on the ecosystem dynamics and possible methods for evaluating these processes. Ecological Forest Management is another important area of research with spatial implications for the forested areas of Canada (Larocque, 2016). For example, the Alberta Forest Conservation Strategy is an ecologically based management plan that combines a system for monitoring biodiversity levels, and the sustainable production of forest services and goods. Impact assessments are not fully developed (using measurable parameters) but are an area of potential geomatics application (both for short and long-term applications) (Schneider, 2000; Luckert et al., 2011).

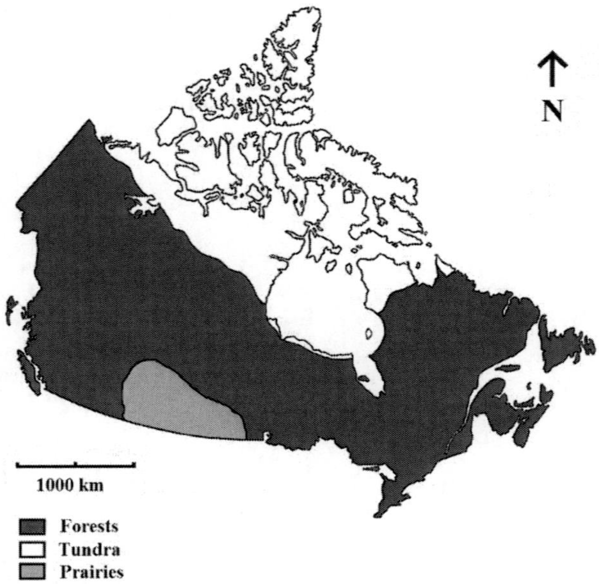

Figure 4b. Vegetation zones (forest and non-forest zones of Canada. The forest includes boreal forest or taiga, temperate rainforest and deciduous forest.

In the past decades, management in many Canadian parks did not conduct dense, comprehensive monitoring, although there was some use of indicator species to ascertain environmental impacts (Campbell & Wiersima, 2000). However, recently more monitoring is planned or undertaken (Parks Canada, 2017). These have acknowledged the need for environmental and social issues for monitoring, and analytical, synthetic and feedback components (*ibid.*). It is also acknowledged that conservation methods must be applied to the entire landscape, rather than only in parks and protected areas, because there are intrinsic links between the protected areas and proximate lands (Normand & Toule, 2000). In a case study, geomatics-based research methods were demonstrated as useful within an integrated research project developed by the Toronto and Region Conservation Authority for application to watershed-based natural heritage strategies. Indicators included patch and landscape configuration change (ibid.).

Greater proximity between people and the conserved areas, increases the requirement for effective assessment systems. For the examination of such contexts, southern Canada is a good case study, with many interesting emphases according to region (Campbell, 2013a). As pointed out by Beechey et al., (2000, 21) because of the high human population:

> "the southern settled reaches of Canada present a formidable challenge to those engaged in biodiversity conservation and protected areas. With almost one third of Canada's population, and many of its most endangered species and ecosystems, southern Ontario stands out as a region where innovation and adaptive approaches are needed to establish formally protected areas."

Similar issues have been detected in British Columbia (Figure 4), where the densely populated urban areas near protected areas record substantial conflict between people and wildlife (Campbell, 2012, 2015, 2016, 2017a). Supporting actions include the creation of more protected areas and safeguarding existing parks. An example of a policy action is the Ontario Parks Legacy, which includes fund raising and land acquisition (Beechey et al., 2000). These actions must be supported by improved assessments before the benefits of policy actions may be fully determined (Campbell, 2013a).

Wildlife

Wildlife data management is also important in Canada (Campbell, 2016). Important spatial parameters include time series population change, landscape/wildlife relations, selection of indicator species for environmental change and social indicators. The selection of these parameters with recognition of scale and time factors provides the bases for geomatics-based analyses. This would be based on parameter quantification and analyses (Campbell, 2015).

In terms of the assessment of human impacts on wildlife, Brooks (2000) describes the Sasajwun Wildlife Research Station as a case study of the possibilities for such a database and the associated issues. This study examined research programs on small mammals, frogs and turtles within long term projects, which derived knowledge that obtained from short-term projects. However, Brooks (2000) acknowledges that a strong commitment and a strong organizational structure are necessary for the development of such information bases, as the input is more difficult than in short term studies. Consequently, more shorter-term projects are undertaken (Campbell, 2013a).

Where there is a change in the configuration of the landscape, it is particularly important to develop information bases that document spatial and temporal variations in wildlife populations (Campbell, 2009a, 2012, 2016, 2017). Birds may be used as indicator species when landcover is modified by social action (Dobson, 2000). This study, based on Jasper National Park demonstrates an integration of ecological (numbers of breeding birds) and social indicators (human landuse intensity). The results enabled the prediction of serious impacts of landcover change on bird presence and habitats. Another example is the 3-year study by Cheskey (2000) on the impact of residential development on forest birds near Waterloo, Ontario (Figure 4a). DeSante and Kaschube (2006) also describe the Canadian Migration Monitoring Network (CMMN) and the Monitoring for Avian Productivity and Survivorship (MAPS) in ecosystem monitoring across Canada.

Danby's (2000) study explored the St. Elias Region which lies between British Colombia, Alaska and the Yukon Territory, a large contiguous region of protected areas. The study demonstrated how a spatial analysis of the biological connectivity of the protected areas, using certain large mammal species as indicators (mountain goat *Oreamnos americanus*, moose *Alces alces,* caribou *Rangifer tarandus*, grey wolf *Canis lupus*, grizzly *Ursus arctos* and Dall sheep *Ovis dalli*). Materials produced included distribution and habitat maps, synthesized with field information, for the assessment of population size, movements and the existence of corridors. The conclusion was that such studies are useful for regional scale assessment of conservation areas and cross boundary studies.

Bears have been used as indicator species for human impacts (see Hodgins & Schrenek 2000 on Grizzly Bears *Ursus arctos* in a two-year study in Jasper National Park; and Sutton et al., 2000 on Black Bears *Ursus americanus* in a three-year study in Bruce Peninsula National Park (Figure 4a), under the Ecological Integrity Monitoring Program (EIMP)). Both studies accumulated data on the movements, population size, and habitat needs of the bears, for strategic, conservation planning. Radio collars for bear tracking and the participation of stakeholders were key research themes. Results included contributions to the development of a monitoring system for the Ecological Integrity Monitoring System.

Tourism and Conservation

Strong environmental awareness in Canada encourages a balance between human needs and environmental protection (Campbell, 2012, 2017). Both Federal and Provincial parks emphasize environmental protection and public use as integral to their mandates (Porter, 2000). In addition, human use of areas within and proximate to national parks is increasing (Eagles, 2000; Campbell, 2016; Parks Canada, 2017). Public awareness of this trend is crucial, given the fragility of some ecosystems. Curthoys and Clark (2000; 32) argue that "an ecologically literate citizenry that understands the concept of landscape health and thus is supportive of conservation initiatives both within and outside of protected area boundaries" determines "the long-term viability of protected areas."

Monitoring is one method devised to cope with these developments (Parks Canada, 2017). This must be supported by data collection methods, which are accurate and standardized, and these methods must be used in all parks and protected areas to allow comparison and cross-referencing (Sinclair, 2000). Monitoring, supported by such data collection methods (necessarily including geomatics) may also evaluate the status of new policies. For example, eco-tourism, (a generic term for ecologically sensitive tourism planning, which emphasizes environmental responsibility in visitor landuse) has been touted as a step forward in environmental management (Matheus & Raimundo, 2016). According to Taylor et al., (2000) several agencies managing Ontario Parks and within the Ministry of Natural Resources follow guidelines related to the tenets of eco-tourism (see also Penny, 2014). Upton (2000) describes the development of Integrated Ecological

Integrity Monitoring Programs, which are included in park Ecological Integrity Statements.

There is the need for indicators of human use, parallel to those of animal use outlined above, to allow pro-active planning. An experimental framework developed for Georgian Bay Islands National Park (Figure 4a) is a step in this direction (Upton, 2000). Human landuse parameters (numbers, vehicular traffic paths, trials, frequency of passage, patch level habitat modification) must be examined in relation to environmental parameters (species movements, density, breeding sites) for the assessment of the appropriate balance between human needs and conservation. From such studies evaluative designations may be derived.

The gradual modification or even degradation of the vegetated habitats requires the measurement of ecological integrity (EI) (Leung et al., 2000). EI was stipulated as a priority by the National Parks Act (Campbell & Wiersima, 2000), but assessment of this complex issue is difficult without measurement. Wilkinson (2000) mentions the balance between use and protection that must exist before ecological integrity can be maintained, this being achieved by monitoring and critical control. Without this type of management, intensive use of ecological resources may create serious problems (see for example Gertsh (2000), for a description of the impacts of campsites on vegetation in Quetico Provincial Park, in Northwestern Ontario, and Vasseur (2000) on campsites in Kejimkujik National Park; Figure 4.a).

Important, measurable human activities in protected areas include snowmobiling, skiing and walking (Huebel, 2012). For example, as described in a study by Cassie et al., (2000) in the Gros Morne National Park of Western Newfoundland (Figure 4a), a critical study of snow mobiling intensity showed degradation of vulnerable plant species, especially phanerophytes. Giles (2000) examined the effects of walking on ecosystems near trails, and the possibilities for the creation of trail surfaces. Due to the wide occurrence of hiking in protected areas, the derivation of measured parameters and associated geomatics methods would be useful. Huebel (2012) found that heli-skiing had impacts on the behavior and spatial distribution of mountain caribou (*Rangifer tarandus caribou*).

Hunting affects ecosystems, dependent on the species to be hunted, the period of hunting and the effects on the wider ecosystem (Campbell, 2017a). Forbes et al., (2000) argue that (using case studies of the black bear *Ursus americanus* in Fundy National Park and wolves *Canis lupus* in Algonquin Park, Ontario, Figure 4a) hunting has had negative impacts on conservation, and more knowledge is required on ecological impacts before hunting can responsibly be combined with the objective of the maintenance of "ecological integrity" (*ibid.* 40). Conflicts between wild animals and people exist even where there is little hunting (Campbell, 2017a). Mercer (2000) gives an apt title to the article "The Biker and the Bear: Linking Human Use and Wildlife Movement." Crucially for some conservation areas, conflicts are especially likely in the valley bottoms of mountainous national parks, as such areas are likely to be preferred by both people and wildlife. The case study is the Three Valley Confluence, where human presence cuts through wildlife corridors, affecting wildlife access to some parts of the ecosystem (Mercer, 2000). Studies of these issues require detailed collection of data on the environmental (animal species density migration, reproduction, numbers, landscape ecology) and social parameters (human movements and numbers). Broader assessments, e.g., corridors and other landscape configuration, may also be needed (Mercer, 2000; Campbell, 2017a).

Ghana

In Ghana (Figure 5), environmental management is also controlled by the government, but unlike Canada, Ghana is a unitary rather than federal state. Hence, the Regions (the approximate equivalents of the Provinces in Canada) do not have the same high level of policy power as the Canadian provinces. The main environmental/ conservation institutions in Ghana are the Environmental Protection Agency (EPA), the Ministry of Environment, Science and Technology and the Ministry of Lands and Forestry. The Survey Department, Lands Department and Environmental Protection Agency have limited geomatics capability. The Institute of Renewable Natural Resources (at the Kwame Nkrumah University of Science and Technology) and the Department of Geography and Resource Development (at the University of Ghana) also have specialists and laboratories for geomatics applications (Campbell, 1998, 2017b.c).

Materials are mostly aerial photographs and literature sources (satellite and radar images are not easily available). These institutions study serious environmental problems, but there is little collaboration between them, although this is improving markedly. Funding is also limited, hampering logistical and equipment support. This contributes to the general lack of information on environmental problems, a problem considering the governments prioritization of local environmental and social planning (Campbell, 2013b).

Relevant, spatial environmental and conservation issues for geomatics in Ghana include wide ranging bush fires, landuse conflicts, urbanization and resource management (mainly forests, rivers and wildlife). The lack of information on environmental issues has been one of the factors behind the prevalence of some problems (in addition to infrastructural and financial problems, and lack of political support). For example, several policies that were to be implemented by the EPA were unsuccessful (Campbell, 2013b).

For example, the National Anti-Bush Fire Committee (NABFC) was established to prevent bushfires (Benneh & Agyepong 1990; Campbell, 2013b), but was unsuccessful due to limited financial resources allocated by the central government, the absence of an effective implementing machinery and the lack of incentives for the people who were involved in bush-fire-fighting.

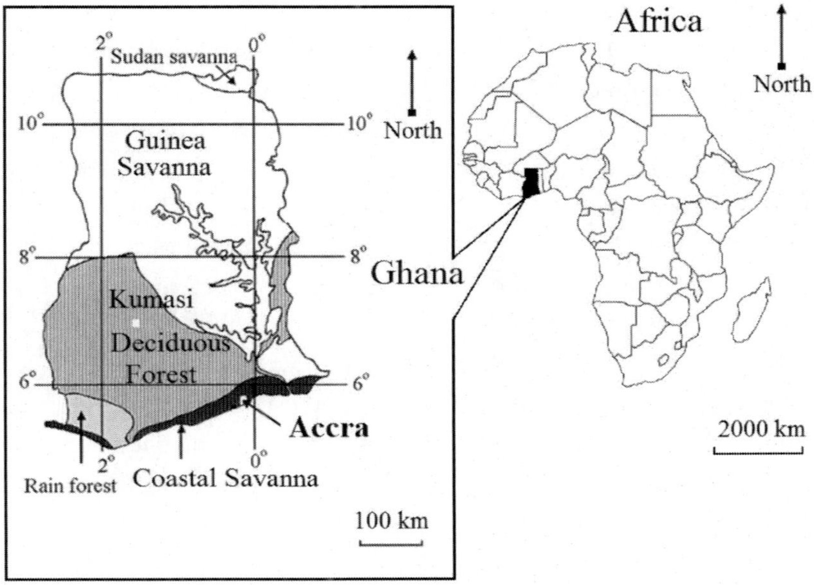

Figure 5. Map of the vegetation zones of Ghana, showing the study sites of Kumasi and the coastal savanna.

The related National Plan of Action to Combat Desertification (NPACD), the second major policy action of the 1980s, was set up to deal with the increased threat of desertification during and after the disastrous 1983/84 drought. Spatial analysis estimated the areas under threat from desertification as 83,487 square kilometers, or 35 percent of the national area (Benneh & Agyepong, 1990). The Environmental Action Plan (1988), the most comprehensive document on environmental policy ever compiled in Ghana (Dorm-Adzobu, 1995) formed the basis of a National Environmental Policy (NEP) currently in operation (Campbell, 2013b). However, the implementation of this plan was ineffective as information availability and management competence were serious problems (Campbell 1993, 2013b).

Currently, geomatics is relevant, through its integrative and analytical capabilities (Campbell, 2017b, c). There is insufficient information available for informed judgments on landuse change. For example, for deforestation, "dramatic" forest loss has been described for this region over the last century, with increased losses in the "last few decades" (Fairhead & Leach, 1998, 3). When however, remotely sensed data is examined, it is frequently noticed that there is a gap between reported change and the image results (*ibid.*). Leach and Fairhead (2000) note that the satellite-based surveys of Ghanaian forest reserves do not necessarily support studies that claim rapid rates of deforestation are occurring on reserve boundaries. This evidence which underscores the need for geomatics based integrated research (using field surveys, social information, and remotely sensed images).

Due to the comparative rarity of geomatics-based studies in Ghana, only two cases will be cited. Both studies were partly based on my participation in the British Development for International Development (DFID) projects in Ghana (1999 to 2000). The first (Campbell & Palmer-Jones, 1999) was based in the coastal savanna of Ghana (Figure 5), with the objective of landscape analysis. Here common perceptions of landscape change hold that there has been steady deforestation. However, examination of remotely sensed images and extensive fieldwork revealed both deforestation and multidirectional change. The second study (DFID, 2000) is based on water resources change in Kumasi (Ghana's second city, Figure 5), the objective being the study of the processes of water management. Here a geomatics-based study of spatial factors for water pollution shed light on the problems inherent in water resources management in the area. It must be pointed out that despite the merit of these studies, they illustrate the point mentioned above, in that they do not have data of sufficient detail and frequency of measurement to be useful for local planning (local planning is currently recognized as vital for development in Ghana) (Campbell, 2013b).

In the case study based on the coastal savanna (DFID Project No. 7219 (GIS Analysis of Aerial Photographs to Provide Quantitative Assessment of Environmental Dynamics in a West Africa Savanna)), Campbell and Palmer-Jones (1999) examined the socioenvironmental landscape dynamics described in a field survey (Campbell, 1998) and demonstrated how these parameters could be measured and analyzed with geomatics. The materials and methods were time series images, and intensive vegetation and social surveys. This methodology examined parameters of landscape change as areas under grass/farms/bare-ground/forest and combinations of these vegetation types. It was argued that geomatics-based research within an integrated research framework enables more effective documentation of the multi-directional vegetation change described above. It also provides beneficial information that cannot be easily collated and analysed through other research methods (Campbell & Palmer-Jones, 1999).

Two areas within the study area were selected for more detailed analyses: the environs of the villages of Aplaku and those of Kokrobite/Oshiyie (Figure 5). These areas were selected due to the total overlap in coverage for 1960, 1986 (aerial photographs) and 1998 (satellite) images, and the forms of vegetation variation, which typify both common and extreme forms of change in the wider area of the Tuba/Aplaku plains of coastal Ghana. The borders of the two case studies were according to the coverage of the aerial photographs.

In the area around Aplaku, the quantification of multi-directional change showed general deforestation, grassland expansion and retraction, farmland expansion and retraction and the relationship between topography, landuse and vegetation. In the Kokrobite/Oshiyie area, multi-directional change was also evident. Changes in the areas of tree/stands were multi-directional, but less extreme than in the Aplaku area. Cultivated areas also underwent multi-directional change and grass dominated areas declined. The relationships between the topography, landuse and vegetation form was also complex but less extreme than that of the Aplaku area. The results of the research therefore supported the contention that geomatics methods provide beneficial information that cannot easily be collated through other research methods (Campbell & Palmer-Jones, 1999; Campbell, 2009b,c,d).

In the Kumasi case study (DFID Project No. 7330: Peri-urban Natural Resources Management at the Watershed Level in Kumasi, Ghana) water resource management in Kumasi, inclusive of biodiversity, landuse change and social problems were examined. "There have been many studies on water pollution and externalities in the Kumasi area, but none embracing a holistic watershed perspective" (DFID Proposal, 2000, 2). The project included information collected under a previous DFID project (R6799: Kumasi Natural Resources Management): land management systems, biogeographical issues, village livelihood surveys and coping strategies, soil fertility improvement resources and soil fertility. The former project included geomatics for documentation, analytical and modeling purposes. The assessment used four approaches: (1) rapid field assessment and conventional impact checklists; (2) conceptual models and geomatics-based analysis; (3) lumped parameter statistical models; and (4) distributed, process-based simulation models.

According to the DFID Proposal (2000, 4) "the main aim of the modeling effort is to develop diagnostic frameworks and geomatics applications which define sensitive locations within the watersheds that are vulnerable to site degradation or pollution or are likely sources of offsite propagation of pollutants; and which assess the risk of deterioration in the condition of the watersheds in the future." The work was based on the theoretical framework of pollution risk by Cairney (1995). This included three elements that influence pollution risk: hazard location, pathway and target. The parameters in these categories in Kumasi were: the location of peri-urban activities within mixed urban/forest and savanna landcover, landuse and waste output, the proximity of these areas to hydrological pathways (areas where water flows, overland or channeled), the frequency and magnitude of surface runoff from the collection areas via the hydrological pathways, and changes in the flow quantity through time. The geomatics-based software would enable spatial displays of these parameters to document dynamic change in pollution risk.

The geomatics aspect of the project had limited impact and sustainability during the time of its inception (McGregor et al., 1999). It was to be income generating, with potential clients purchasing maps and analytical papers produced by the system, based on their requested needs. However, at the time of the field survey, (mid-2000) this was not apparent. Feedback from stakeholders attending numerous seminars pointed to this failure. Problems included the failure to involve all stakeholders in the formative period of the project, which therefore resulted in a lack of knowledge about the strength of the geomatics base of the project. Excessive technical jargon, excluded many less informed stakeholders, who otherwise might have understood the geomatics software. The difficulties in disseminating geomatics technology in environmental projects have been noted in other studies (Campbell, 2013a). For example, electricity and other logistical supports were insufficiently reliable to allow the spread of geomatics-based technology to rural areas. Social and cultural issues were also evident: the media was not fully involved in disseminating information about the usefulness of the project products; additionally, costs were too high to be funded by the earnings from orders from any clients, as the income levels in the country were not fully considered. This is still a basic issue for geomatics applications in low income nations, especially in rural areas (Campbell, 2017a).

This case illustrated the problems embedded in geomatics use in Ghana, as indicative of other applications in less developed nations (Campbell, 2009c,d, 2017a,c). There are several other possible applications. The following sections describe the general issues of forestry, wildlife and rural landuse. The main problem in the analysis of these parameters is the lack of sufficient data (remotely sensed images, field measurements, and social surveys) for analysis, limited geomatics facilities and rural poverty. This problem is compounded by the realization that decentralization (devolution of decision making power, supporting locally relevant information) is the first step in development. The information currently available is too general to allow such detailed local planning applications (Campbell, 2013a).

Forestry

Phytogeographical change and the application of mapping techniques have been issues in Ghana from the pre-colonial and colonial periods (pre-1957), from the records of travelers and administrators, and is still an important topic (Dickson & Benneh, 1988; Benneh & Agyepong, 1990; Parren & Graaf, 1995; Palo et al., 1996; Campbell & Palmer-Jones, 1999). Today the issue of spatial deforestation is still very important (Agbesinyale, 1992; Gyasi et al., 1995; Hawthorne & Abu-Juam, 1995; Satti, 1995; Campbell, 2013a, 2017a,b,c). The loss of forest stands in Ghana in the last 100 years is argued to be one of the highest in Africa (Dixon et al., 1996), but some authors also question the extent (as there is little authoritative, geomatics-based evidence, and some researchers examine 'positive' issues such as increased open savanna land for savanna species, reforestation and effective forest management (Parren & Graaf 1995; Richards 1995; Campbell, 1998). Fairhead and Leach (1998) support a more critical look at the assumption of historical linear deforestation, noting that in many cases this belief was based on insufficient image-based evidence, poor judgment and generalization of contextual findings (see also Campbell 2013b,c,d).

The spatial documentation of human resource use and landcover change is a key issue. Several human activities influence the spatial distribution of vegetation (Carson, 1985; Campbell & Palmer-Jones 1999; Campbell 2017b,c). Food and cash cropping is the dominant activity in rural Ghana. Bush burning is the main method of clearance for farming, hunting and building, and firewood cutting is practiced for income and energy. Firewood is the main energy source, due to a lack of electricity (Jampoh, 1989; Campbell, 2009a,b,c). Carson (1985, 7) notes that in the savannas of coastal Ghana (Figure 5), firewood cutting "has received little attention in the literature ... and its effects are probably under estimated."

Desertification is also an important problem, with few supporting spatial information sources and geomatics-based analyses (Campbell, 2017c). Benneh and Agyepong (1990) describe desertification as the "ultimate form of land degradation" where "the biological, productive capacity of the land or water is more or less completely destroyed." Measurable, spatial parameters include area of bare ground, visible surface erosion, changes in canopy cover, grass density and plant biodiversity, ground water depletion and the numbers, depths and lengths of gullies (Campbell, 2017c). Despite the lack of permanent deserts, as noted above, the total area at risk from desertification has been estimated at 35 percent of the total area of Ghana (Benneh & Agyepong, 1990, Figure 5). This figure may have expanded recently (Campbell, 2017c). Factors for this include the incidence of droughts and bush fires during the dry season (November to May), intensive agriculture, high population densities in small local areas, dry climates, cattle herding, lack of rural financial investments and resultant soil erosion (Campbell, 2017c).

Wildlife and Rural Landuse

In Ghana, forest reserves (established in the early 20th century) and national parks/game reserves exist (for example, the Mole Game Reserve in the northern savanna; the Shai Hills National Park in the coastal savanna and the Kade Reserve in the central forest). However, wildlife ecology and management are generally neglected in the literature, despite the existence of remnants of the past dense wildlife populations (Campbell, 2017b). In the northern savanna, leopards (*Panthera pardus*), lions *Panthera leo* and elephants (*Loxodonta africana*) are recorded. The forest region contains duikers (for example, crowned duiker (*Sylvicopia grimmia*), black duiker (*Cephalophus niger*). red flanked duiker *Cephalophus rufilatus.* The coastal savanna is currently low in mammal species. Mammal species still in existence, which are the main hunted animals, include the Maxwell's duiker (*Cephalophus maxwelli*), bushbuck (*Tragelaphus scriptus*) and the African civet (*Viverra civetta*). Several studies document the habitat ranges of these mammals, but none give a full picture of the spatial changes in distribution over the years and the possible role of conservation areas (Campbell, 1998).

In some cases, there are differences in location perceptions of zoogeographical change. This is important, as official, 'scientific' assessments may be scanty, and local knowledge may be the main information source, even for data input for geomatics analysis (Campbell, 2017b). There may be competing sources of information for wildlife. For example, in one study of the coastal savanna (Campbell 1998), the people in the area differed in their perception of the indicators of antelope species change. The hunters measured animal species change by the number of animals caught, length of hunting periods, distance walked in hunting, sightings and footprints, especially after rain. Non-hunters rather observed species reduction, due to reduced crop losses to antelope grazing. Both hunters and non-hunters agreed that reduced animal numbers were due to excessive hunting, drought, deforestation and the increasing proximity of settlements. Deforestation in relation to hunting was perceived to be a negative phenomenon (Campbell 1998). This local information is vital for local planning but is generally not documented (Campbell, 2017a,b,c).

Rural Landuse and Conservation

Unlike Canada, in Ghana there is little documentation of tourism's role in conservation and landscape change (Campbell, 2917b). Tourist areas relevant to conservation biology topics include the national parks, forests and beaches, and to a lesser extent, historical cultural activities (traditional farming, firewood cutting and hunting). There is more focus on how these activities may change the landcover (Carson, 1985; Campbell, 1998, 2017b,c; Campbell & Palmer-Jones, 1999). For example, in the coastal savanna, farming and firewood cutting were common in shrub/grass mosaics, and hunting was common in forest clumps. In forested areas urban sprawl, road building, mining and farming have depleted stands (Benneh & Agyepong, 1990; Fairhead & Leach, 1998). Measurable parameters in images and field surveys include visible urban infrastructure and vegetation classes, but the complexity of the landscape and sparse data may hamper identification of landcover change.

CONCLUSION

This chapter has explored the possible links of geomatics and conservation biology in furthering the development of environmental and managerial complexes. The examples, though not exhaustive, exemplify the complexity of the issues involved. The concepts of space, pattern, place and integrated analyses were evident through all the comparisons. The disciplines of geomatics and conservation biology are both navigating transformative periods, with inputs largely geared towards the adaptation to new problems, greater complexities of old problems, more effective and less costly management, and the new disciplinary issues. As both disciplines are strongly interlinked, their integrated applications offer a whole that is greater than the sum of the parts. The management of the emphases on methods within a larger flexible methodology would enable the improvement of both short-term tactical management and longer term strategic planning. The derivation of quantifiable parameters for measurement using geomatics is essential for such improvement. The numerous theoretical and applied studies that have emerged to enable such syntheses illustrate the complexity of effective integrated methodologies. The next chapters will examine the relationships between geomatics and conservation biology in more detail.

Regarding the case studies, points concerned contrasts in technological development and differing landcover issues. Canada's advancement was significant, but there were development possibilities for Ghana. Geomatics applications in Canada contributed significantly to policy making and (to a lesser degree) environmental monitoring. In Ghana, possibilities provide the basis for more detailed studies and solutions to problems with compatibility and reliability of data sources. Such cross-country research, currently rare, can produce comparisons of value to future work, as contextual parameters are good methods of comparisons. Such comparisons must acknowledge the social context of the technology, the rarity and limitations of spatially and temporally detailed data, the requirement for broad sets of parameters (covering social and environmental indicators), and the involvement of the stakeholders (inclusive of local people). Such awareness acknowledges the complexities involved in the integration of geomatics and conservation biology, both for theoretical advancement and practical application.

REFERENCES

Agbesinyale, P. K., (1992). *Agricultural Expansion and the State of Ghana's Environment: The Need for Sustainable Development.* Cape Coast: University of Cape Coast.

Agrawal, A. (2000). Adaptive management in transboundary protected areas: The Bialowieza National Park and Biosphere Reserve as a case study. *Environmental Conservation* 27(4): 326-333.

Aikman, J. Neil, G. Cannons, A. & Milton, J. (1998). *So, where are the stones and where are the swamps anyways?* A resource assessment of Stony Swamp, Nepean, Ontario (Unpublished paper). Ottawa: Carleton University.

Barry, D. & Oelschlaeger, M. (1996). A science for survival: values and conservation biology. *Conservation Biology* 10: 905–911.

Beechey, T. J., Duncan, J. E., Turner, S. J. & Powell, D. M. (2000). Ontario Parks Legacy 2000: Building Nature Reserve for the New Millenium. In *Challenges to*

National Parks and Protected Areas: Learning from the Past, Looking to the Future. The Fourth International Conference of Science and the Management of Protected Areas (SAMPAA IV), the Third Annual Parks Research Forum of Ontario (PRFO III). May 14-19, 2000. Waterloo: University of Waterloo.

Bengston, D. N., Fan, D. P & Celarier, D. N. (1999). A new approach to monitoring the social environment for natural resource management and policy: The case of US National forest benefits and values. *Journal of Environmental Management* 56 (3): 181-193.

Benneh, G. & Agyepong, G. A. (1990). *Land Degradation in Ghana*. London: Commonwealth Secretariat.

Brennan, J., Bender, D. J., Contreras, D. A. & Fahrig, L. (2001). Focal patch landscape studies for wildlife management: Optimizing sampling effort across scales. In Liu, J. & Taylor, W. W. (Eds.), *Integrating Landscape Ecology into Natural Resource Management,* 68 – 91. Cambridge: Cambridge University Press.

British Columbia Institute of Technology (2018) *Geomatics*. Retrieved from https://www.bcit.ca/study/programs/geomatics.

Brooks, R.J. (2000). The Value and Maintenance of Long -Term Research Programs in Protected Areas. In *Challenges to National Parks and Protected Areas: Learning from the Past, Looking to the Future. The Fourth International Conference of Science and the Management of Protected Areas (SAMPAA IV), the Third Annual Parks Research Forum of Ontario (PRFO III).* May 14-19, 2000. Waterloo, ON: University of Waterloo.

Brunner, R. D. & Clark, T. W. (1997). A practice-based approach to ecosystem management. *Conservation Biology* 11(1): 48–58.

Burley, J. (2002). *Forest biological diversity: an overview*. Unasylva 53(209): 3–9.

Cairney, T. (1995). *The Re-Use of Contaminated Land: A Handbook of Risk Assessment.* Chichester: Wiley.

Caissie, A., Frego, K. A. and Flemming, S. (2000). Effects of Various Amounts of Snow Mobile Traffic on the Plant Species Composition of a Coastal Plain Sphagnum Bog in Gros Morne National Park, Western Newfoundland. In *Challenges to National Parks and Protected Areas: Learning from the Past, Looking to the Future. The Fourth International Conference of Science and the Management of Protected Areas (SAMPAA IV), the Third Annual Parks Research Forum of Ontario (PRFO III).* May 14-19, 2000. Waterloo, ON: University of Waterloo.

Callicott, J. B. (1990). Whither conservation ethics? *Conservation Biology*, 4:15–20.

Campbell, M. (1993) *Implementing the Environmental Action Plan of the Environmental Protection Council: Problems and Prospects*. M. P. A. Thesis, University of Ghana, Accra.

Campbell, M., (1998). *Interactions between Biogeography and Rural Livelihoods in the Coastal Savanna of Ghana*. PhD Thesis, University of London (Imperial at Wye College).

Campbell, M. and Palmer-Jones, R. (1999). *The Utility and Problems of GIS Methods in the Analysis of Social Imprints on the Landscape of the Coastal Savanna of Ghana: A Report to ESCOR*. DFID Project No. 7219. GIS Analysis of Aerial Photographs to Provide Quantitative Assessment of Environmental Dynamics in a West Africa Savanna. Norwich: University of East Anglia.

Campbell, M. (2002a). *The Assessment of Biogeographical Change in the Western Greenbelt of Ottawa Using Geomatics*. Master of Arts Thesis. Ottawa: Carleton University.

Campbell, M. (2002b). *Vegetation Change and Urbanization in the Western Greenbelt of Ottawa (1934- 1999)*. Ottawa: National Capital Commission Report.

Campbell, M. (2005). The ecological and social context of mammal hunting in the coastal savanna of Ghana. *Geoforum* 36: 667–680. doi: 10.1016/j.geoforum.2005.04.001.

Campbell, M. (2009). The impact of habitat characteristics on bird presence and the implications for wildlife management in the environs of Ottawa, Canada. *Urban Forestry and Urban Greening* 8: 87–95.

Campbell, M. (2009b). Repositioning zoogeography within the nature–culture borderlands: An animal geography of reptiles in southern Ghana. *Applied Geography* 29: 260–268.

Campbell, M. (2009c). A new zoogeography of domestication and agricultural planning in Southern Ghana. *Area* 41(2), 139–148.

Campbell, M. (2009d). Proximity in a Ghanaian savanna: human reactions to the African palm civet *Nandinia binotata*. *Singapore Journal of Tropical Geography* 30: 220–231.

Campbell, M. (2012). The relevance of age and gender for public attitudes to brown bears (*Ursus arctos*), black bears (*Ursus americanus*) and cougars (*Puma concolor*) in Kamloops, British Columbia. *Society and Animals* 21 (4): 341-359.

Campbell, M. (2013a). Biogeographic change and environmental policy in the western greenbelt of Ottawa, Canada. In Creighton, E. & Danovich, P. (Eds.) *Environmental Policy,* 148 – 164. New York, NY: Nova Science Publishers.

Campbell, M. (2013b). *The Political Ecology of Agricultural History in Ghana*. New York, NY: Nova Science Publishers.

Campbell, M. (2015). *Vultures: Their Evolution, Ecology and Conservation*. New York, NY: Taylor and Francis, CRC Press.

Campbell, M. (2016). Avian reactions towards human approaches in different urban greenery structures in Nanaimo. *Urban Forestry & Urban Greening* 19: 47–55.

Campbell, M. (2017). Integrated research methods and geomatics in protected area management. In Campbell, M. (Ed.), 165 – 189. *Biological Conservation in the 21st*

Century: A Conservation Biology of Large Wildlife. New York, NY: Nova Science Publishers.

Campbell, M. (2017a). *Biological Conservation for the 21st Century: A Conservation Biology of Large Wildlife*. New York, NY: Nova Science Publishers.

Campbell, M. (2017b). The ecology and conservation of the West African lion. In Jenkins, O. P. (Ed.), *Advances in Animal Science and Zoology*. Volume 11. New York, NY: Nova Science Publishers.

Campbell, M. (2017c). An applied biogeography of forest/savanna land cover change in Ghana. In Daniels, J., *Advances in Environmental Research*. Volume 61. New York, NY: Nova Science Publishers.

Campbell, M. & Wiersima, Y. (2000). A Monitoring Framework for National Parks: Assessing Integrity Across a System. In *PRFO, Challenges to National Parks and Protected Areas: Learning from the Past, Looking to the Future. The Fourth International Conference of Science and the Management of Protected Areas (SAMPAA IV), the Third Annual Parks Research Forum of Ontario (PRFO III)*. May 14-19, 2000. Waterloo, ON: University of Waterloo.

Carson, W. P. (1985). State of Knowledge Report: The Ecology of the Accra-Winneba Plains with Some Aspects of Related Savanna Ecosystems. In *Man and the Biosphere Project 3: Impact of Human Activities on the Structure and Productivity of the Savanna Ecosystem in Ghana Phase 1: Following and Rotation in the Accra-Winnable Plains*. Report on Preliminary Activities, 4-74, Accra: Environmental Protection Agency.

Conservation Halton (2017). *Ecological Monitoring Protocols*. Retrieved from https://www.conservationhalton.ca/uploads/ch_monitoring_protocols_v1_final_june2017.pdf

Craig, B. Lamb, J. & Vagnan, H. (2000) The Ecological Monitoring and Assessment Network (EMAN) Volunteer Monitoring Program: An effective tool for community outreach. In *PRFO, Challenges to National Parks and Protected Areas: Learning from the Past, Looking to the Future. The Fourth International Conference of Science and the Management of Protected Areas (SAMPAA IV), the Third Annual Parks Research Forum of Ontario (PRFO III)*. May 14-19, 2000. Waterloo, ON: University of Waterloo.

Craighead, J. J. (1998). *Thinking like a mountain. Wildlife Society Bulletin* 26: 908-910.

Crisci, J. V., Katinas, L. & Posadas, P. (2003). *Historical Biogeography: An Introduction*. Cambridge: Harvard University Press.

Curthoys, L. P. & Clark, J. (2000). Interpretative Planning at the Landscape Level: A Tool for Giving People a Voice in the Future of Their Communities. In *PRFO, Challenges to National Parks and Protected Areas: Learning from the Past, Looking to the Future. The Fourth International Conference of Science and the Management of Protected Areas (SAMPAA IV), the Third Annual Parks Research Forum. (PRFO III)*. May 14-19, 2000. Waterloo, ON: University of Waterloo.

Danby, R. (2000). *PRFO, Challenges to National Parks and Protected Areas: Learning from the Past, Looking to the Future. The Fourth International Conference of Science and the Management of Protected Areas (SAMPAA IV), the Third Annual Parks Research Forum. (PRFO III)*. May 14-19, 2000. Waterloo, ON: University of Waterloo.

Decker, D. J., Riley, S. J. & Siemer, W. F. (2012). *Human Dimensions of Wildlife Management*. Second edition. Baltimore: John Hopkins University Press.

Dempsey, C. (2011). Conservation Biology and GIS. *GIS Lounge*. Retrieved from https://www.gislounge.com/conservation-biology-and-gis/.

DeSante, D. F. & KasChube, D. R. (2006). The monitoring avian productivity and survivorship (maps) program 1999, 2000, and 2001 report. *Bird Populations* 7: 23-89.

Department for International Development (DFID) (1999). *DFID project (R6799: Kumasi Natural Resources Management)*. London: University of Greenwich.

Dickson, K. B. & Benneh, G. (1988). *A New Geography of Ghana*. London: Longman.

Diduck, A. (1999). Critical education in resource and environmental management: learning and empowerment for a sustainable future. *Journal of Environmental Management* 57 (2), 85-97.

Dixon, R. K., Perry, J. A., Vanderlein, E. L. & Hiol, F. H. (1996). Vulnerability of forest resources to global climate change: Case study of Cameroon and Ghana. *Climate-Research* Vol. 6 (2), 127-133.

Dobson, T. (2000). *PRFO, Challenges to National Parks and Protected Areas: Learning from the Past, Looking to the Future. The Fourth International Conference of Science and the Management of Protected Areas (SAMPAA IV), the Third Annual Parks Research Forum. (PRFO III)*. May 14-19, 2000. Waterloo, ON: University of Waterloo.

Dorm-Adzobu, C., (1995). *New Roots: Institutionalizing Environmental Management in Africa*. Washington DC: World Resources Institute.

Dubuisson, B., 1975. *Practique de la photogrammétrie et de moyens cartographiques derivés des ordinateurs*. Paris: Editions Eyrolles. [*Practical photogrammetry and cartographic means derived from computers*]

Eagles, P. F. J. (2000) International Trends in Park Tourism. In *PRFO, Challenges to National Parks and Protected Areas: Learning from the Past, Looking to the Future. The Fourth International Conference of Science and the Management of Protected Areas (SAMPAA IV), the Third Annual Parks Research Forum. (PRFO III)*. May 14-19, 2000. Waterloo, ON: University of Waterloo.

Ehlers, M. (2008). Geoinformatics and digital earth initiatives: a German perspective. *International Journal of Digital Earth*, 1(1): 17-30, DOI: 10.1080/1753894070 1781975.

Ehlers, M. (1993). Integration of GIS, remote sensing, photogrammetry and cartography: the geoinformatics approach. *Geo-Informations-Systeme* (GIS) 6 (5): 18-23.

Ehrenfeld, D. W. (1970). *Biological Conservation*. New York, NY: Holt, Rinehart and Winston.

Ehrenfeld, D. W. (1992). Conservation biology: Its origins and definitions. *Science* 255: 1625 – 1626.

Elliott, J. A. & Campbell, M. (2002). The environmental imprints and complexes of social dynamics in rural Africa: Cases from Zimbabwe and Ghana. *Geoforum* 33: 221–237.

Fagan, A. (2000). The Characteristics of the Older Camper: How to Attract the Aging Camper. In *PRFO, Challenges to National Parks and Protected Areas: Learning from the Past, Looking to the Future. The Fourth International Conference of Science and the Management of Protected Areas (SAMPAA IV), the Third Annual Parks Research Forum. (PRFO III).* May 14-19, 2000. Waterloo, ON: University of Waterloo.

Fairhead, J. & Leach, M. (1998). *Reforming Deforestation: Global Analysis and Local Realities - Studies in West Africa*. London: Routledge.

Forbes, G. Theberge, J. and Woodley, S. (2000). Hunting in Canadian National Parks. In *PRFO, Challenges to National Parks and Protected Areas: Learning from the Past, Looking to the Future. The Fourth International Conference of Science and the Management of Protected Areas (SAMPAA IV), the Third Annual Parks Research Forum. (PRFO III).* May 14-19, 2000. Waterloo, ON: University of Waterloo.

Gagnon, P. & Coleman, D. J., 1990. Geomatics: an integrated, systemic approach to meet the needs for spatial information. *Canadian Institute of Surveying and Mapping Journal* 44 (4): 377-382.

Geomatica (2001). *Geomatics*. Canadian Institute of Geomatics, Ottawa.

Gerlach, L. P. & Bengston, D. (1994). If ecosystem management is the solution, what is the problem? *Journal of Forestry* 92(8): 18–21.

Gertsch, F. (2000). The Effect of Education on Compliance with Wilderness Campsite Closures. In *PRFO, Challenges to National Parks and Protected Areas: Learning from the Past, Looking to the Future. The Fourth International Conference of Science and the Management of Protected Areas (SAMPAA IV), the Third Annual Parks Research Forum (PRFO III).* May 14-19, 2000. University of Waterloo, Ontario, Canada.

Gibeau, M. L. (2000). *A Conservation Biology Approach to Management of Grizzly Bears in Banff National Park*, Alberta. PhD Thesis. University of Calgary, Calgary.

Giles, A. (2000) Integrating the Environmental, Social and Economic Aspects of Trail Surfacing Decisions. In *PRFO, Challenges to National Parks and Protected Areas: Learning from the Past, Looking to the Future. The Fourth International Conference of Science and the Management of Protected Areas (SAMPAA IV), the Third Annual Parks Research Forum. (PRFO III).* May 14-19, 2000. University of Waterloo, Ontario, Canada.

Gomarasca, M. (2010) Basics of Geomatics. *Applied Geomatics*, 2, 137-146. https://doi.org/10.1007/s12518-010-0029-6.

Goodchild, M., (1992). Geographical information science. *International Journal of Geographical Information Systems* 6 (1), 31-45.

Goodfellow, D. (2014). How GIS is Being Used in Conservation Biology. *GIS Lounge.* Retrieved from https://www.gislounge.com/gis-used-conservation-biology/.

Gunderson, L. (1999). Resilience, flexibility and adaptive management antidotes for spurious certitude? *Conservation Ecology* 3 (1), 7. Retrieved from http://www.consecol.org/vol3/iss1/art7.

Gyasi, E., Agyepong, G. T., Ardayfio-Schandorf, E., Enu-Kwesi, L., Nabila, J. S. & Owusu-Bennoah E., (1995). Production pressure and environmental change in the forest-savanna zone of southern Ghana. *Global Environmental Change* 5(4): 355-366.

Hales, D. F. (1991). The Pinelands National Reserve: An approach to cooperative conservation. In West, P. & Brechin, S. (Eds.), Resident Peoples and National Parks: Social Dilemmas and Strategies in International *Conservation,* 239-250. Tucson: University of Arizona.

Hawthorne, W. D. & Abu-Juam, M. (1995) *Forest Project in Ghana: With Particular Reference to Vegetation and Plant Species.* Gland: International Union of Nature and Natural Resources (IUCN).

Hays, S. (1959). *Conservation and the Gospel of Efficiency: The Progressive Conservation, 1890–1929.* Cambridge: Harvard University Press.

Heywood, I., Cornelius, S. & Carver, S. (2006) *An Introduction to Geographical Information Systems*. Third Edition. Harlow: Pearson Education.

Hodgins, A. & Schrenek, V. (2000). *PRFO, Challenges to National Parks and Protected Areas: Learning from the Past, Looking to the Future. The Fourth International Conference of Science and the Management of Protected Areas (SAMPAA IV), the Third Annual Parks Research Forum. (PRFO III).* May 14-19, 2000. University of Waterloo, Ontario, Canada.

Holling, C. S. (1978). *Adaptive Environmental Assessment and Management.* New York, NY: John Wiley.

Huebel, K. (2012). *Assessing the Impacts of Heli-Skiing on the Behavior and Spatial Distribution of Mountain Caribou (Rangifer tarandus caribou).* MSc. Thesis, Thompson Rivers University, Kamloops.

Jackson, H. B. & Fahrig, L. (2015). Are ecologists conducting research at the optimal scale? *Global Ecology and Biogeography* 24:52–63.

Jampoh, E. L. (1989). *The Organization of Fuelwood and Charcoal Trade in Ghanaian Urban Centres: Case Studies in Accra and Kumasi*. BSc Dissertation, Institute of Renewable Natural Resources, University of Science and Technology, Kumasi.

Jensen, J. R. (2007). *Remote Sensing of the Environment: An Earth Resource Perspective*, 2nd Edition. Upper Saddle River: Pearson Education Inc.

Jensen, M. E. & Bourgeron, P. S. (Eds.). (1994). *Volume II: Ecosystem Management: Principles and Applications.* PNW-GTR-318. Portland, OR: U. S. Department of Agriculture.

Johnson, B. L. (1999). The role of adaptive management as an operational approach for resource management agencies. *Conservation Ecology* 3(2), 8. Retrieved from http://www.consecol.org/vol3/iss2/art8/.

Journal of Remote Sensing & GIS (2016). Geomatics. Retrieved from https://www.omicsonline.org/scholarly/geomatics-journals-articles-ppts-list.php.

Kolko, G. (1963). *The Triumph of Conservatism: A Re-Interpretation of American History, 1900–1916.* New York, NY: Free Press.

Larocque, G. R. (Ed.). (2016). *Ecological Forest Management Handbook.* Boca Raton: CRC Press.

Larson, B. M., Riley, J. L., Snell, E. A. & Godschalk, H. G. (1999). *The Woodland Heritage of South Ontario: A Study of the Ecological Change, Distribution and Significance*. Don Mills: Federation of Ontario Naturalists.

Leach, M. & Fairhead, J. (2000). Challenging Neo-Malthusian deforestation analyses in West Africa's dynamic forest landscapes. *Population and Development Review* 26(1): 17-43.

Lee, K. N. (1999). Appraising adaptive management. *Conservation Ecology* 3(2): 3. Retrieved from http://www.consecol.org/vol3/iss2/art3.

Leonard, T. C. (2009). American economic reform in the progressive era: Its foundational beliefs and their relation to eugenics. *History of Political Economy* 41(1), 109 – 141. DOI 10.1215/00182702-2008-040.

Leong, K. M., Decker, D. J. & Lauber, T. B. (2012). Stakeholders as beneficiaries of wildlife management. In Decker, D. J., Riley, S. J. & Siemer, W. F. (Eds.) *Human Dimensions of Wildlife Management,* 26–40. Baltimore: JHU Press.

Leung, Y. F., Marion J. L. & Cole, D. N. (2000). Visitor Impact Assessment and Monitoring Systems for Protected Areas: A Methodological Review. In *PRFO, Challenges to National Parks and Protected Areas: Learning from the Past, Looking to the Future. The Fourth International Conference of Science and the Management of Protected Areas (SAMPAA IV), the Third Annual Parks Research Forum. (PRFO III).* May 14-19, 2000. University of Waterloo, Ontario, Canada.

Lomolino, M. V. & Heaney, L. R. (Eds.) (2004). *Frontiers of Biogeography*. Sunderland: Sinauer Associates.

Lomolino, M. V., Sax, D. F. & Brown, J. H. (Eds.) (2004). *Foundations of Biogeography*. Chicago: University of Chicago Press.

Lomolino, M. V., Riddle, B. R. & Brown, J. H. (2006). *Biogeography*, 3rd Edition. Sunderland: Sinauer Associates.

Lourie, S. A. & Vincent, A. C. J. (2004). Using biogeography to help set priorities in marine conservation. *Conservation Biology* 18:1004–1020.

Luckert, M. L., Haley, D. & Hoberg, G. (2011). *Policies for Sustainably Managing Canada's Forests: Tenure, Stumpage Fees, and Forest Practices.* Vancouver: UBC Press.

MacArthur, R. H. & Wilson, E. O. (1967). *The Theory of Island Biogeography.* Princeton: Princeton University Press.

Machlis, G. E & Force, J. E. (1997). The Human Ecosystem: Part I: The Human Ecosystem as an Organizing Concept in Ecosystem Management. *Society and Natural Resources* 10(4): 347-367.

Maini, J. S. & Carlise, J. A. (2000). Conservation. *The Canadian Encyclopedia*. 2-4. Ottawa: Historical Foundation of Canada.

Matheus, F. S. & Raimundo, S. (2016). Public use and ecotourism policies in Brazilian and Canadian protected areas. *Études caribéennes*, 33-34. DOI: 10.4000/etudes caribeennes.9344. [*Caribbean Studies*]

Mazzocchi, F. (2006). Western science and traditional knowledge. *EMBO Reports* 7(5): 463 – 466.

McGregor, D., Simon, D. &Thompson, D. (2000) Proposal. DFID Project No. 7330: *Peri-urban Natural Resources Management at the Watershed Level in Kumasi Ghana.* Egham: University of London (Royal Holloway and Bedford New College.

Meffe, G. K., & Carroll. C. R. (1994). *Principles of Conservation Biology.* Sunderland: Sinauer Associates.

Meffe, G. K., & Viederman. S. (1995). Combining science and policy in conservation biology. *Wildlife Society Bulletin* 23: 327-332.

Meine, C. (2004). *Correction Lines: Essays on Land, Leopold, and Conservation.* Washington, DC: Island Press.

Meine, C., Soule, M. & Noss, R. F. (2006). "A mission-driven discipline'': The growth of conservation biology. *Conservation Biology* 20(3): 631–651.

Meine, C. (2010). Conservation Biology: Past and Present. In N. S. Sodhi & P. S. Ehrlich (Eds.), *Conservation Biology for All*, 7 – 26. Oxford: Oxford University Press.

Mercer, G. (2000). The Biker and the Bear: Linking Human Use Wildlife Movement. In *PRFO, Challenges to National Parks and Protected Areas: Learning from the Past, Looking to the Future. The Fourth International Conference of Science and the*

Management of Protected Areas (SAMPAA IV), the Third Annual Parks Research Forum. (PRFO III). May 14-19, 2000. University of Waterloo, Ontario, Canada.

Mitchell, M. S. & Hebblewhite, M. (2012). Carnivore habitat ecology: integrating theory and application. In Boitani, L. & Powell. R. A. (Eds.), *Carnivore Ecology and Conservation,* 218–255. New York, NY: Oxford University Press.

Morzillo, A. T., De Beurs, K. M. & Martin-Mikle, C. J. (2014). A conceptual framework to evaluate human-wildlife interactions within coupled human and natural systems. *Ecology and Society* 19(3), 44. Retrieved from http://dx.doi.org/10.5751/ES-06883-190344.

Natural Resources Canada (2018). *Geomatics.* Retrieved from https://www.nrcan.gc.ca/earth-sciences/geomatics/10776.

Norgaard, R. B, (1994). *Development Betrayed: The End of Progress and a Coevolutionary Revisioning of the Future*. London: Routledge.

Normand, L. A. & Towle, K. G. (2000). An Innovative Approach to Biodiversity Conservation and Restoration Across Settled Landscape. In *PRFO, Challenges to National Parks and Protected Areas: Learning from the Past, Looking to the Future. The Fourth International Conference of Science and the Management of Protected Areas (SAMPAA IV), the Third Annual Parks Research Forum. (PRFO III).* May 14-19, 2000. Waterloo, ON: University of Waterloo.

Odenbaugh, J. (2003). Values, advocacy, and conservation biology. *Environmental Values* 12: 55–69.

Orr, D. W. (2004). *The Nature of Design: Ecology, Culture, and Human Intention.* New York, NY: Oxford University Press.

Page, R., Bayley, S., Cook, J. D., Green, J. E. & Ritchie, J. R. B. (1996). Banff-Vallee dela Bow: a l'Heure des Choix. Ottawa: Ministere des Travaux Publics et Services Gouvernmentaux. [*Banff Valley Bow: At Choice Time*. Ottawa: Ministry of Public Works and Government Services.]

Palo, M., Yirdaw, E. and Mery, G., (1996). Deforestation and sustainable forestry challenge in Ghan. In Palo, M., (Ed.) *Sustainability Forestry Challenges for Developing Countries*, 343-357. Dordrecht: Kluwer Academic Publishers.

Parks Canada (2017). *Science and Conservation: Ecosystem Monitoring*. Retrieved from https://www.pc.gc.ca/en/nature/science/controle-monitoring/ecosysteme-ecosystem.

Parren, M. P. E. & Graaf, N. R. (1995). *The Quest for Natural Forest Management in Ghana, Coted'Ivoire and Liberia.* Wageningen, Stichting Tropenbos.

Penny, E. (2014). *In the Eye of the Beholder: Perceptions of Ecotourism in Algonquin Provincial Park.* Masters Thesis, University of Waterloo, Waterloo.

Pickett, S. T. A., Parker, V. T. & Fiedler, P. L. (1992). The new paradigm in ecology: Implications for conservation biology above the species level. In Fiedler, P. L. & Jain, K. (Eds.), *Conservation Biology: The Theory and Practice of Nature*

Conservation, Preservation and Management, 66-87. New York, NY: Chapman and Hall.

Pollock-Ellwand, N. & Nelson, G. (1999). Landuse, Landscape Change and the Role of Parks and Protected Areas. In *Parks and Protected Areas Research in Ontario* (PPRFO) Parks and Protected Areas Research Forum of Ontario. 11-15. Ottawa: PPRFO.

Porter, J. (2000). Improving the role of recreationists in the conservation of parks and protected areas. In *PRFO, Challenges to National Parks and Protected Areas: Learning from the Past, Looking to the Future. The Fourth International Conference of Science and the Management of Protected Areas (SAMPAA IV), the Third Annual Parks Research Forum. (PRFO III).* May 14-19, 2000. Waterloo, ON: University of Waterloo.

Posadasa, P., Crisci, J. V. & Katinas, L. (2006). Historical biogeography: A review of its basic concepts and critical issues. *Journal of Arid Environments* 66: 389–403.

Primack, R. B. (1993). *Essentials of Conservation Biology*. Sunderland, MA: Sinauer Associates.

Primack, R. B. (2002). *Essentials of Conservation Biology*. Third Edition. Sunderland, MA: Sinauer Associates.

Proulx, R., Massicotte, P. & Pepino, M. (2014). Googling trends in conservation biology. *Conservation Biology* 28: 44–51.

Rhemtulla, J. & Higgs, E. (2000). Eighty Years of Change: The Montane Vegetation of Jasper National Park. In *PRFO, Challenges to National Parks and Protected Areas: Learning from the Past, Looking to the Future. The Fourth International Conference of Science and the Management of Protected Areas (SAMPAA IV), the Third Annual Parks Research Forum of Ontario (PRFO III).* May 14-19, 2000. Waterloo, ON: University of Waterloo.

Richards, P. (1985) *Indigenous Agricultural Revolution*. London: Hutchinson.

Riley, J. L. & Mohr, P. (1994). *The Natural Heritage of Southern Ontarios Settled Landscapes: A Review of Conservation and Restoration Ecology for Land-use and Landscape Planning*. Peterborough: Ontario Ministry of Natural Resources.

Roebuck, P. & Phifer, P. (1999). The persistence of positivism in conservation biology. *Conservation Biology* 13(2): 444-446.

Samuelsen, R. (2008). *The Field of Geomatics: Scientific, Technological, Business and Infrastructural Aspects.* NF-working paper 1027/2008.

Satti, A. M. (1995). *Country Environmental Profile: Ghana*. Abidjan: African Development Bank.

Schneider, R. (2000). Integration of Protected Areas and Ecological Forest Management: A Case Study from Alberta, Canada. In *PRFO, Challenges to National Parks and Protected Areas: Learning from the Past, Looking to the Future. The Fourth International Conference of Science and the Management of Protected Areas*

(SAMPAA IV), the Third Annual Parks Research Forum. (PRFO III). May 14-19, 2000. Waterloo: University of Waterloo.

Scoones, I. (1997). Landscape, fields and soils: Understanding the history of soil fertility management in Southern Zimbabwe. *Journal of Southern African States* 2: 615-634.

Shellford, V. E. (1932) Ecological Society of America: A Nature Sanctuary Plan unanimously adopted by the Society. December 28, 1932. *Ecology* 14(2): 240-245.

Sinclair, T. (2000) Visitor Statistics: A Need for Accurate and Reliable Data; A Case Example of Ontario. In *PRFO, Challenges to National Parks and Protected Areas: Learning from the Past, Looking to the Future. The Fourth International Conference of Science and the Management of Protected Areas (SAMPAA IV), the Third Annual Parks Research Forum. (PRFO III).* May 14-19, 2000. Waterloo. ON: University of Waterloo.

Skibicki, A. A., Stadel, T., Welch, D. & Nelson, J. G. (1994) *Ecological Monitoring and National Park Working Paper No. 7.* Heritage Resources Centre. Waterloo, ON: University of Waterloo.

Sodhi, N. S. & Ehrlich, P. S. (Eds.). (2010). *Conservation Biology for All.* Oxford: Oxford University Press.

Solecki, W. D. (1994). Putting the biosphere reserve concept into practice: Some evidence of impacts in rural communities in the United States. *Environmental Conservation* 21 (3): 242-247.

Soule, M. E. (1985). What is conservation biology? *Bioscience* 35: 727-734.

Soulé, M. E. (1986a). *Science of Scarcity and Diversity.* Sunderland: Sinauer Associates.

Soulé, M. E. (1986b). Conservation biology and the real world. In Soulé, M. E. (Ed.), *Conservation biology: The Science of Scarcity and Diversity*, 1-12. Sunderland: Sinauer Associates.

Soulé, M. E. & Orians, G. H. (Eds.). (2001). *Conservation Biology: Research Priorities for the Next Decade.* Washington, DC: Island Press.

Soulé, M. E. & Wilcox, B. A. (1980). *Conservation Biology: An Evolutionary-Ecological Perspective.* Sunderland: Sinauer Associates.

Stanley, T. R., Jr. (1995). Ecosystem management and the arrogance of humanism. *Conservation Biology* 9: 255-262.

Statistics Canada (2011). *Forest Land by Province and Territory.* Retrieved from http://www.statcan.gc.ca/tables-tableaux/sum-som/l01/cst01/envi34a-eng.htm.

Sutton, G. (2000). *PRFO (2000) Challenges to National Parks and Protected Areas: Learning from the Past, Looking to the Future. The Fourth International Conference of Science and the Management of Protected Areas (SAMPAA IV), the Third Annual Parks Research Forum. (PRFO III).* May 14-19, 2000. Waterloo, ON: University of Waterloo.

Taylor, J. R. Mlair, H. & Reid, D. (2000). Best Practices for Ecotourism in Parks and Protected Areas. In *PRFO, Challenges to National Parks and Protected Areas:*

Learning from the Past, Looking to the Future. The Fourth International Conference of Science and the Management of Protected Areas (SAMPAA IV), the Third Annual Parks Research Forum. (PRFO III). May 14-19, 2000. Waterloo, ON: University of Waterloo.

Temple, S. A., Brussard, P. F., Bolen, E. G., Salwasser, H., Soule, M. E. & Teer, J. G. (1988). What's So New About Conservation Biology? *Transactions of the North American Wildlife and Natural Resources Conference* 53: 609-612.

Upton, D. (2000). The Development of a Human Use Indicators Screening Framework for Georgian Bay Islands National Park. In *PRFO, Challenges to National Parks and Protected Areas: Learning from the Past, Looking to the Future. The Fourth International Conference of Science and the Management of Protected Areas (SAMPAA IV), the Third Annual Parks Research Forum. (PRFO III).* May 14-19, 2000. Waterloo, ON: University of Waterloo.

Vasseur, L. (2000). Connection Between Ecosystem Health and Park Management: the Impact on Camp Grounds. In *PRFO (2000) Challenges to National Parks and Protected Areas: Learning from the Past, Looking to the Future. The Fourth International Conference of Science and the Management of Protected Areas (SAMPAA IV), the Third Annual Parks Research Forum. (PRFO III).* May 14-19, 2000. Waterloo, ON: University of Waterloo.

Wedding, L. M., Lepczyk, C. A., Pittman, S. J., Friedlander, A. M. & Jorgensen, S. (2011). Quantifying seascape structure: Extending terrestrial spatial pattern metrics to the marine realm. *Marine Ecology Progress Series* 427: 219 - 232.

Wells, M., Brandon, K. & Hannah, L., (1992). *People and Parks, Linking Protected Area Management with Local Communities*. Washington, DC: The World Bank.

Whittaker R. J. & Ladle, R. J. (2011). The roots of conservation biogeography. In Ladle, R. J. & Whittaker, R. J. (Eds.) *Conservation Biogeography*, 1 – 12. Hoboken: Blackwell Publishing Ltd.

Whittaker, R. J., Araújo, M. B., Jepson, P., Ladle, R. J., Watson, J. E. M. & Willis, K. J. (2005). *Conservation biogeography: Assessment and prospect. Diversity and Distributions* 11:3 – 23.

Wilknison, P. F. (2000). Protecting for Ecological Integrity in Canada's National Parks. Allowable and Appropriate Visitor Use. In *PRFO, Challenges to National Parks and Protected Areas: Learning from the Past, Looking to the Future. The Fourth International Conference of Science and the Management of Protected Areas (SAMPAA IV), the Third Annual Parks Research Forum. (PRFO III).* May 14-19, 2000. Waterloo, ON: University of Waterloo.

Williams, P. H. & Araújo, M. B. (2000). Integrating species and ecosystem monitoring for identifying conservation priorities. *European Conservation* 4: 17–18.

Wilson, E. O. & Peter, F. M. (Eds.) (1988). *Biodiversity.* Washington DC: National Academy Press.

Woodley, S. & Carruthers, J. A. (2000). *National Parks. The Canadian Ecyclopaedia*, 5-9. Ottawa: Historical Foundation of Canada.

In: Geomatics and Conservation Biology
Editor: Michael O'Neal Campbell

ISBN: 978-1-53614-468-0

Chapter 2

MAPPING AND MANAGEMENT OF VULTURES IN AN INDIAN STRONGHOLD

Kaushalendra Kumar Jha*
Tropical Forest Ecosystem and Environment Management Unit,
Technical Forestry Faculty Area,
Indian Institute of Forest Management, Bhopal, India

ABSTRACT

Vulture populations in India have been reduced by the occurrence of diclofenac poisoning, habitat degradation and climatic variation. An effective assessment of the status of Indian vultures requires a methodology including literature study, field research and the mapping of the changing ranges and habitat distributions of the various species. This chapter investigates the evidence for the decline of vulture populations in India, taking a case study of Central India, a stronghold for vulture populations. The study is based on a survey of the literature on vulture population counts, ranging and habitat mapping and predictive models of vulture presence. It investigates the status of endangered vulture species, their conservation requirements, the data collection methodologies and technology required for such assessments. Maps of vulture distribution and ecological relations are presented and discussed. Vulture conservation policies are also explored.

INTRODUCTION

Vultures as scavengers provide one of the most important yet underappreciated and little-studied ecosystem services of any avian group (Ruxton & Houston, 2004;

* Corresponding Author Email: Jhakk1959@gmail.com.

Sekercioglu et al., 2004; Sekercioglu, 2006; Markandya et al., 2008; Inger et al., 2016). Some species of vultures are among the world's largest flying birds, and all species are considered at least medium sized. The twenty- three species of vultures are grouped into two families, the New World or Cathartid vultures and the Old World or Accipitrid vultures. The Cathartid vultures comprise seven species, while the Accipitrid vultures include 16 species (Ogada et al., 2012a). All the Accipitrid vultures are either migratory or resident in Asia, except for the White-headed Vulture *Trignoceps occipitalis*, Hooded Vulture *Necrosyrtes monachus*, White-backed Vulture *Gyps africanus*, Ruppell's Vulture *Gyps rueppelli*, Cape Vulture *Gyps coprotheres*, and Lappet-faced Vulture *Torgos tracheliotos*. In India, the Long-billed Vulture *Gyps indicus*, White-rumped Vulture *Gyps benghalensis*, Red-headed Vulture *Sarcogyps calvus*, Slender-billed Vulture *Gyps tenuirostris*, and Egyptian Vulture *Neophron percnopterus* are resident vultures. The Himalayan Griffon *Gyps himalayensis*, Cinereous Vulture *Aegypius monachus*, Eurasian Griffon *Gyps fulvus* and Bearded Vulture *Gypaetus barbatus* are the migratory vultures.

Vultures, as obligate scavengers, provide important ecological and economic services through their consumption of carrion and the consequent regulation of disease and pests. They reduce the exposure time of rotting matter, which strongly contributes to reducing the rate of transmission of infectious diseases (Ogada et al., 2012b). They also assist industrial services by quickly removing soft tissue from the bones of carcasses, which can then be used for fertilizer production. Cultural and recreational services are also associated with this group of species (DeVault, et al., 2016). However, in the absence of vultures, the populations of facultative scavengers like feral dogs, rats and red fox may increase (Pain et al., 2003; Prakash et al., 2003; Markandaya et al., 2008; Moleon et al., 2014), though this has not been conclusively established. The scavenging efficiency of facultative scavengers is inferior to that of vultures, hence micro-organisms may flourish and contribute to diseases (Morales-Reyes et al., 2017). Hill et al. (2018) suggest that under the warm and humid conditions, facultative scavengers do not compensate for the absence of vultures, as carcasses remain for longer periods and decomposers replace scavengers. The resultant proliferation of zoonotic pathogens and other micro-organism like rabies, etc. may spread diseases among wild and domestic populations of animals and even humans (Stern & Wentzel, 1949 and Reed & Rocke, 1992 in Inger et al., 2016; Markandaya et al., 2008).

Vultures and Management

Populations of obligate scavengers (Africa, Europe - except Western Europe, and Asia) have significantly declined during past few decades, due to threats from a variety of anthropogenic factors including unintentional poisoning (Diclofenac and baits) and energy infrastructure (collision and electrocution) (DeVault et al., 2016; Botha et al.,

2017; Campbell, 2017). Around 61% of obligate scavenger species of the world are currently threatened with extinction (Ogada et al., 2012a). The highest rates of decline have been reported in South Asia and particularly in India; at one point only one in a thousand Long-billed vultures (*Gyps indicus*) remained alive (Prakash et al., 2007). This unnatural death rate was largely the result of human actions. The leading contributory factor was unintentional poisoning (Diclofenac use among cattle in the subcontinent), scarcity of food and loss of habitat. Other threats arise from energy infrastructure like windmills and electricity power lines, disease causing pathogens, etc. As a result, in India all the resident *Gyps* vultures (Long-billed, White-rumped and Slender-billed vultures) and the Red-headed vulture are critically endangered, the Himalayan griffon is near threatened and the Egyptian vulture is endangered. Migratory or winter visitors, such as the Cinereous and Bearded vultures are near threatened, and the Eurasian griffon is Least Concern (IUCN, 2017).

Cortes-Avizanda et al. (2018) argue that global vulture population decline is due to poisoning, persecution, habitat destruction, food scarcity, changed farming practices, etc. However, on the Indian subcontinent there are a few positive signs of population stabilization or recovery (Cuthbert et al., 2011; Bowden, 2012; Chaudhary et al., 2012; Ruperal, 2016 and Kishorel, 2016 in Campbell, 2017). To restore the viable population first and then make it grow on a sustainable basis, actions must include conservation management (Oppel et al., 2016), provision of safe nesting sites, supplementary food, reduction of disturbances and protection of nests against poaching and persecution (Lindsey, 1992; Liberatori & Penteriani, 2001; Vaughan et al., 2005; Gonzalez et al., 2006; Oro et al., 2008; Cortes-Avizanda et al., 2010; Altwegg et al., 2014; Demerdzhiev et al., 2014; Zuberogoitia et al., 2014). The management of vulture populations is important to avoid the permanent degradation of ecosystem functioning and associated services (Ogada et al., 2012a; Mateo-Tomas et al., 2017).

Vulture Measurement

Management expert Peter Drucker is often quoted as saying that "you can't manage what you can't measure." This proposition is primarily related to business projects; however, this can be taken as a thumb-rule for most of the wildlife management projects. James Harrington also argues that "measurement is the first step that leads to control and eventually to improvement. If you can't measure something, you can't understand it. If you can't understand it, you can't control it. If you can't control it, you can't improve it." Such improvement in the case of wildlife would be the sustainable management and conservation of a species population.

Measurement of a biological entity for its management may require ecological parameters such as richness and abundance of the species, spatial and temporal distribution, habitat requirement e.g., availability of inhabiting sites, food and water, existing threats to survival, etc. In the case of vultures, this would initiate the discovery of vulture species presence, numbers and spatial and temporal distributions. Other important parameters would include tree species and structure, roosting and nesting locations on cliffs and monuments, availability of food sources from ungulate populations, threats such as power structures, use of poison, etc. These inform conservation efforts directed towards vultures.

Measuring Biological Systems

Measurement of vulture related parameters over extensive areas enables extensive and intensive data generation for the analysis of the strengths and weaknesses in the system. Standardized protocols are required for the collection of quality data. Many statistical models are available for the analysis of biological populations, but some models are ineffective in producing the required results. For example, models of species distribution may not be applicable to diverse regions. Use of models to inform management decisions in regions other than that used for model development should be undertaken with caution (Bamford et al., 2009a). Pavokivic and Susic (2005) averred that model predictions are not facts, rather they are hypothetic scenarios of what could happen in the future, based upon the present knowledge on the species. It cannot be said for sure whether the future will resemble the past due to possible changes in the system. Probably for this reason, Chitty (1996) asserted that "mathematical modelers have done more harm than good through possessing a medieval faith in the power of logic to unravel secrets of the natural world."

However, in view of the recent evolution and efficacy of statistical models, they could be used for understanding the changing dynamics of an ecosystem. For such a purpose, a census or total count of individuals could be more useful than using a transect count, etc. provided the constraints of census are removed. For example, in the case of heterogeneous distribution, a larger area would cover higher variations in terms of population. However, the main difficulty of total count or census over large areas is that it requires a lot of resources (man power and wages) and may entail a lot of uncertainty. With available manpower and dovetailing of their works as well as funds, a total count of vultures may be done by institutions such as the Forest Departments (FD) of different states, provided a suitable protocol is available.

Basic data collection through a census may be insufficient for managing a species, as an index-based maps with weighted parameters may be required. Therefore, one must rely on modelling for research-based results. For example, in a search for unlisted nesting or roosting sites in forest vegetation, it is necessary to identify potential landcover using known habitat attributes. Using known parameters of observed habitats, a habitat suitability index may be created and used for a modelling exercise to predict vulture presence. The resultant map could be used to support conservation policy and management for vulture species. Parameters for nesting site suitability, such as tree size and species, optimum road, human settlement and waterbody distances, open or dense canopied vegetation, ungulate and cattle populations, etc. could be analyzed and an index (probability of occurrence or suitability of sites or potential sites) could be prepared.

VULTURE STRONGHOLD

Accipitrid vultures are found in Africa, Europe and Asia. Ten species are present in Africa (Botha et al., 2017), while only four species are present in Europe (Bearded vulture *Gypaetus barbatus*, Cinereous vulture *Aegypius monachus*, Egyptian vulture *Neophron percnopterus*, and the Eurasian griffon *Gyps fulvus*) (Moreno-Oppo et al., 2015). Sixteen species occur in Asia, with nine occurring in the Indian subcontinent, including some European species (Botha et al., 2017). Vulture populations were reported to be in steep decline during the 1990s and the first decade of the 21st century, but this appears to have slowed in recent years and the populations may have stabilized (Prakash et al., 2003, 2007; Green et al., 2004; 2007 & 2008; Cuthbert et al., 2011; Martinez-Abrain et al., 2012; Ogada et al., 2012a; Prakash et al., 2012). Therefore, recent estimation of vulture populations in different states of India can be taken as the minimum surviving population (Figure 1).

Within India, species counts have mostly relied on scattered, isolated populations within states rather than total counts of the whole state (Sivakumar & Manakadan, 2005; Subramanya et al., 2006; Chhangani & Mohnot, 2008; Das et al., 2011; Pande et al., 2013; Ramakrishnan et al., 2014; Venkitachalam & Senthilnathan, 2015; Saran, 2017, etc.). Few studies have attempted statewide surveys. Thakur and Narang (2012) reported around 150 White-rumped vultures in Himachal Pradesh. Chhangani (2009) recorded 5080 temporally distributed vultures (1086 Long-billed, 325 White-rumped, 84 Red-headed, 2418 Egyptian, and 217 Cinereous vultures, and 240 Himalayan and 643 Eurasian griffon vultures) over a period of four years (2004-2007) in Rajasthan. Jha (2015) recorded 2084 vultures (932 Egyptian, 516 Slender-billed, 341 Long-billed, 209 White-rumped, 16 Red-headed, and 14 Himalayan Griffon vultures) in Uttar Pradesh. Kamboj et al. (2016) reported 999 vultures of four species (458 White-rumped, 385 Long-billed, 132 Egyptian and 34 Red-headed vultures) in Gujarat.

As per Jha (2017) Madhya Pradesh has 7028 vultures of seven species (3351 Long-billed, 1874 Egyptian, 1522 White-rumped, 93 Red-headed, 53 Cinereous, 95 Eurasian and 42 Himalayan griffon vultures). In a recent report of Jhala et al. (2017) out of 6828 reported vultures (227 Griffon, 1424 Egyptian, 2346 White-rumped, 2830 Long-billed and one Slender-billed vulture), 92% were in the Central Indian landscape. This landscape comprises Rajasthan, Madhya Pradesh, Odisha, Jharkhand, Maharashtra, Telangana and Andhra Pradesh. This report may be an underestimation (considering the reports of Jha, 2015 and Jha, 2017), but it indicates Madhya Pradesh as a stronghold for vultures in India and for Critically Endangered *Gyps* vultures (White-rumped and Long-billed vultures). This may therefore be numerous enough for a population recovery provided that appropriate conservation measures or management efforts are taken up immediately. The habitat usage and breeding status of these vulture species are depicted in the following chart, derived from field data (Figure 2).

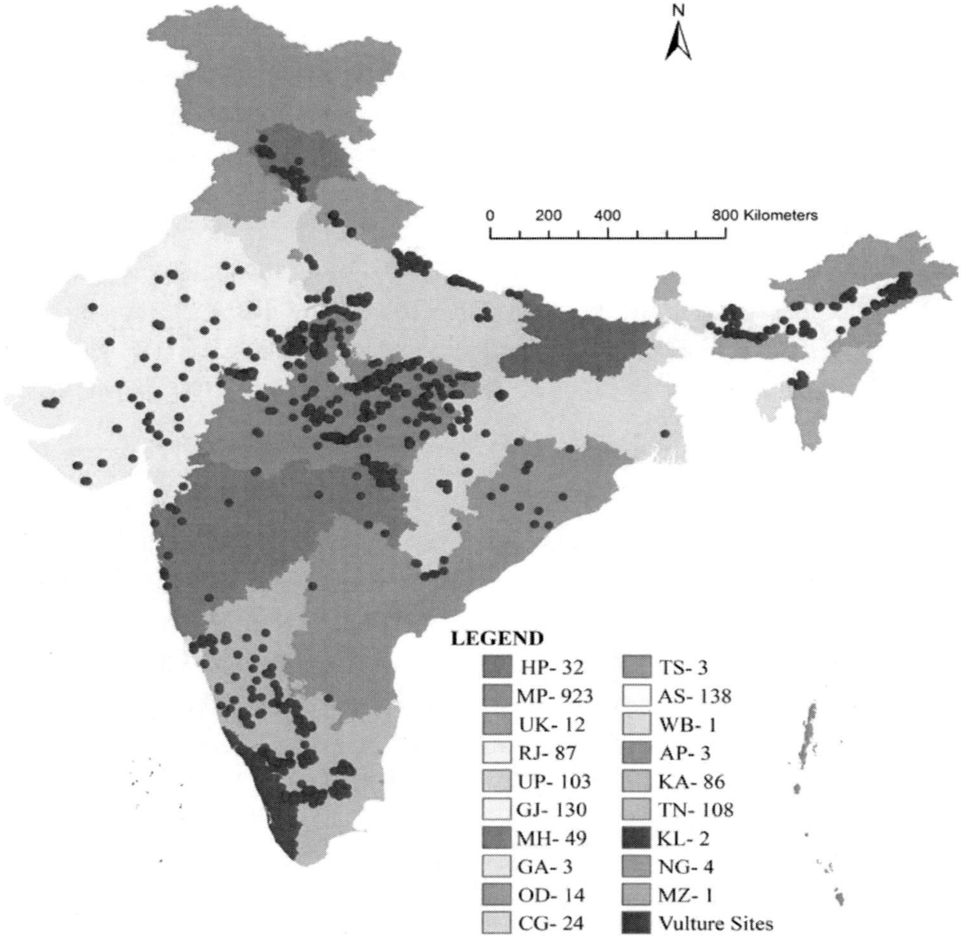

Figure 1. Base map of India showing vulture sites extracted from different studies. The latest reports from states have been compiled with assumption that the reported sites still exist and support at least the same number of vultures. Please note that in comparison to other states Madhya Pradesh supports maximum sites (and vultures), making it appear a vulture stronghold. HP – Himachal Pradesh, MP – Madhya Pradesh, UK – Uttarakhand, RJ – Rajasthan, UP – Uttar Pradesh, GJ – Gujarat, MH – Maharashtra, GA - Goa, OD – Odisha, CG – Chhattisgarh, TS – Telangana, AS – Assam, WB – West Bengal, AP – Andhra Pradesh, KA – Karnataka, TN – Tamil Nadu, KL – Kerala, NG – Nagaland, MZ – Mizoram.

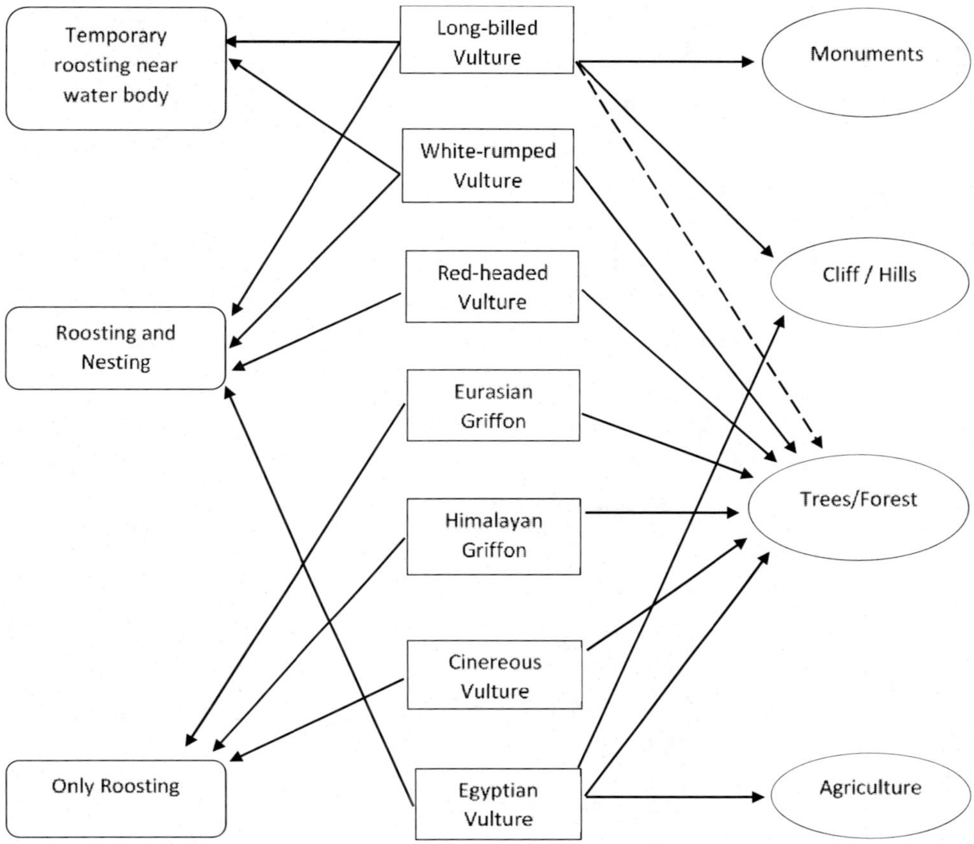

Figure 2. Chart depicting the relationship (arrows) among habitat use (ovals) and breeding status (rectangles with rounded corners) of different vulture species (rectangles) in Central India-MP. Broken arrow indicates occasional incidence.

METHODOLOGY

Protocol Assumption

The road transect method is useful for estimating vulture richness and abundance (Prakash et al., 2007) in large areas. This method requires less expenditure in time, human and monetary resources. A limitation is that road transects do not cover areas distant from roads. Census or complete enumeration methods (especially when maximum vultures are confined in forested area), avoid this problem provided the data from census is accurate. Census methods for population assessment has been adopted in the case of raptors (Di Vittorio et al., 2012), although in smaller areas and for small populations. Hyde et al. (2006) suggest that the traditional methods using field surveys are time consuming, but more accurate than the modern modelling methods as far as the structural measurement of forest (canopy volume, basal area, species diversity, etc.) for wildlife management is concerned.

In the present case study of the Indian vulture stronghold, the total enumeration method was adopted since huge manpower was available in the FD and undertook the survey alongside their regular duty. Monetary and other resources were also dovetailed in census work. The first assumption of the count protocol was that vultures leave roosts in the morning and return in the evening, and reuse roost and nest sites; migratory vultures remain in the destination country for at least a season; and vultures show seasonal and local movements in reaction to environmental conditions (Newton, 1979; Dodge et al., 2014; Saran, 2017). The second assumption was that people involved in counting and collecting other data were cognizant of the subject and objective of the survey and financial resource were sufficient to cover a large area. The third assumption was that the equipment was adequate for the data collection.

Data Collection

Vultures have attracted a strong interest from the point of view of conservation biology because of their recognized role in ecosystems as obligate scavengers and flagship species (Donazar et al., 2016). In India, their status as important actors in ecosystems was not recognized until their sudden decline in the first decade of current century and the observation of the environmental impacts of this decline. Even at this stage, the perception of the problem existed only at higher levels in the FD, while the execution and frontline staff were less aware of the issues. Therefore, it was a challenge for unskilled workers to collect the specialized data necessary for information development. Partly due to this situation, six regional workshops were conducted for training the trainers (ca. 240 resource persons) by experts (foresters, vulture biologists and veterinarians) to impart the knowledge of vulture biology, habitat, diseases and identification; and data collection in a predefined pro forma. Ten more workshops were organized at local levels to train the frontline staff by the trained trainers. The participants were instructed on the methodology of the single day census along following lines:

(i) Identification of potential and actual vulture sites in all the beats (smallest territorial unit of the FD) by gathering current evidence and historical records. Registering these sites by noting their longitude and latitude, and name of the beat.

(ii) Recording other important features, such as the nearest human settlements, waterbodies, high tension power towers, windmills, metaled or seasonal roads, etc.

(iii) Composing the count team of a minimum of two persons (one observer and another data recorder) and involving, preferably, one resource person in the team from outside the FD.

(iv) Fixing of a single count date for the state (the largest possible area) to avoid repeat counts and scheduling counts for winter and summer to capture vulture residency proportions.

(v) Issuing the count guidelines, including the time of the count (between 6:00 am to 10:00 am, to capture all the individuals roosting or nesting before late morning departure), counting only resting vultures, as the counting of flying birds may result in multiple counting by neighboring teams.

(vi) Providing essential instruments to all the teams, for example, binoculars, GPS sets, cameras and predesigned pro forma for data recording, etc.

(vii) Appointing the team of observers as 'check and balances' during counts in very important areas, which are those with high or low vulture occurrence.

(viii) Fixing the very next day of the census for returning the data electronically from field to the headquarters via controlling offices.

Mapping and Modelling

Predicting the spatial distribution of animals is an important and widely used tool, with applications in wildlife management, conservation, and population health (Morris et al., 2006). Rodriguez et al. (2007) reviewed that predictive models have been successfully used in a variety of conservation biology studies, with the main purpose of addressing pressing conservation problems (Brotons et al., 2004; Linkie et al., 2006). However, confronted with extremely limited data on the distribution, abundance, and dynamics of most species on earth (Rodríguez, 2003; Mace et al., 2005), models allow for the extrapolation of relatively few field samples and museum records to the entire potential range of a species, thus creating predicted distribution maps (Guisan & Zimmermann, 2000; Heglund, 2002; Anderson et al., 2003; Guisan & Thuiller, 2005).

Spatial modelling approaches have been successfully demonstrated in ecology, biogeography, evolution and conservation ecology (Guerrero-Casado et al., 2013). Spatial modelling of raptors is important for their conservation, especially when the spatiotemporal distribution of species is not documented sufficiently over large landscapes. Further, understanding the relationships between habitat and the distribution of raptors may become important for the successful management of forests and other landscapes (Liminana et al., 2011; Di Vittorio et al., 2012). Better knowledge of species distributions at regional extents may be of great value in several disciplines, such as the design of nature reserve networks or the assessment of environmental impacts (Rodrıguez et al., 2007).

For the development of successful conservation programs, species distribution modelling has been increasingly used to identify suitable habitat and to predict potential distribution (Austin, 2002; Rushton et al., 2004). Species distribution models can provide critical baseline distribution information for the conservation of poorly understood species (Coxen et al., 2017). Predictive species distribution models have been used in a wide range of applications including regional biodiversity assessments, conservation biology, wildlife management and conservation planning (Rodriguez et al., 2007). On the other hand, habitat suitability models can use environmental variables to predict the presence/absence or the abundance of a species throughout a study area (Hirzel et al., 2008). Habitat modelling based on statistical techniques and Geographic Information System (GIS) tools have important potential applications, such as forecasting species occurrence in poorly documented areas, the identification of adequate target areas for species reintroduction or conservation programs or predicting the climatic change impact (Rodriguez et al., 2007; Munoz et al., 2013).

Technology Used

The rapid development of spatial technologies in recent decades has made available new tools and capabilities, especially the evolution of geographic information systems (GIS), the global positioning system (GPS), and remote sensing (RS) technologies, which increased the effectiveness and efficiency of field data collection and analysis (Milla et al., 2005). In the current project, a GIS-GPS-RS combination was used to analyze natural resource data and generate different types of information and management maps. GIS applications were used for the storage, management, and analysis of spatially distributed data. GPS is a major advance in mapping technique for wildlife biology because it is simple to use, cheap, lightweight and can give instantaneous, very accurate, locational information in real time (Hulbert & French, 2001).

Different categories of maps were generated during the study, for the presentation of results and promotion of the methodology of the study. These included base maps, such as the state map with district boundaries, which was prepared using Survey of India topo-sheets. These were further used in understanding the richness and abundance of the species in different regions and zones categorized by agriculture or ecological features. There were also landcover maps. These were thematic maps showing features such as forest cover, agriculture areas, and water bodies (Figure 3) useful for species presence prediction. These were generated from remote sensing data by visual image interpretation and ground truthing (Reddy et al., 2016). A thematic map focuses in a specific idea or theme and illustrates a specific subject unlike the general map, in which the variety of geological and geographical phenomena appear together. Thematic maps also emphasize spatial variations of one or a small number of geographic distributions (Suryawanshi and

Bhutada, 2010). Prediction maps were created, which generally help in predicting the suitability of habitat for species. Such maps are model-based using variables such as rainfall, temperature, vegetation, elevation, food availability, reduced threats, etc. relating to target species, to predict the suitability of, for example, habitat for a specific species in terms of abundance, roosting, nesting, etc. (Lauria et al., 2015). ArcGIS, a computer program, was used in the current project for its high efficiency in creating and using maps, compiling geographic data, analyzing mapped information, sharing and discovering geographic information, using maps and geographic information in a range of applications, and managing geographic information in a database.

Selected Study Variables

The distribution of vulture species with respect to various habitat parameters was considered for the spatial modelling. Based on the field survey, certain attributes were collected, and some variables were also retrieved from spatial data analysis. As per the pre-survey knowledge on the vulture habitats, visits were made to all the 51 districts of the state to collect the geo-coordinates of vulture sightings. The observations were made during two different seasons, namely summer and winter. The geo-coordinates collected were stored in computers along with habitat parameters like vegetation composition, distance of road/waterbody, slope and aspect, etc.

The landuse/land cover map of the state was generated to document the habitat, using satellite remote sensing data as pre-defined land cover class, i.e., forest cover, agriculture land and water body, including perennial and ephemeral rivers. The forests types mapped included tropical dry deciduous forests, tropical moist deciduous forests and tropical thorn forests. About 5% of the forest area in the State has been earmarked as Protected Areas for wildlife conservation. Although the flagship species in different PAs is the tiger, vulture populations are also supported in considerable numbers. Therefore, boundary/extent of selected PAs was also identified for further use. Since the climatic conditions influence the vulture populations, all of the 11 agro-climatic regions of the State were delineated digitally.

Mapping Challenges

Around 3000 persons, eighty percent of them frontline staff and other lower level workers, were engaged and more than 6000 data sheets, each containing 15-20 parametric fields, were used for information collection during the pre-census survey, questionnaire survey and two censuses. The management of this huge data, collected by a large survey group over the vast forest area (77,700 km^2 spread over whole state) was a - herculean exercise. However, this was possible due to the involvement of the whole forest force headed by the Principal Chief Conservator of Forests, of Madhya Pradesh.

The use of the fixed pro forma did not collect only uniform field data, but heterogeneous data were collected creating classification and analytical problems. For example, vernacular names of the nesting and roosting trees varied regionally, creating problems for the surveyors. For example, *Terminalia arjuna* has regionally variable local names, such as *Arjun*, *Kahua*, *Kahu*, *Kahwar*, *Koha*, *Kau,* etc.). Equalization to *Terminalia arjuna* was necessary for the software-based analysis. Coordinates (longitude and latitude) of points such as roosting or nesting sites were also recorded variably like "degree, minutes and seconds" and "degree decimal" since the latter is the requirement of processing software – ArcGIS. Vulture species identification, including the distinction between adults and juveniles was sometimes difficult, due to distance and natural barriers such as high cliffs and tall trees with dense canopies. Repeat visits were sometimes required to correct errors.

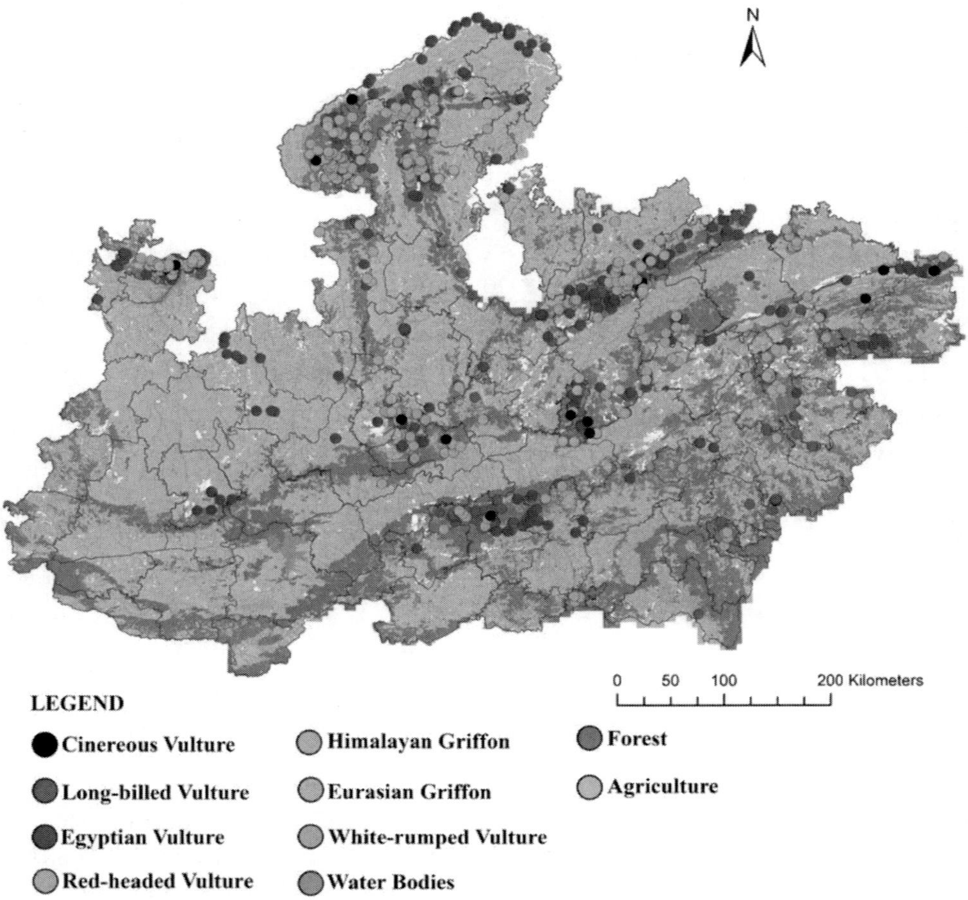

Figure 3. A landcover map of Indian vulture stronghold (MP) with vulture sites of different species. Please note the presence of maximum sites in the forests and a few in the agriculture field.

MAPPING OUTPUT

The following maps and information bases were created from the analysis of the field and supporting data.

(i) Vulture richness and abundance and their spatial as well as seasonal distribution for Central India-Madhya Pradesh, useful for future referencing during any study indicating changes or trends in the status of vultures.

(ii) Forest and non-forest distributions of vultures; out of 7028 vultures, 87% were recorded in forest area and 13% in agricultural or other non-forest landcover, contrary to the common belief, based on anecdotal narratives, that vultures are non-forest species.

(iii) Forty five percent of the vultures recorded in forest were found in Protected Areas (PAs) which represent 5% of the forest area of the state, indicating that PAs are effective in vulture conservation.

(iv) Vultures were mostly recorded near water bodies, while nesting and roosting on trees or cliffs; the minimum distance between vulture sites and waterbodies and human settlements was between a few meters to 8 km and 15 km, respectively.

(v) Adult and juvenile distributions, varied in different ecozones, indicating the recruitment status of the species. The adult and juvenile proportion (juveniles per hundred adults: jpha) or fresh recruitment (juvenile percentage in total population: (jptp) for all species in the state averaged 12 jpha and 11 jptp, respectively.

(vi) The distribution of active, inactive and abandoned nests provided information on breeding and non-breeding population. The average number of these nests was 459, 641 and 225, respectively.

(vii) The abundance of vulture species in different eco-zones indicated the possible influence of different forest structures and climatic factors. Very dense forests and high rainfall areas had relatively lower numbers of vultures.

(viii) Changes in the number of sites during winter and summer counts indicated vulture movements in summer with reference to winter sites. While 832 sites were recorded in winter, 91 sites were documented as new sites in summer and 120 winter sites were found deserted.

(ix) Cattle population maps of districts surrounding the PAs indicated additional availability of food. Neelgai (*Boselaphus tragocamelus*), Wild pig (*Sus scrofa*), Langur (*Semnopithecus antellus*), Cheetal (*Axis axis*), Sambar (*Rusa unicolor*) and chinkara (Gazella bennettii) were common in different PAs with varying density (2.1 km^{-2} to 75.2 km^{-2}). Cattle populations in the neighboring districts varied from 234,159 to 1,327,461 (DAHD 2012).

(x) Maps of threat structures, like windmills and high- tension power towers in the districts suggested the need for management intervention for bird collision reduction.

Use of Maps

Velevsky et al. (2014) analyzing species habits and habitats, argued that generally, resident species are affected by direct human induced mortality, and migratory species are affected by changes in productivity, degradation of habitats, livestock and ungulate carcass availability, persecution and electrocution. Therefore, mapping of vultures, their habit and habitat, is very useful for knowledge of ecosystem relations and the bases of vulture management and conservation. Indian forests are mostly managed for forest products and wildlife diversity conservation, guided by well-prepared termed working plans (for territorial divisions) and management plans (for wildlife divisions). The management plan, generally, revolves around the flagship species, for example, tiger, rhino, elephant, lion, etc. But, minor species like vultures are less mentioned by conservation planners, may be for want of awareness or lack of data on species management. Therefore, information derived from vulture censuses or otherwise could be included in the management plans for recovery and further conservation efforts of vultures. Useful maps include some of the following.

Threat Maps

Mapping vulture locations supports knowledge of their proximity to wind farms, roads and other developmental activities. Although wind farms are the preferred option for alternate energy generation (due to the reduction of air pollution and fossil fuel consumption), they contribute to raptor population reduction (Leddy et al., 1999; Laiolo & Tella 2006; Kuvlesky et al., 2007; Garvin et al., 2011). This is especially problematic in areas with high vulture population densities (Carrete et al., 2012; Reid et al., 2015) and during the migratory period (de Leucas et al., 2012). Power-lines with tall towers also contribute to vulture injuries and deaths in some parts of India and are the leading cause of injuries in some areas (Naidoo et al., 2011; Saran & Purohit, 2012). Therefore, it is apparent that there is need of applying the precautionary principle during windfarm planning to minimize the impact on long lived species (Carrete et al., 2010, 2012). Suggested measures include the establishment of rescue and rehabilitation centers. Although the rehabilitation rate reported is not very high, the true value of rehabilitation efforts is their ability to protect the species, as even a single bird saved contributes to the continued maintenance of genetic diversity within the threatened population (Naidoo et al., 2011). Selective stopping of turbines during nearby vulture aggregation could also be done with minimal loss (de Leucas et al., 2012). Bird-flight diverters could also be established to reduce collision rates, as attempted in South Africa for cranes and bustards (Anderson, 2002 in DeVault et al., 2016) of vultures.

Arroyo and Razin (2006) envisaged that human development and human recreational use of nature are likely to increase in the future, hence, it is important to understand response of wildlife to human activities. Some workers (Layna and Rico, 1991; Donazar et al., 1993; Bridgeford & Bridgeford, 2003) argue that human-created disturbances have been influencing the breeding success of vultures. Cinereous vultures are affected by anthropogenic disturbances; hence they select breeding areas far away from roads and villages (Moran-Lopez et al., 2006; Margalida et al., 2011; Guerrero-Casado et al., 2013). Sources of such human-created disturbances include heavy traffic on nearby roads. However, breeding bearded vultures were unaffected by noisy activity beyond 500-700 m. Therefore, management interventions may concentrate on keeping breeding colonies away from disturbances, at least during breeding season, by diverting the vehicular traffic.

Spatial distributions of vultures indicate the favorable range for the species where management intervention could be intensified. Recently, the biggest threat for Indian vultures has been the drug, Diclofenac, administered on the cattle, which causes the unintentional poisoning of vultures (Saini et al., 2012). The current research indicates that most of the vultures are confined to forest areas where there are few domestic cattle. Diclofenac use could be banned in such areas (including action against its illegal use, as veterinary use of this drug is prohibited in India, Taggart et al., 2007; DeVault et al., 2016) and in nearby areas (livestock grazing inside forest) and the villages (established or not relocated) within the forests.

Effective nature conservation in human dominated landscapes require a deep understanding of human behavior, perceptions and values. Human-wildlife conflict is a relatively well studied global scale challenge (Santangeli et al., 2016). Therefore, maps are needed of the unintentional use of poison, other than diclofenac, used for killing wild animals and affecting vulture populations. This is especially the case for areas near Protected (Wildlife) Areas or forests, and in areas of wild-animal and human conflict. Unintentional poisoning occurs when vultures consume poison baits set out for other species or when they consume carcasses of animals that have died from poisoning. Poisons may be laid to kill animals that eat crops, such as elephants, buffalo and other herbivores (Botha et al., 2017). Such a study can play a key role in delivering the knowledge and evidence for the implementation of effective, timely conservation actions Ogada et al., 2016).

Resource Maps

Resource mapping covers the resources used by vultures in their life activities. For example, some tree species or tree associations are frequented for roosting and nesting; maps recording these would constitute vulture resource maps. The level of detail may vary and include tree size and associated vegetation. Common nesting and roosting trees are *Ficus religiosa*, *Boswellia serrata* and *Terminalia arjuna* in Central India and *Pinus roxburghii* in the temperate Himalayan region (Thakur and Narang, 2012), *Bombax ceiba* in the Himalayan *Tarai* region (Baral et al., 2005) and *Prosopis cineraria* in the arid regions (Chhangani, 2007; Murn et al., 2014). These are large trees that offer safety and security to vultures. Large trees in roosts likely enhance microclimate conditions for vultures by causing temperature inversions in upper branches on calm nights. Well-spaced branches allow vultures mobility in close quarters (Geiger, 1965 and Wright et al., 1986 in Thompson et al., 1990). Potential nesting trees near vulture colonies could be given the same protection as those already utilized by vultures. Poirazidis et al. (2004) suggest that optimal habitat for nesting is mature trees surrounded by opening or smaller trees. Such areas should be monitored, and canopy closure must be regulated.

Conservation of vultures depends not only on the protection of the colony and mature trees for nesting, but also the surrounding habitat, in terms of food availability, like dehesas (an agro-pastoral ecosystem in Mediterranean Spain) for adequate foraging (Carrete & Donazar, 2005). Mapping of other resources such as potential food (e.g., the ungulates within and outside the forest) by using secondary data is also important. Such data is readily available from the Forest and Agriculture Department. Studies of the mortality rates of local animals could also help in the construction of food availability maps. Such maps could be used to locate areas with scanty food supplies, to determine potential locations for food supply centers or feeding stations popularly known as vulture restaurants. Cortes-Avizanda et al. (2016) has also recorded that one of the efforts and investments towards reversing or reducing the decline of vultures worldwide has been supplementary feeding in the area where natural food is in shortage. Other secondary data, like location of windmills, power towers, etc. outside the forest could be collected from the Energy Department to use in identification of rescue and rehabilitation centers.

Vulture abundance, breeding sites and carcass shortages can be correlated using area mapping, and control measures could be taken up by establishing Vulture Restaurants (Becker, et al., 2009; Deygout et al., 2009; Garcia-Heras et al., 2013; Kane et al., 2015; Lieury et al., 2015; Moreno-Opo et al., 2015; Cortes-Avizanda et al., 2016 etc.). Although vulture restaurants could be an ecological trap (Fluhr et al., 2017), a, supply of a small quantity of food to mimic the original condition of temporal and spatial unpredictability is useful (Cortes-Avizanda et al., 2010) and supplementary food supply has been a very effective and important tool for conservation and reintroduction of avian scavengers (Cortes-Avizanda et al., 2016). However, the risk of a vulture consuming contaminated food will have to be eliminated, as a single carcass containing a high diclofenac residue level could completely decimate a wild (captive or reintroduced) *Gyps* colony (Saini et al., 2012). Mapping could be also useful in deciding the location of breeding areas, as the probability of visiting a feeding station is inversely related to the distance from breeding sites (Garcia-Heras et al., 2013).

Other Maps

In a recent study in the vulture stronghold of India (Jha, 2017), only direct observations of vulture-inhabited sites were used in the preparation of maps, which could be treated as representative of suitable sites for the vultures in Central India. Such sites may have biophysical and environmental features which is favored by specific species. These natural and anthropogenic variables (vegetation, food, disturbance, topography, etc.) along with climatic factors (rainfall, temperature, etc.) could be measured resulting into a set of attributes (Grinnell, 1917 in Hirzel et al., 2008; Guisan et al., 2000; Bridgeford & Bridgeford, 2003; Virani et al., 2012). By using the habitat characteristics of the small area and putting them in to an appropriate model, new suitable areas could be identified for focused conservation efforts. This would be helpful in expanding the vulture conservation area to be used by future populations after facilitative intervention, to increase the population from the current level. The species distribution modelling (a tool for conservation planning and resource management), owing to the broad use of geographical information systems in recent years, could be used for identifying suitable habitat and to predict potential distributions of vulture populations (Robertson et al., 2003, Rushton et al., 2004; Vaz et al., 2008).

The recovery of vulture populations under current circumstances is therefore a mammoth task. Increasing the population is a further challenge, as the rate of reproduction of vultures is very slow, compounded by the major environmental threats. Therefore, facilitation of high quality habitat conditions will be required, to create new territory for the recovering populations or reintroduced birds. Required supportive resources will include prospective management maps of suitable nesting and roosting areas, including large trees in open landcover removed from roads and human settlements. The human activities, if required, would have to be evaluated for negative impacts. For example, logging activities, if any, may have to be shifted to non-breeding seasons and selective logging would have to be done instead of clear felling (Mihoub et al., 2013).

Climate is one of the main factors for the geographical range of species (Walther et al., 2001 & 2002; Pearson and Dawson, 2003; Marquez et al., 2011). Birds may respond to climate change by changing wintering areas, migration routes and breeding grounds (Ahola et al., 2004; Gordo et al., 2005), undergoing changes in their phenology (Ptaszyk et al., 2003; Sanz et al., 2003; Tryjanowski et al., 2005; Zalakevicius et al., 2006; Gordo and Sanz, 2006; Halupka et al., 2008). Their local abundances and overall distributions may also change (Huges, 2000; McCarty, 2001; Lemoine et al., 2007; Virkkala et al., 2008; Barbet-Massin et al., 2009). Climate change may pose an additional direct threat to vultures and mammal distributions (reducing the availability of carrion and suitable foraging habitat), with indirect impacts through climate driven changes in vegetation or habitat, in addition to those threats mentioned in this chapter (Thuiller et al., 2006a,b; Chamaille-Jammes & Fritz, 2009; Phipps et al., 2017). In the case of the Cape and Bearded vulture, climate change may have played a role in their extinction in the southern African region based on long term increases of temperatures and changing rainfall patterns (Simmons & Jenkins, 2007; IPCC, 2014).

There is evidence that rainfall patterns influence vulture breeding success (Bridgeford & Bridgeford, 2003; Virani et al., 2012). In terms of temperature change, breeding Cape vultures suffer increased levels of heat stress in higher temperatures with longer exposure to sunlight (Chaudhry, 2007 in Phipps et al., 2017). Cape vulture nest sites at higher elevations are more likely to be occupied and breeding success is higher when they are more sheltered from extreme weather conditions (Pfeiffer et al., 2017). Increased temperatures and carbon dioxide levels enhance woody vegetation cover, inhibiting the visual foraging of vultures by obscuring carcasses (Schultz, 2007 in Phipps et al., 2017; Bamford et al., 2009b; Midgley and Bond, 2015). Model-based maps of climatic and habitat variables are therefore important for monitoring and management of the vulture habitats.

CONCLUSION

This chapter has examined the issues concerning the status of vultures in Central India and looked at a research project that studied these issues. It argued for the vital importance of GIS derived maps in the portrayal and evaluation of vulture ecology and management strategies for vulture conservation. Central India was selected as a case study, due to its importance in the recent declines of vulture populations and the ameliorative actions that have since been employed. The evidence presented indicates that there are certain hypotheses related to vulture habits and habitat. Therefore, reference maps are useful for the comparison of past records with the present status of vulture populations and testing of hypotheses.

One hypothesis is that vultures repair the older nests and reuse them if required. Therefore, recording of such nests over time may allow the refutation or acceptance of the hypothesis that vultures reuse nesting sites. A second hypothesis is that during the summer season, vultures move to areas near large, permanent waterbodies. Reference maps of distribution may allow the acceptance or refutation of this hypothesis. A third hypothesis is that migratory vultures return to their home during summer. Cinereous Vultures were seen during summer count of 2016. Therefore, further monitoring and mapping may detail and possibly explain the factors for the longer presence in the destination country. A fourth hypothesis is that Himalayan Griffon vultures stay confined in the adjacent landscapes of Himalayan *Tarai* states of India. Their presence in central India (up to 23°N) further south needs to be assessed, as to whether the species is expanding its territory southwards, where the average temperature is higher than that of its native range.

Certain recommendations also emerge from the materials presented in this study. The most important issue concerns the balancing of the ecosystem and the maintenance of the vultures' economic services. To ensure this outcome, their conservation and management guidelines and efforts must aim for certain objectives. Reference maps must be created that serve as the bases for management tools and indicators supportive of policy decisions on vulture conservation and monitoring. Old, large trees in open forests must be conserved to enable the current and future expansion of vulture ranges. The protection of vulture colonies must be upgraded to Wildlife Sanctuary status. Disturbance of breeding vultures must be minimalized or prevented, and these policies extended to potential vulture habitats. There must also be continuous monitoring of Diclofenac and other poisons, and the promotion of a zero-tolerance policy. Vulture restaurants for supplementary feeding may be established as a last resort. Traditional pastoralism may be encouraged (Mateo-Tomas and Olea, 2009), without violating the conservation of Protected Areas (Pykala, 2000 and Steinfeld et al., 2010 in Mateo-Tomas and Olea, 2015). Abandonment of such practices negatively influences the species distribution (Mateo-Tomas and Olea, 2015). All stakeholders will have to be organized for vulture centric development. Finally, the management will have to prioritize the conservation of vulture species in the PAs on the above lines. More concerted efforts are also needed in the forests outside the PAs, in acknowledgment that the potential habitat of the species is much larger in area (> 90% area) than the area within the PAs, and future populations may require this external landcover.

ACKNOWLEDGMENTS

I sincerely acknowledge the discussion input from Drs Advait Edgaonkar, Ashish David and Ompraksh Maidguni, and the mapping input of Shri Ankit Sinha, all from IIFM Bhopal. Academic input from Ms Radhika Jha in building up the structure of this review is highly appreciated. Quality editing of the manuscript by Dr Campbell is duly admired.

REFERENCES

Ahola, M., Laaksonen, T., Sippola, K., Eeva, T., Rainio, K. & Lehikoinen, E. (2004). Variation in climate warming along the migration route uncouples arrival and breeding dates. *Global Change Biology*, *10*, 1-8.

Altwegg, R., Jenkins, A. & Abadi, F. (2014). Nest boxes and immigration drive the growth of an urban Peregrine Falcon *Falco peregrinus* population. *Ibis*, *156*, 107-115.

Anderson, R., Lew, D. & Peterson, A. (2003). Evaluating predictive models of species' distributions: criteria for selecting optimal models. *Ecological Modelling*, *162*, 211–232.

Arroyo, B. & Razin, M. (2006). Effect of human activities on bearded vulture behavior and breeding success in the French Pyrenees. *Biological Conservation*, *128*, 276-284.

Austin, M. P. (2002). Spatial prediction of species distribution: an interface between ecological theory and statistical modelling. *Ecological Modelling*, *157*, 101–118.

Bamford, A. J., Monadjem, A., Anderson, M. D., Anthony, A., Borello, W. D., Bridgeford, M., Bridgeford, P., Hancock, P., Howells, B., Wakelin, J. & Hardy, I. C. W. (2009a). Trade-offs between specificity and regional generality in habitat association models: a case study of two species of African vulture. *Journal of Applied Ecology*, 46, 852–860. doi: 10.1111/j.1365-2664.2009.01669.x.

Bamford, A. J., Monadjem, A. & Hardy, I. C. W. (2009b). An effect of vegetation structure on carcass exploitation by vultures in an African savanna. *Ostrich*, *80*, 135–137.

Baral, N., Gautam, R. & Tamang, B. (2005). Population status and breeding ecology of White-rumped Vulture *Gyps bengalensis* in Rampur Valley, Nepal. *Forktail*, *21*, 87–91.

Barbet-Massin, M., Walther, B. A., Thuiller, W., Rahbek, C. & Jiguet, F. (2009). Potential impacts of climate change on the winter distribution of Afro-Palaearctic migrant passerines. *Biology Letters*, *5*, 248-251.

Becker, N., Choresh, Y., Bahat, O. & Inbar, M. (2009). Economic analysis of feeding stations as a means to preserve an endangered species: The case of Griffon Vulture (*Gyps fulvus*) in Israel. *Journal for Nature Conservation*, *17*, 199-211.

Botha, A. J., Andevski, J., Bowden C. G. R., Gudka, M., Safford, R. J., Tavares, J. & Williams, N. P. (2017). *Multi-species Action Plan to Conserve African-Eurasian Vultures*. CMS Raptors MOU, Abu Dhabi, United Arab Emirates.

Bowden, C. (2012). Vulture population trends and current status in the wild – a global and regional perspective. *Symposium on Developing a Regional Response to the Conservation of South Asia's Critically Endangered Vulture Species*, 3rd – 4th May 2012, New Delhi.

Bridgeford, P. & Bridgeford, M. (2003). Ten years of monitoring breeding Lappet-faced Vultures *Torgos tracheliotos* in the Namib-Naukluft Park, Namibia. *Vulture News*, *48*, 3-11.

Brotons, L., Manosa, S. & Estrada, J. (2004). Modelling the effects of irrigation schemes on the distribution of steppe birds in Mediterranean farmland. *Biodiversity and Conservation*, *13*, 1039–1058.

Campbell, M. O. (2017). The conservation biology of vultures. In: *Biological Conservation in 21st Century: A Conservation Biology of Large Wildlife.* (Ed) MO Campbell. Nova Science Publishers New York.

Carrete, M. & Donazar, J. A. (2005). Application of central-place foraging theory shows the importance of Mediterranean dehesas for the conservation of the cinereous vulture, *Aegypius monachus*. *Biological Conservation*, *126*, 582–590.

Carrete, M., Sanchez-Zapata, J. A., Benitez, J. R., Lobon, M., Camina, A., Lekuona, J. M., Montelio, E. & Donazar, J. A. (2010). The precautionary principle and wind-farm planning: data scarcity does not imply absence of effects. *Biological Conservation*, *143*, 1829-1830.

Carrete, M., Sanchez-Zapata, J. A., Benitez, J. R., Lobon, M., Montoya, F. & Donazar, J. A. (2012). Mortality at wind-farms is positively related to large-scale distribution and aggregation in griffon vultures. *Biological Conservation*, *145*, 102–108.

Chamaille-Jammes, S. & Fritz, H. (2009). Precipitation-NDVI relationships in eastern and southern African savannas vary along a precipitation gradient. *International Journal of Remote Sensing*, *30*, 3409-3422.

Chaudhry, M. J. I. (2007). *Are Cape Vultues (Gyps coprotheres) Feeling the Heat? Behavioural Differences at North and South Facing Colonies in South Africa.* University of Cape Town, Cape Town, South Africa.

Chaudhry, M. J. I., Ogada, D. L., Mallik, R. N., Virani, Z. & Giovani, D. (2012). First evidence that populations of critically endangered long-billed vulture *Gyps indicus* have increased following the ban of the toxic veterinary drug diclofenac in south Asia. *Bird conservation International*, *22*, 389–397. doi: 10.1017/S095927091 2000445.

Chhangani, A. K. (2007). Sightings and nesting sites of Red-headed Vulture *Sarcogyps calvus* in Rajasthan, India. *Indian Birds*, *3*(6), 218–221.

Chhangani, A. K. (2009). Status of vulture population in Rajasthan, India. *Indian Forester*, *135*(2), 239-251.

Chhangani, A. K. & Mohnot, S. M. (2008). Demography of migratory vultures in and around Jodhpur, India. *Vulture News*, *58*, 23-34.

Cortes-Avizanda, A., Carrete, M. & Donazar, J. A. (2010). Managing supplementary feeding for avian scavengers: Guidelines for optimal design using ecological criteria. *Biological Conservation*, *143*(7), 1707-1715. DOI 10.1016/j.biocon.2010.04.016.

Cortes-Avizanda, A., Blanco, G., DeVault, T. L., Markandya, A. Virani, M. Z., Brandt, J. & Donazar, J. A. (2016). Supplementary feeding and endangered avian scavengers: benefits, caveats, and controversies. *Frontiers in Ecology and the Environment*, *14*(4), 191–199, doi:10.1002/fee.1257.

Cortes-Avizanda, A., Martin-Lopez, B., Ceballos, O. & Pereira, H. M. (2018). Stakeholders perceptions of the endangered Egyptian vulture: Insights for conservation. *Biological Conservation*, *218*, 173-180. DOI 10.1016/j.biocon.2017. 09.028.

Coxen, C. L., Frey, J. K., Carleton, S. A. & Collins, D. P. (2017). Species distribution models for a migratory bird based on citizen science and satellite tracking data. *Global Ecology and Conservation*, *11*, 298-311.

DAHD. (2012). Department of Animal Husbandry, Dairying and Fisheries. *19th Livestock Census District wise Report 2012*. http://dahd.nic.in/documents/statistics/livestock-census. Retrieval date: 22/03/2018.

Das, S. K., Dashahare, A., Marathe, S., Kundu, N. & Kesharwani, R. (2011). Status of Raptors with Special Reference to Vultures in and around Rajaji National Park, India. *World Journal of Zoology*, *6* (4), 350-356.

de Lucas, M., Ferrer, M., Bechard, M. J. & Munoz, A. R. (2012). Griffon vulture mortality at wind farms in southern Spain: Distribution of fatalities and active mitigation measures. *Biological Conservation*, *147*(1), 184-189.

Demerdzhiev, D., Stoychev, S., Dobrev, D., Spasov, S. & Terziev, N. (2014). Conservation measures undertaken to improve the population status of eastern imperial eagle (*Aquila heliaca*) in Bulgaria. *Slovak Raptor Journal*, *8*, 27–39.

Deygout, C., Gault, A., Sarrazin, F. & Bessa-Gomes, C. (2009). Modeling the impact of feeding stations on vulture scavenging service efficiency. *Ecological Modelling*, *220*, 1826–1835.

DeVault, T. L., Beasley, J. C., Olson, Z. H., Moleon, M., Carrete, M., Margalida, A. & Sanchez-Zapata, J. A. (2016). Ecosystem Services Provided by Avian Scavengers. In: *Why Birds Matter: Avian Ecological Function and Ecosystem Services*. C. H. Şekercioğlu, D. G. Wenny, and C. J. Whelan, eds. University of Chicago Press, Chicago, IL, USA., 2016, p. 235-270.

Di Vittorio, M., Maurizio, S. & Lopez-Lopez, P. (2012). Habitat preferences of Bonelli's Eagles *Aquila fasciata* in Sicily. *Bird Study*, *59*, 207–217.

Dodge, S., Gil, B., Keith, B., Davidson, S. C., Weinzierl, R., Bechard, M. J., Barber, David., Kays, R., Brandes, D., Han, J. & Wikelski, M. (2014). Environmental drivers of variability in the movement ecology of turkey vultures (*Cathartes aura*) in North and South America. *Philosophical Transactions of the Royal Society*, *B*, *369*, 20130195.

Donazar, J. A., Hiraldo, F. & Bustamante, J. (1993). Factors influencing nest site selection, breeding density and breeding success in the bearded vulture (*Gypaetus barbatus*). *Journal of Applied Ecology*, *30*, 504-514.

Fluhr, J., Benhamou, S., Riotte-Lambert, L. & Duriez, O. (2017). Assessing the risk for an obligate scavenger to be dependent on predictable feeding sources. *Biological Conservation*, *215*, 92–98.

Garcia-Heras, M. S., Cortes-Avizanda, A. & Donazar, J. A. (2013). Who Are We Feeding? Asymmetric Individual Use of Surplus Food Resources in an Insular Population of the Endangered Egyptian Vulture *Neophron percnopterus*. *PLOS ONE*, *8*(11), e80523. doi: 10.1371/journal.pone.0080523.

Garvin, J. C., Jennelle, C. S., Drake, D. & Grodsky, S. M. (2011). Response of raptors to a windfarm. *Journal of Applied Ecology*, *48*, 199–209.

Geiger, R. (1965). *The climate near the ground.* Harvard Univ. Press, Cambridge, Mass. 611 pp.

Gonzalez, L. M., Margalida, A., Sanchez, R. & Oria, J. (2006). Supplementary feeding as an effective tool for improving breeding success in the Spanish imperial eagle (*Aquila adalberti*). *Biological Conservation*, *129*, 477–486.

Gordo, O., Brotons, L., Ferrer, X. & Comas, P. (2005). Do changes in climate patterns in wintering areas affect the timing of the spring arrival of trans-Saharan migrant birds? *Global Change Biology*, *11*, 12-21.

Gordo, O. & Sanz, J. J. (2006). Climate change and bird phenology: a long-term study in the Iberian Peninsula. *Global Change Biology*, *12*, 1993–2004.

Green, R. E., Newton, I., Shultz, S., Cunningham, A. A., Gilbert, M., Pain, D. J. & Prakash, V. (2004). Diclofenac poisoning as a cause of vulture population declines across the Indian subcontinent. *Journal of Applied Ecology*, *41*(5), 793–800.

Green, R. E., Taggart, M. A., Senacha, K. R., Raghavan, B., Pain, D. J. & Cuthbert, R. (2007). Rate of decline of the oriental white-backed vulture population in India estimated from a survey of diclofenac residues in carcasses of ungulates. *PLOS ONE*, *2*(8), 686.

Green, R. E., Hunt, W. G., Parish, C. N. & Newton, I. (2008). Effectiveness of Action to Reduce Exposure of Free Ranging California Condors in Arizona and Utah to Lead from Spent Ammunition. *PLoS One*, *3*, e4022.

Grinnell, J. (1917). Field tests of theories concerning distributional control. *The American Naturalist*, *51*(602), 115–128.

Guerrero-Casado, J., Arenas, R. & Tortosa, F. S. (2013). Modelling the nesting-habitat of the Cinereous Vulture *Aegypius monachus* on a fine scale for conservation Purposes. *Bird Study*, *60*, 533–538.

Guisan, A. & Thuiller, W. (2005). Predicting species distribution: offering more than simple habitat models. *Ecology Letters*, *8*, 993–1009.

Guisan, A. & Zimmermann, N. E. (2000). Predictive habitat distribution models in ecology. *Ecological Modelling*, *135*, 147–186.

Halupka, L., Dyrcz, A. & Borowiec, M. (2008). Climate change affects breeding of reed warblers *Acrocephalus scirpaceus*. *Journal of Avian Biology*, *39*, 95–100.

Heglund, P. J. (2002). *Foundations of species-environment relations. Predicting species occurrences: issues of accuracy and scale* (ed. by J. M. Scott, P. J. Heglund, M. L. Morrison, J. B. Haufler, M. G. Raphael, W. A. Wall and F. B. Samson), pp. 35–41. Island Press, Washington, D.C.

Hill, J. E., DeVault, T. L., Beasley, J. C., Rhodes, Jr. O. E. & Belant, J. L. (2018). Effects of vulture exclusion on carrion consumption by facultative scavengers. *Ecology and Evolution*, 1-9. DOI: 10.1002/ece3.3840.

Hirzel, A. H. & Lay, G. L. (2008). Habitat suitability modelling and niche theory. *Journal of Applied Ecology*, *45*, 1372–1381 doi: 10.1111/j.1365-2664.2008.01524.x.

Hulbert, I. A. R. & French, J. (2001). The accuracy of GPS for wildlife telemetry and habitat mapping. *Journal of Applied Ecology*, *38*, 869–878.

Hughes, L. (2000). Biological consequences of global warming: Is the signal already apparent? Trends in *Ecology and Evolution*, *15*, 56–61.

Hyde, P., Dubayah, R., Walker, W., Blair, J. B., Hofton, M. & Hunsaker, C. (2006). Mapping forest structure for wildlife habitat analysis using multi-sensor (LiDAR, SAR/InSAR, ETM+, Quickbird) synergy. *Remote Sensing of Environment*, *102*, 63–73.

Inger, R., Cox, D. T. C., Per, E., Norton, B. A. & Gaston, K. J. (2016). Ecological role of vertebrate scavengers in urban ecosystems in the UK. *Ecology and Evolution*, *6*, 7015-7023. DOI: 10.1002/ece3.2414.

IPCC. (2014). *Climate Change 2014: Impacts, Adaptation, and Vulnerability. Part B: Regional Aspects.* Contribution of Working Group II to the Fifth Assessment Report of the Intergovernmental Panel on Climate Change. Barros et al., Eds. (Cambridge, UK.).

IUCN. (2017). *IUCN Red list of Threatened Species*. http://www.iucnredlist.org/search

Jha, K. K. (2015). Distribution of vultures in Uttar Pradesh, India. *Journal of Threatened Taxa*, *7*(1), 6750–6763.

Jha, K. K. (2017). *Vulture Atlas Central India* – MP. Indian Institute of Forest Management, Bhopal, India.

Jhala, Y. V., Qamar, Q. & Saini, S. (2017). *Status of vultures in tiger landscapes of India*. National Tiger Conservation Authority, New Delhi and Wildlife Institute of India, Dehradun.

Kane, A., Jackson, A. L., Monadjem, A., Colomer, M. A. & Margalida, A. (2015). Carrion ecology modelling for vulture conservation: are vulture restaurants needed to sustain the densest breeding population of the African white-backed vulture? *Animal Conservation*, *18*(3), 279–286. DOI: 10.1111/acv.12169.

Kuvlesky, W. P., Brennan, L. A., Morrison, M. L., Boydston, K. K., Ballard, B. M. & Bryant, F. C. (2007). Wind energy development and wildlife conservation: challenges and opportunities. *Journal of Wildlife Management*, *71*, 2487–2498.

Laiolo, P. & Tella, J. L. (2006). Fate of unproductive and unattractive habitats: recent changes in Iberian steppes and their effects on endangered avifauna. *Environmental Conservation*, *33*, 223–232.

Layna, J. R. & Rico, M. (1991). Incidencia de molestias humanas sobre territories de nidificacion de quebrantahuesos: vigilancia de nidos (Incidence of human discomfort on nesting grounds of bearded vultures: nest monitoring). In: Heredia, R., Heredia, B., El Quebrantahuesos (*Gypaetus barbatus*) en los Pirineos (The Lammergeier (*Gypaetus barbatus*) in the Pyrenees). *Coleccion Tecnica ICONA*, Madrid, pp. 109–115.

Leddy, K. L., Higgins, K. F. & Naugle, D. E. (1999). Effects of wind turbines on upland nesting birds in conservation reserve program grasslands. *Wilson Bulletin*, *11*, 100–104.

Lemoine, N., Bauer, H. G., Peintinger, M. & Bohning-Gaese, K. (2007). Effects of climate and land-use change on species abundance in a central European bird community. *Conservation Biology*, *21*, 495–503.

Lieury, N., Gallardo, M., Ponchon, C., Besnard, A. & Millon, A. (2015). Relative contribution of local demography and immigration in the recovery of a geographically-isolated population of the endangered Egyptian vulture. *Biological Conservation*, *191*, 349-356.

Liminana, R., Arroyo, B. E., Surroca, M., Urios, V. & Reig-Ferrer, A. (2011). Influence of habitat on nest location and reproductive output of Montagu's Harriers breeding in natural vegetation. *Journal of Ornithology*, *152*, 557–565.

Lindsey, G. (1992). Nest guarding from observation blinds: strategy for improving Puerto Rican Parrot nest success. *Journal of Field Ornithology*, *63*, 466–472.

Linkie, M., Chapron, G., Martyr, D. J., Holden, J. & Leader-Williams, N. (2006). Assessing the viability of tiger subpopulations in a fragmented landscape. *Journal of Applied Ecology*, *43*, 576–586.

Mace, G., Masundire, H., Baillie, J., Ricketts, T., Brooks, T., Hoffmann, M., Stuart, S., Balmford, A., Purvis, A., Reyers, B., Wang, J., Revenga, C., Kennedy, E., Naeem, S., Alkemade, R., Allnutt, T., Bakarr, M., Bond, W., Chanson, J., Cox, N., Fonseca, G., Hilton-Taylor, C., Loucks, C., Rodrigues, A., Sechrest, W., Stattersfield, Janse van Rensburg, B. & Whiteman, C. (2005). Biodiversity. *Ecosystems and human well-being: current state and trends*, *Vol. 1. Findings of the condition and trends working group* (ed. by R. Hassan, R. Scholes and N. Ash), pp. 77–122. Island Press, Washington, D.C.

Mateo-Tomas, P. & Olea, P. P. (2009). Combining scales in habitat models to improve conservation planning in an endangered vulture. *Acta Oecologica*, *35*, 489–498.

Mateo-Tomas, P. & Olea, P. P. (2015). Livestock-driven land use change to model species distributions: Egyptian vulture as a case study. *Ecological Indicators*, *57*, 331–340.

Margalida, A., Moreno-Opo, R., Arroyo, B. E. & Arredondo, A. (2011). Reconciling the conservation of endangered species with economically important anthropogenic activities: interactions between cork exploitation and the cinereous vulture in Spain. *Animal Conservation*, *14*, 167–174.

Markandya, A., Taylor, T., Longo, A., Murty, M. N., Murty, S. & Dhavala, K. (2008). Counting the cost of vulture decline—An appraisal of the human health and other benefits of vultures in India. *Ecological Economics*, *67*, 194-204.

Marquez, A. L., Real, R., Olivero, J. & Estrada, A. (2011). Combining climate with other influential factors for modelling climate change impact on species distribution. *Climatic Change*, *108*, 135–157.

Martinez-Abrain, A., Tavecchia, G., Regan, H. M., Jimenez, J., Surroca, M. & Oro, D. (2012). Effects of wind farms and food scarcity on a large scavenging bird species following an epidemic of bovine spongiform encephalopathy. *Journal of Applied Ecology*, *49*, 109–117.

Mateo-Tomas, P., Olea, P. P., Sanchez-Barbudo, I. S. & Mateo, R. (2012). Alleviating human–wildlife conflicts: Identifying the causes and mapping the risk of illegal poisoning of wild fauna. *Journal of Applied Ecology*, *49*, 376–385.

Mateo-Tomas, P., Olea, P. P., Moleon, M., Selva, N. & Sanchez-Zapata, J. A. (2017). Both rare and common species support ecosystem services in scavenger communities. *Global Ecology and Biogeography*, DOI: 10.1111/geb.12673.

McCarty, J. P. (2001). Ecological consequences of recent climate change. *Conservation Biology*, *15*, 320–331.

Midgley, G. F. & Bond, W. J. (2015). Future of African terrestrial biodiversity and ecosystems under anthropogenic climate change. *Nature Climate Change*, *5*, 823–829.

Mihoub, J. B., Jiguet, F., Lecuyer, P., Eliotout, B. & Sarazine F. (2013). Modelling nesting site suitability in a population of reintroduced Eurasian black vultures *Aegypius monachus* in the Grands Causses, France. *Oryx*, *48*(1), 116–124 doi: 10. 1017/S0030605312000634.

Milla, K. A., Lorenzo, A. & Brown, C. (2005). GIS, GPS, and Remote Sensing Technologies in Extension Services: Where to Start, What to Know. *Journal of Extension*, *43*(3), *16*. Feature Articles, 3FEA6. http://joe.org/joe/2005june/a6.php.

Morales-Reyes, Z., Sanchez-Zapata, J. A., Sebastian-Gonzalez, E., Botella, F., Carrete, M. & Moleon, M. (2017). Scavenging efficiency and red fox abundance in Mediterranean mountains with and without vultures. *Acta Oecologica*, *79*, 81-88.

Moran-Lopez, R., Sanchez, J. M., Costillo, E., Corbacho, C. & Villegas, A. (2006). Spatial variation in anthropic and natural factors regulating the breeding success of the cinereous vulture (*Aegypius monachus*) in the SW Iberian Peninsula. *Biological Conservation*, *130*, 169–182.

Moreno-Opo, R., Trujillano, A., Arredondo, A., Gonzalez, L. M. & Margalida, A. (2015). Manipulating size, amount and appearance of food inputs to optimize supplementary feeding programs for European vultures. *Biological Conservation*, *181*, 27–35.

Morris, L. R., Proffitt, K. M. & Blackburn, J. K. (2016). Mapping resource selection functions in wildlife studies: Concerns and recommendations. *Applied Geography*, *76*, 173-183.

Munoz, A. R., Marquez, A. L. & Real, R. (2013). Updating known distribution models for forecasting climate change impact on endangered species. *PLOS ONE*, *8*, e65462.

Murn, C., Saeed, U., Khan, U. & Iqbal, S. (2014). Population and spatial breeding dynamics of a Critically Endangered Oriental White-backed Vulture *Gyps bengalensis* colony in Sindh Province, Pakistan. *Bird Conservation International*, pp 1 – 11. DOI: 10.1017/S0959270914000483.

Naidoo, V., Wolter, K., Espie, I. & Kotze, A. (2011). Vulture rescue and rehabilitation in South Africa: An urban perspective. *Journal of the South African Veterinary Association*, *82*(1), 24–31.

Newton, I. (1979). *Population Ecology of Raptors*. Buteo Books. Vermillion, South Dakota 57069 USA.

Ogada, D. L., Keesing, F. & Virani, M. Z. (2012a). Dropping dead: causes and consequences of vulture population declines worldwide. *Annals of The New York Academy of Sciences*, *1249*(1), 57–71.

Ogada, D. L., Torchin, M. E., Kinnaird, M. F. & Ezenwa, V. O. (2012b). Effects of vulture declines on facultative scavengers and potential implications for mammalian disease transmission. *Conservation Biology*, *26*, 453–60.

Ogada, D., Shaw, P., Beyers, R. L., Buij, R., Murn, C., Thiollay, J. M., Beale, C. M., Holdo, R. M., Pomeroy, D., Baker, N., Kruger, S. C., Botha, A., Virani, M. Z., Monadjem, A. & Sinclair, A. R. E. (2016). Another Continental Vulture Crisis: Africa's Vultures Collapsing toward Extinction. *Conservation Letters*, *9*, 89-97.

Oro, D., Margalida, A., Carrete, M., Heredia, R. & Donazar, J. A. (2008). Testing the goodness of supplementary feeding to enhance population viability in an endangered vulture. *PLoS One*, *3*(12), e4084. https://doi.org/10.1371/journal.pone.0004084.

Pain, D. J., Cunningham, A. A., Donald, P. F., Duckworth, J. W., Houstan, D. C., Katzner, T., Pary-Jones, J., Poole, C., Prakash, V., Round, P. & Timmins, R. (2003). Causes and effects of temporospatial declines of *Gyps* vultures in Asia. *Conservation Biology*, *17*, 661–671.

Pande, S., Mestri, P., Deshpande, P., Warange, A. & Mahabal, A. (2013). Promising trend of in situ breeding of Oriental White-rumped Vulture *Gyps bengalensis* in Raigad District, Maharashtra, India: conservation implications for re-introduction of ex situ populations. *Journal of Threatened Taxa*, *5*(7), 4106–4109.

Pearson, R. G. & Dawson, T. P. (2003). Predicting the impacts of climate change on the distribution of species: Are bioclimate envelope models useful? *Global Ecology and Biogeography*, *12*, 361–371.

Pfeiffer, M. B., Venter, J. A. & Downs, C. T. (2017). Cliff characteristics, neighbor requirements and breeding success of the colonial Cape Vulture *Gyps coprotheres. Ibis*, *159*, 26-37.

Phipps, W. L., Diekmann, M., MacTavish, L. M., Mendelsohn, J. M., Naidoo, V., Wolter, K. & Yarnell, R. W. (2017). Due South: A first assessment of the potential impacts of climate change on Cape vulture occurrence. *Biological Conservation*, *210*, 16–25.

Poirazidis, K., Goutner, V., Skartsi, T. & Stamou, G. (2004). Modelling nesting habitat as a conservation tool for the Eurasian black vulture (*Aegypius monachus*) in Dadia Nature Reserve, northeastern Greece. *Biological Conservation*, *118*, 235–248.

Prakash, V., Pain, D. J., Cunningham, A. A., Donald, P. F., Prakash, N., Verma, A., Gargi, R., Sivakumar, S. & Ramani A. R. (2003). Catastrophic collapse of Indian White Backed (*Gyps bengalensis*) and Long Billed (*Gyps indicus*) Vultures population. *Biological Conservation*, *109*(3), 381-390.

Prakash, V., Green, R. E., Pain, D. J., Ranade, S. P., Saravanan, S., Prakash, N., Venkitachalam, R., Cuthbert, R., Rahmani, A. R. & Cunningham, A. A. (2007). Recent changes in populations of resident Gyps vultures in India. *Journal of the Bombay Natural History Society*, *104* (2), 129-135.

Prakash, V., Bishwakarma, M. C., Chaudhary, A., Cuthbert, R., Dave, R., Kulkarni, M., Kumar, S., Paudel, K., Ranade S., Shringarpure, R. & Green, R. E. (2012). The Population Decline of *Gyps* Vultures in India and Nepal Has Slowed Since Veterinary Use of Diclofenac was Banned. *PLoS One*, *7*(11), e49118. https://doi.org/10.1371/journal.pone.0049118.

Ptaszyk, J., Kosicki, J., Sparks, T. H. & Tryjanowski, P. (2003). Changes in the timing and pattern of arrival of the White Stork (*Ciconia ciconia*) in western Poland. *Journal of Ornithology*, *144*, 323–329.

Pykala, J. (2000). Mitigating human effects on European biodiversity through traditional animal husbandry. *Conservation Biology*, *14*, 705–712, http://dx.doi.org/10.1046/j.1523-1739.2000.99119.x.

Ramakrishnan, B., Kannan, G., Samson, A., Ramkumar, K. & Ramasubramanian, S. (2014). Nesting of White-Rumped Vulture (*Gyps bengalensis*) in the Segur Plateau of the Nilgiri North Forest Division, Tamilnadu, India. *Indian Forester*, *140*(10), 1014-1018.

Reddy, K. Y., Krishna, K. V. S. G. M. & Reddy, I. S. (2016). Remote Sensing and GIS Based Topograghicial Mapping For Environmental Management Study. *International Journal of Applied Engineering Research*, *11*(2), 1325-1334

Reed, T. M. & Rocke, T. E. (1992). The role of avian carcasses in botulism epizootics. *Wildlife Society Bulletin*, *20*, 175-182.

Reid, T., Kruger, S., Whitfield, D. P. & Amar, A. (2015). Using spatial analyses of bearded vulture movements in southern Africa to inform wind turbine placement. *Journal of Applied Ecology*, *52*, 881–892. doi: 10.1111/1365-2664.12468.

Robertson, M. P., Peter, C. I., Villet, M. H. & Ripley, B. S. (2003). Comparing models for predicting species' potential distribution: A case study using correlative and mechanistic predictive modelling techniques. *Ecological Modelling*, *164*, 153–167.

Rodriguez, J. P. (2003). Challenges and opportunities for surveying and monitoring tropical biodiversity — a response to Danielsen et al. *Oryx*, *37*, 411–411.

Rodriguez, J. P., Brotons, L., Bustamante, J. & Seoane, J. (2007). The application of predictive modelling of species distribution to biodiversity conservation. *Diversity and Distribution*, *13*, 243–251.

Rushton, S. P., Ormerod, S. J. & Kerby, G. (2004). New paradigms for modelling species distributions? *Journal of Applied Ecology*, *41*, 193–200.

Ruxton, G. D. & Houston, D. C. (2004). Obligate vertebrate scavengers must be large soaring fliers. *Journal of Theoretical Biology*, *228*, 431-436. DOI: 10.1016/j.jtbi.2004.02.005.

Saini, M., Taggart, M. A., Knopp, D., Upreti, S., Swarup, D., Das, A., Gupta, P. K., Niessner, R., Prakash, V., Mateo, R. & Cuthbert, R. J. (2012). Detecting diclofenac in livestock carcasses in India with an ELISA: A tool to prevent widespread vulture poisoning. *Environmental Pollution*, *160*, 11-16.

Santangeli, A., Arkumarev, V., Rust, N. & Girardello, M. (2016). Understanding, quantifying and mapping the use of poison by commercial farmers in Namibia – Implications for scavengers' conservation and ecosystem health. *Biological Conservation*, *204*(B), 205-211.

Sanz, J. J., Potti, J., Moreno, J., Merino, S. & Frias, O. (2003). Climate change and fitness components of migratory bird in the Mediterranean region. *Global Change Biology*, *9*, 461–472.

Saran, R. (2017). Population monitoring and annual population fluctuation of migratory and resident species of vultures in and around Jodhpur, Rajasthan. *Journal of Asia-Pacific Biodiversity*, *10*, 342-348.

Saran, R. P. & Purohit, A. (2012). Eco-transformation and electrocution: A major concern for the decline in vulture population in and around Jodhpur. *International Journal of Conservation Science*, *3*(2), 111-118.

Schultz, P. (2007). *Does Bush Encroachment Impact Foraging Success of the Critically Endangered Namibian Population of the Cape Vulture Gyps coprotheres?* Masters Project. University of Cape Town.

Sekercioglu, C. H. (2006). Increasing awareness of avian ecological function. *Trends in Ecology and Evolution*, *21*, 464-471.

Sekercioglu, C. H., Daily, G. C. & Ehrlich, P. R. (2004). Ecosystem consequences of bird declines. *Proceedings of the National Academy of Sciences of the United States of America, 101*, 18042-18047.

Simmons, R. E. & Jenkins, A. R. (2007). Is climate change influencing the decline of Cape and Bearded Vulures in southern Africa? *Vulture News*, *56*, 41-51.

Sivakumar, S. & Manakadan, R. (2005). Decline of vultures in Sriharikota, Nellore District, Andhra Pradesh, India. *Indian Birds*, *1* (1), 14.

Steinfeld, H., Mooney, H. A., Schneider, F. & Neville, L. E. (2010). *Livestock in a Changing Landscape. Drivers, Consequences and Responses.* Island Press, Washington.

Sterne, M. & Wentzel, M. (1949). Botulism in Animals in South Africa. *Report on the 14th International Veterinary Congress, London 8th to 13th August 1949*, 3, 329-331.

Subramanya, S. & Naveein, O. C. (2006). Breeding of Long-billed Vulture *Gyps indicus* at Ramanagaram hills, Karnataka, India. *Indian Birds*, *2*(2), 32-34.

Suryawanshi, S. L. & Bhutada, S. H. (2010). Development of thematic maps for a watershed using remote sensing and geographic information system. *International Journal of Agricultural Engineering*, *3* (1), 147-152.

Taggart, M. A., Senacha, K. R., Green, R. E., Jhala, Y. V., Raghavan, B., Rahmani, A. R., Cuthbert, R., Pain, D. J. & Meharg, A. A. (2007). Diclofenac residues in carcasses of domestic ungulates available to vultures in India. *Environment International*, *33*, 759–765.

Thakur, M. L. & Narang, S. K. (2012). Population status and habitat-use pattern of Indian white-backed vulture (*Gyps bengalensis*) in Himachal Pradesh, India. *Journal of Ecology and the Natural Environment*, *4*(7), 173-180.

Thompson, W. L., Yahner, R. H. & Storm, G. L. (1990). Winter Use and Habitat Characteristics of Vulture Communal Roosts. *The Journal of Wildlife Management*, *54*(1), 77-83.

Thuiller, W., Broennimann, O., Hughes, G., Alkemade, J. R. M., Midgley, G. F. & Corsi, F. (2006a). Vulnerability of African mammals to anthropogenic climate change under conservative land transformation assumptions. *Global Change Biology*, *12*, 424-440.

Thuiller, W., Midgley, G. F., Hughes, G. O., Bomhard, B., Drew, G., Rutherford, M. C. & Woodward, F. I. (2006b). Endemic species and ecosystem sensitivity to climate change in Namibia. *Global Change Biology*, *12*, 759–776.

Tryjanowski, P., Kuzniak, S. & Sparks, T. H. (2005). What affects the magnitude of change in first arrival dates of migrant birds? *Journal of Ornithology*, *146*, 200–205.

Vaz, S., Martin, C. S., Eastwood, P. D., Ernande, B., Carpentier, A., Meaden, G. J. & Coppin, F. (2008). Modelling species distributions using regression quantiles. *Journal of Applied Ecology*, *45*(1), 204-217.

Vaughan, C., Nemeth, N. M., Cary, J. & Temple, S. (2005). Response of a Scarlet Macaw *Ara macao* population to conservation practices in Costa Rica. *Bird Conservation International*, *15*, 119–130.

Velevski, M., Grubac, B. & Tomovic, L. (2014). Population Viability Analysis of the Egyptian Vulture *Neophron percnopterus* in Macedonia and Implications for Its Conservation. *Acta Zoologica Bulgarica*, *66* (1), 43-58.

Venkitachalam, R. & Senthilnathan, S. (2015). Breeding record of Indian vulture (*Gyps indicus*) in Moyar Valley, Tamil Nadu, India. *Current Science*, *109*(258), 258-259.

Virani, M. Z., Monadjem, A., Thomsett, S. & Kendall, C. (2012). Seasonal variation in breeding Ruppell's Vultures *Gyps rueppellii* at Kwenia, southern Kenya and implications for conservation. *Bird Conservation International*, *22*, 260-269.

Virkkala, R., Heikkinen, R. K., Leikola, N. & Luoto, M. (2008). Projected large-scale range reductions of northern-boreal land bird species due to climate change. *Biological Conservation*, *141*, 1343–1353.

Walther, G. R., Post, E., Convey, P., Menzel, A., Parmesan, C., Beebee, T. J. C., Fromentin, J. M., Hoegh-Guldberg, O. & Bairlein, F. (2002). Ecological responses to recent climate change. *Nature*, *416*, 389–395.

Walther, G. R., Burga, C. A. & Edwards, P. J. (2001). *Fingerprints of Climate Change. Adapted Behaviour and Shifting Species Ranges*. New York: Kluwer Academic/ Plenum.

Wright, A. L., Yahner, R. H. & Storm, G. L. (1986). Roost-tree characteristics and abundance of wintering vultures at a communal roost in south central Pennsylvania. *Raptor Research*, *20*, 102- 107.

Zalakevicius, M., Bartkeviciene, G., Raudonikis, L. & Janulaitis, J. (2006). Spring arrival response to climate change in birds: a case study from eastern Europe. *Journal of Ornithology*, *147*, 326–343.

Zuberogoitia, I., Zabala, J., Martinez, J. E., Gonzalez-Oreja, J. A. & Lopez-Lopez, P. (2014). Effective conservation measures to mitigate the impact of human disturbances on the endangered Egyptian vulture. *Animal Conservation*, *17*, 410–418.

In: Geomatics and Conservation Biology
Editor: Michael O'Neal Campbell

ISBN: 978-1-53614-468-0

Chapter 3

GEOSPATIAL MODELLING OF HUMAN-ELEPHANT CONFLICTS IN DALMA WILDLIFE SANCTUARY AND ITS SURROUNDINGS IN INDIA

Somak Das, C. Jeganathan, Saptarshi Mondal and Richa Sharma

Department of Remote Sensing, Birla Institute of Technology (BIT)
Mesra, Ranchi, Jharkhand, India

ABSTRACT

Conflicts between people and elephants are common in some parts of tropical Asia. This may occur when the elephant population is concentrated or increases, in regions with high human population densities or intensive land use. Dalma Wildlife Sanctuary, located near Jamshedpur city in Jharkhand State of India is known for its elephant (*Elephas maximus*, Linnaeus, 1758) habitat and large elephant population. The Dalma elephants migrate between the forested areas located in the border regions of Jharkhand, West Bengal, Odisha as per food and water availability, and climatic conditions. The present study in Dalma Wildlife Sanctuary and surroundings examines the migration patterns, recent migration systems, land use and land cover change for the period 1989 to 2014 and assesses suitable areas for Dalma elephants. The study is based on field surveys, published records and image-based analyses. The objective is to identify potential conflict spots which will support appropriate planning to avoid and reduce human elephant conflict in the study area. Geomatics-based analyses of such issues are vital for the study of the conservation biology of elephants in India and for understanding the general issues of human-large wildlife conflict

INTRODUCTION

Increasing human populations and associated developments and landcover change have led to degradation of natural resources, with negative impacts on the habitats, including food and shelter availability, of wildlife in nearby areas (Ellis & Pontius, 2007). One method for the analysis of such change is the use of geographical information systems or GIS (Diallo et al., 2009). The Asian elephant (*Elephas maximus*, Linnaeus, 1758) has been seriously affected by human populations, land use and land cover change, partly because of the elephants' size, and the vast food and habitat requirements of such large mammals (Calabrese et al., 2017). India has the largest population of Asian elephants and therefore this is an important research area (Baskaran et al., 2011). Human developments, including agriculture and settlements have encroached on the corridors used by elephants for movements, forcing the animals to enter the human populated areas. This leads to conflict, which have led to the deaths of both elephants and people.

As per the recent synchronized elephant population estimation in India (MOEFCC, 2017), the total number of elephants in India is 27312, encompassing 4 regional distributions (South: 11960, North-East: 10139, North-West: 2085, East-Central: 3128). The number of lives lost due to elephant-human conflict is highest in the east-central region where hundreds of elephants and people have been killed annually (Jharkhand Observer, 2011; The Pioneer, 2015). The increasing human populations in this area indicates that such conflicts with fatal results for both people and elephants may increase. GIS is an effective method for the analysis of such issues (Kushwaha et al., 2000; Kushwaha & Roy, 2002; Mongkolsawat & Chanket, 2007; Areendran et al., 2011; Zhao et al., 2014). Such studies have already been used for elephant studies, including India (Bist, 2006; Prasad et al., 2011; Roever et al., 2013) and for modelling habitat suitability for large mammals including elephants (Khanna et al., 2001; Diemer, 2003; Hazarika & Saikia, 2013; Imam & Kushwaha, 2013; Reza et al., 2013; Sitompul et al., 2013; Tikhile et al., 2013). This chapter examines the spatial issues behind this conflict using geo-spatial technology, taking a case study the Dalma Sanctuary of Jharkhand State in India (Government of India, 2012).

In Jharkhand state, at least 2 – 3 human-elephant conflicts are recorded monthly. The main vegetation type in Jharkhand state is tropical forest, with 29.61% of its total geographical area covered with forest. The total number of elephants in Jharkhand is 679 (MOEFCC, 2017). Out of these, more than 150 elephants are in Dalma Wildlife Sanctuary (Jharkhand Forest Department, Census, 2011 - 2012) which is the second most densely population elephant habitat zone in Jharkhand. Elephants migrate during the monsoon and winter seasons, from the higher altitudes and towards warmer regions of south Bengal and Odisha. During the migration, they are attracted to the paddy fields in nearby villages, which naturally results in conflict (Guha, 2017).

Source: MOEFCC, 2017.

Figure 1. Elephant Distribution in India (orange colour).

Study Area, Data and Methods

Jharkhand state was formed out of the southern part of Bihar State in the year 2000. It comprises 28 districts with a total geographical area of 78,714 sq. km and a total population of 32,966,238 (Jharkhand Provisional Result-Census 2011-Data Sheet, 2011). Our study area is Dalma Wildlife Sanctuary and its environs from two districts: East (Purba) Singhbhum and Saraikela Kharsawan districts. It also includes part of Ranchi and West (Paschim) Singhbhum districts (Figure 1). The wildlife sanctuary is situated in the north, approximately 10 km from the steel city - Jamshedpur, which has a total area of

193.22 sq. km (Sambath, 2014). The core area of this forest covers 35 sq. km, while the buffer area covers 158.22 sq. km. The forest is classified as dry deciduous. Dalma Wildlife Sanctuary is surrounded by Dalma Hill (928 meters above sea level). About 10 to 15% of the forest in the Sanctuary is flat land, which also includes undulating terrain with rocky hills and plateaus (Ranjan et al., 2016). The summer temperature varies between 22 °c to 38 °c, and annual rainfall is around 1400 mm. The main water source of this region is the River Subarnarekha. This forest ecology, including abundant water sources is suitable for elephants, and other species such as sloth bear (Melursus ursinus, Shaw, 1791), dhole (*Cuon alpinus*, Pallas, 1811), mouse deer (*Tragulidae* spp., Milne-Edwards, 1864), Indian giant squirrel (*Ratufa indica*, Erxleben, 1777), Indian pangolin *(Manis crassicaudata*, E. Geoffroy, 1803), pythons (Burmese pythons, *Python bivittatus*, Kuhl, 1820, Indian pythons *Python molurus*, Linnaeus, 1758), eagles (*Accipitridae* spp.) and various types of other bird and insect species (Rani et al., 2014; Sambath, 2014; Chatterjee, 2016).

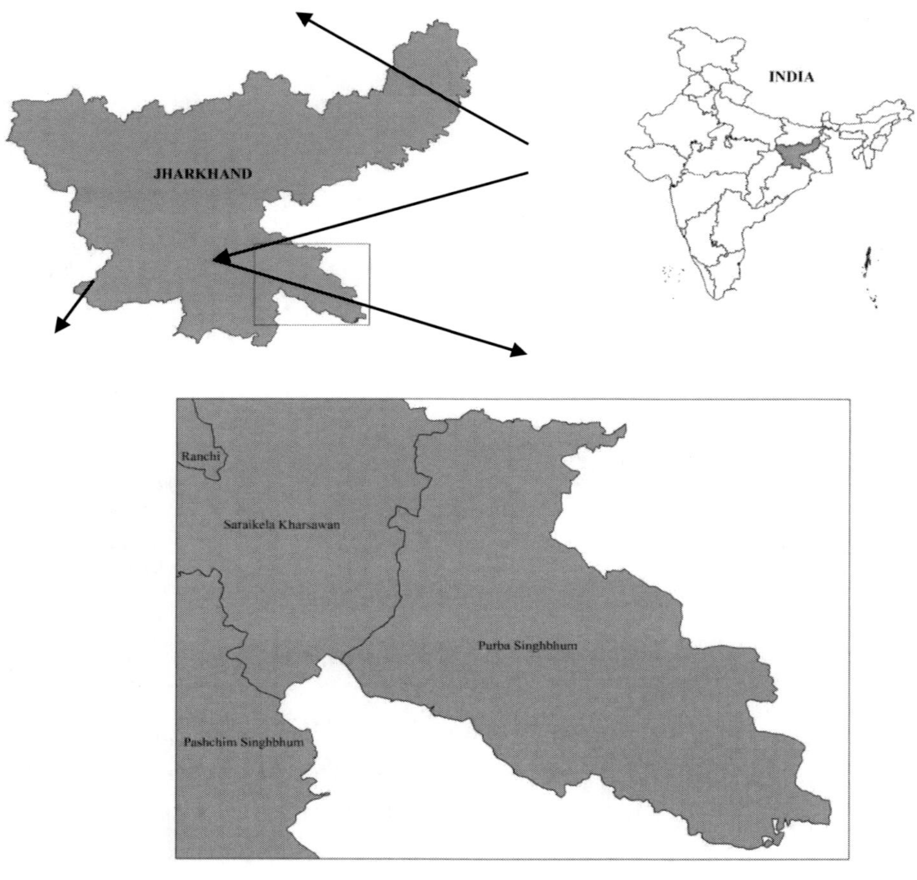

Figure 2. Location of Study Area.

For the study, topographical maps (1:250000), Landsat 4-5 TM and Landsat 8 satellite images (30-meter spatial resolution) and Cartosat DEM (15m resolution) were used. The spatial database about transportation network and water bodies in the study area was created from toposheets. Forest area and settlements were extracted from Google earth. Elevation and slope maps were calculated with the help of Cartosat DEM data. Recent data on human- elephant conflict were collected from Jharkhand Forest Wildlife Department. Elephant migration routes and elephant mortality data were also gathered from forest office. Table 1 below shows the data on the elephant-human conflicts in the study area.

Table 1. Human-Elephant conflicts, Dalma wildlife sanctuary (2009 - 2014)

No.	Reason killed/captured	2009-10	2010-11	2011-12	2012-13	2013-14	<07-2014
1.	Poaching	-	-	1	-	2	-
2.	Poison	-	-	1	-	-	-
3.	Accidental electrocution	-	1	1	3	1	-
4.	Deliberate electrocution	-	-	-	-	-	-
5.	Train accidents	-	-	1	1	1	-
6.	Contagious diseases**	-	-	-	-	3	-
7.	Declared rouge, Sec 11, killed	-	-	-	-	-	-
8.	Declared rouge, Sec 11, captured	-	-	-	-	-	-
9.	Captured under Sec 12	-	-	-	-	-	-
10.	Other unnatural reasons	2	2	1	-	3	-
11.	Natural reasons	11	10	4	4	-	1
12.	Unknown reasons	-	-	-	-	-	-
13.	People killed by elephants	54	69	62	60	56	10
14.	Payments (Rs. Lakh)human deaths	53.93	52.62	67.32	91.17	109.28	10.00
15.	Payments (Rs. Lakh), crops, etc.	123.71	74.94	131.89	120.19	275.61	29.76

Source: Jharkhand forest wildlife department.

Note: Rs. Stands for the Indian rupee, the basic currency unit. The current exchange rate is 1 rupee is 0.0146066USD. $1 = 68.4622.

Elephant Habitat Suitability Analysis

Using information from several existing research works, the current study models the potential suitable habitat locations for elephants. Parameters for this model included factors for elephant presence (food, shelter (in terms of forest cover), water availability), disturbance factors (road and human settlement proximity) and accessibility parameters such as elevation and slope. These were mapped and their contribution towards elephant suitability was modelled. Adult male elephants require 212 liters for water for drinking daily, especially during the peak of summer heat. Elephants frequently wallow in water, which influences their daily range of movement. Therefore, areas with water sources were deemed very suitable for elephant presence. Elephants eat around 150kg of vegetation daily, hence suitable vegetation is deemed a factor for elephant presence. Human settlement affects elephant presence negatively and creates the possibility ofr conflict with people, hence the further locations from settlement the greater the potential reduction of conflict. For elevation, Indian elephants generally migrate up to 300 meters. All the input data layers were aggregated, with their importance as the weighting system. Figure 3 shows the model of the methodology. The parameters or factors for elephant presence are shown in Table 2.

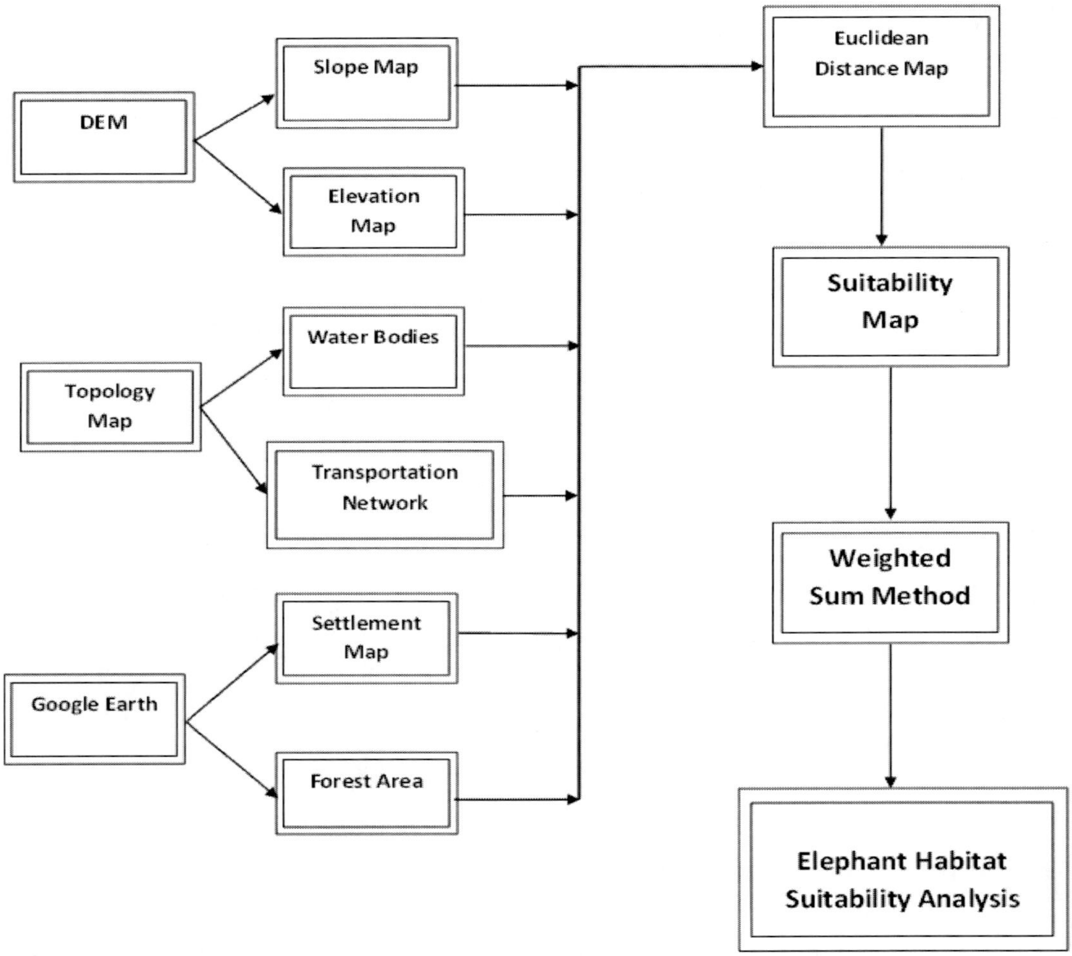

Figure 3. Methodology.

Table 2. Weights of variables and Associated Attributes

Variables	Final weightage
Forest and Settlement Suitability	0.5
Elevation Suitability	0.1
Slope Suitability	0.1
Water bodies Suitability	0.2
Transportation Network Suitability	0.1

Potential Conflict Spot Identification

Land use and land cover maps for 1989 and 2014 were prepared using Landsat images and the change was analysed. Using elephant route data obtained from the forest department, 5 km and 10 km buffer zones were prepared. This was followed by an overlay analysis using a change detection map, elephant route map and elephant attack points data to determine and display the potential conflict spots.

Results and Discussion

The study was limited to the Dalma Wildlife Sanctuary, hence the issues of migration and conflict issues within West Bengal and Odisha states were not examined. Figure 4 shows the elephant sighting area, water holes etc., of Dalma Wildlife Sanctuary Area. The conflict locations of Dalma elephants as observed from the forest field survey report is shown in Figure 5.

Dalma elephants migrate to West Bengal and Odisha during the winter season and return to Dalma in summer. According to the census report of wild animals in Dalma, the elephant migration times vary annually. Table 3 shows the variations in the migration for four years between 2001 and 2014. Dalma elephants migrate continuously in different groups to west Bengal and Orissa in different phases. The different migration routes of elephants in Dalma are shown in Figure 6.

Table 3. Elephant Migration Times, Dalma Wildlife Sanctuary

No.	Year	Time and day of arrival in Dalma	Time to go from Dalma
1.	2001	28.02.2011 – 10.03.2011	I16.08.2011 – 30.08.2011
2.	2012	13.03.2012 – 20.03.2012	29.09.2012
3.	2013	10.05.2013 – 25.05.2013	30.08.2013
4.	2014	19.05.2014 – 03.06.2014	No Data

Source: Forest Range Office

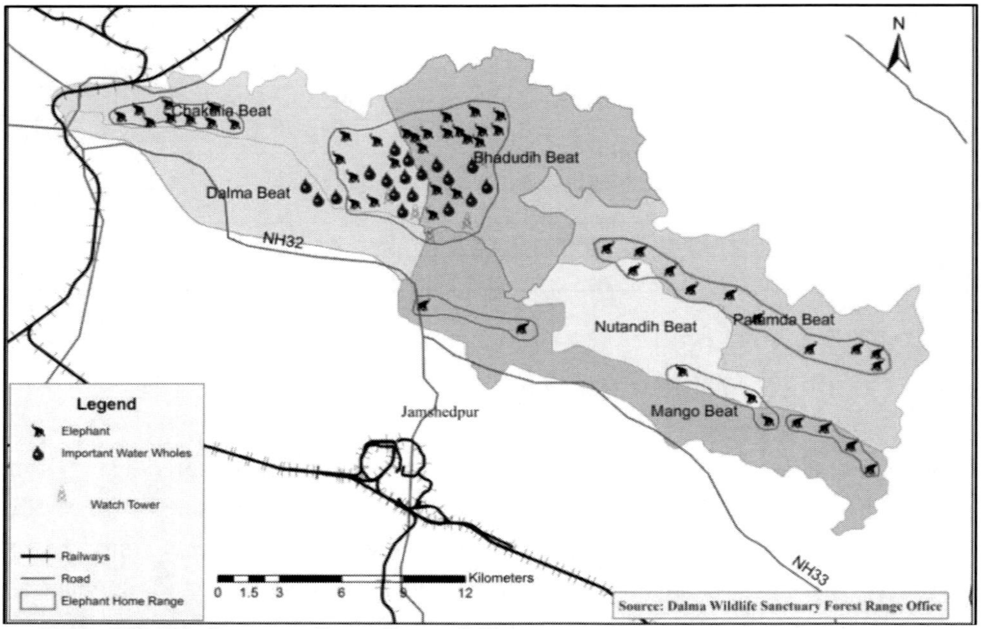

Source: Dalma Wildlife Sanctuary Forest Range Office.

Figure 4. Elephants Sightings, Dalma Wildlife Sanctuary.

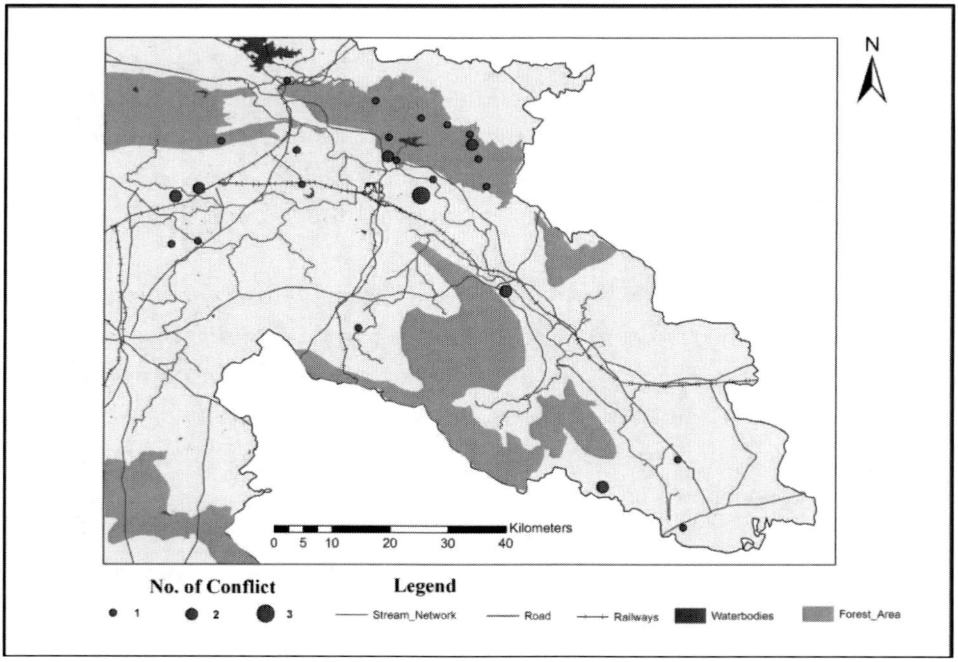

Figure 5. Number of conflicts in different area near Dalma Wildlife Sanctuary and Surroundings.

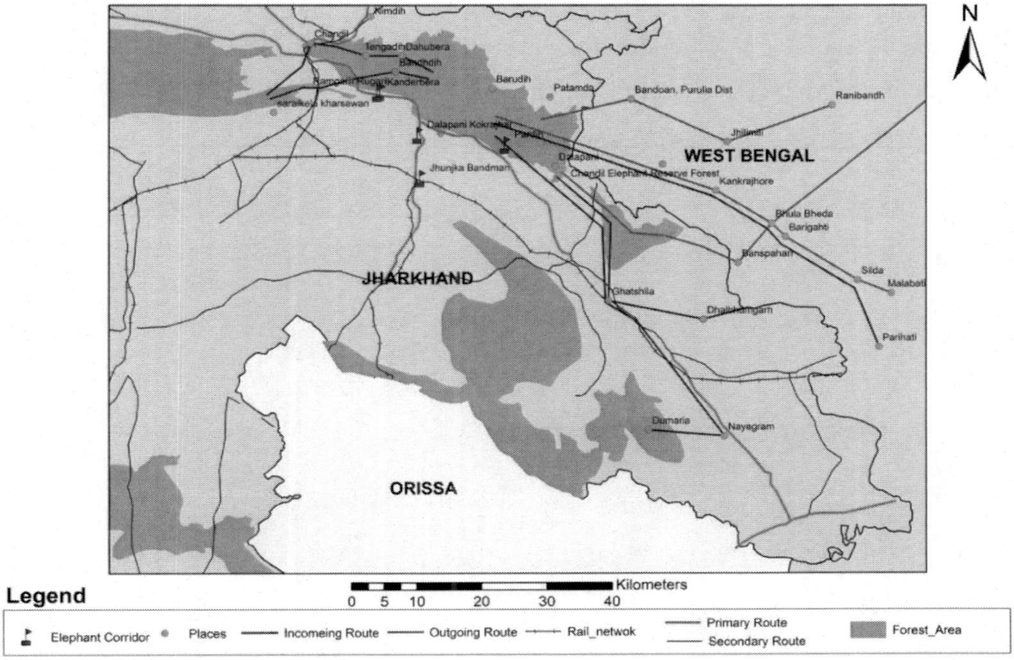

Figure 6. Elephant Migration Route Map.

HABITAT SUITABILITY

Through the analysis of 6 geo-environmental variables as mentioned in the methods section, the suitable areas were classified into 5 groups or levels of habitat suitability, namely: (1) not suitable; (2) less suitable; (3) moderately suitable; (4) suitable; and (5) highly suitable. Suitable areas are mainly spread over those areas where all variables are optimal.

Table 4. Elephant Habitat Suitability Area Statistics

SI No.	Class name	Value range	Actual area (Sq. Km)
1	Not suitable	0-20	580.13
2	Less suitable	20-40	3011.75
3	Moderately suitable	40-60	2480.73
4	Suitable	60-80	425.84
5	Highly suitable	80-100	46.54

The suitable areas should have forest and water bodies, far away from transportation network and settlement along with low elevation and slope. The resultant Elephant Habitat Suitability Map is shown in Figure 7. The statistics are shown in Table 4.

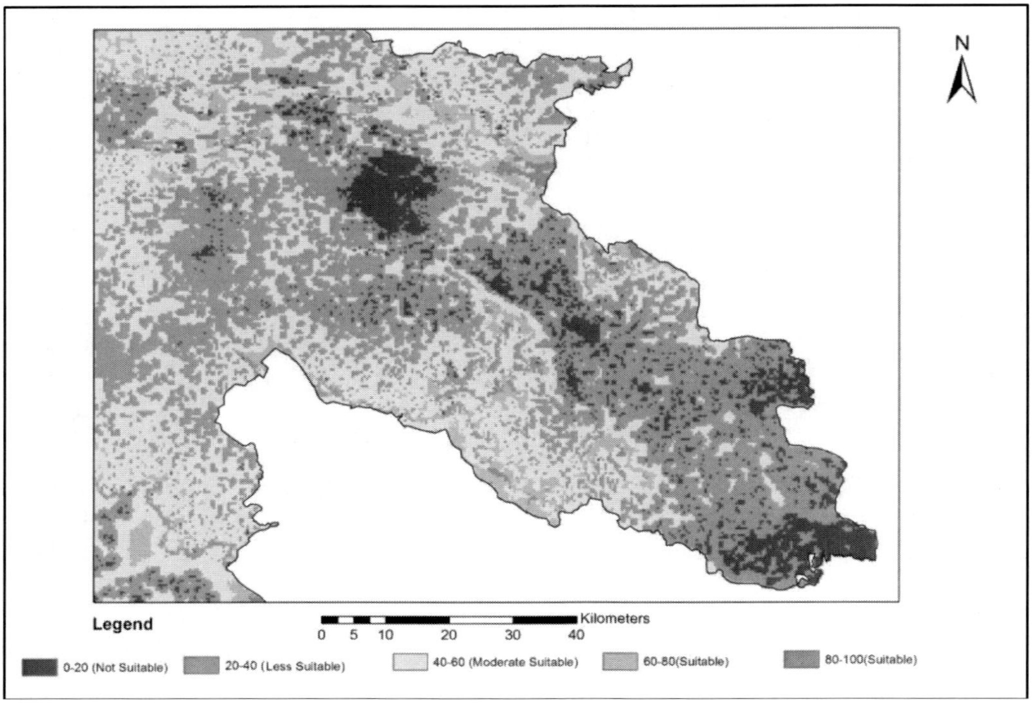

Figure 7. Elephant Habitat Suitability Map.

LAND USE LAND COVER CHANGE ANALYSIS

Land use and landcover change was abstracted from the analysis of satellite images, to show the changes in our study are over last 25 years. These were divided into six classes: (1) built up; (2) water bodies; (3) agriculture land; (4) wet land; (5) waste land; and (6) forest. Human settlement and industrial areas were included in the built up class. The agriculture class included plantation and cropland. These are shown in Figure 8 below. The landcover change statistics are displayed in Table 5. Figure 9 shows the Land Use and Land Cover map in 2014, and Figure 10 shows the distribution of negative (red) and positive (green) change categories from year 1989 to 2014.

The overlay analysis (Figure 11) shows that the maximum human-elephant conflict locations were within a 10-kilometer buffer area. This indicates that the main reason for increased human-elephant conflicts is the expansion of human settlements near the forested areas, and the conversion of forest land to agriculture and human settlements. Some potential conflict spots are exist inside the forest area, which reveals that human activities are also increasing inside the forest area. A clear degradation and fragmentation in the forest area can be seen from the LULC maps of 1989 to 2014.

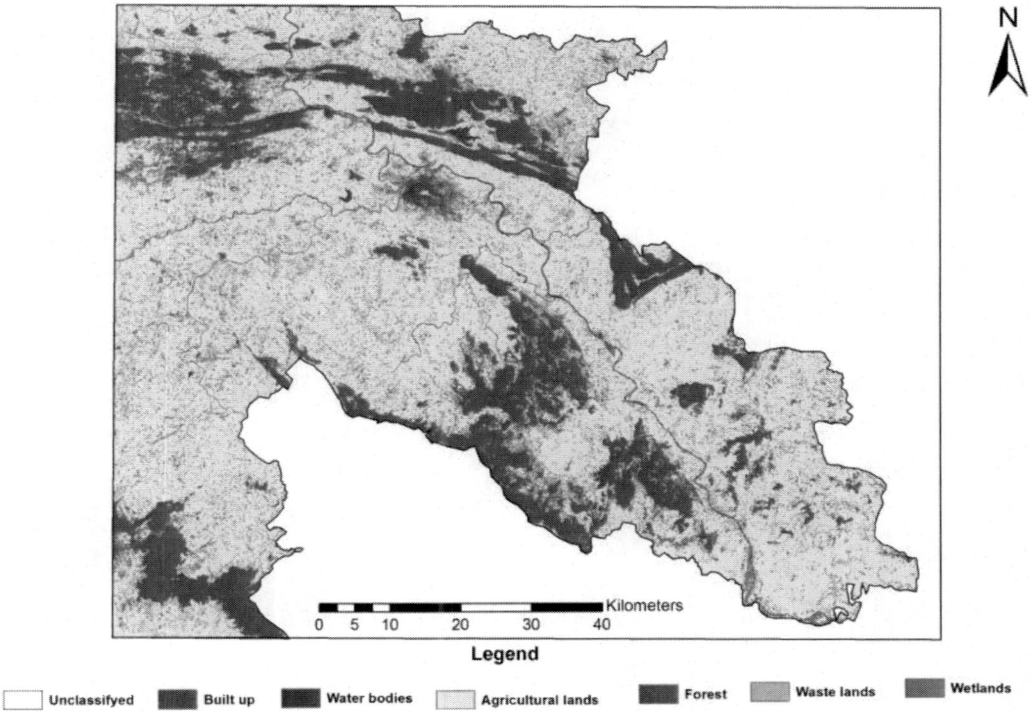

Figure 8. Land Use and Land Cover map in 1989.

Table 5. LULC Change Statistics (in sq. km)

Change Matrix								
		2014						
		Settlement	Agriculture	Forest	Wasteland	Wetland	Waterbody	
1989	Settlement	114.57	0.00	0.00	0.00	0.00	7.09	**121.66**
	Agriculture	149.05	3953.82	0.00	104.34	31.56	60.52	**4299.29**
	Forest	7.32	377.22	1398.55	6.36	1.13	2.43	**1793.01**
	Wasteland	2.99	74.80	0.00	10.02	3.61	7.37	**98.79**
	Wetland	4.03	119.90	0.00	21.05	1.18	7.65	**153.81**
	Waterbody	0.00	18.53	0.00	3.45	0.56	54.04	**76.58**
		277.95	**4544.26**	**1398.55**	**145.23**	**38.05**	**139.10**	

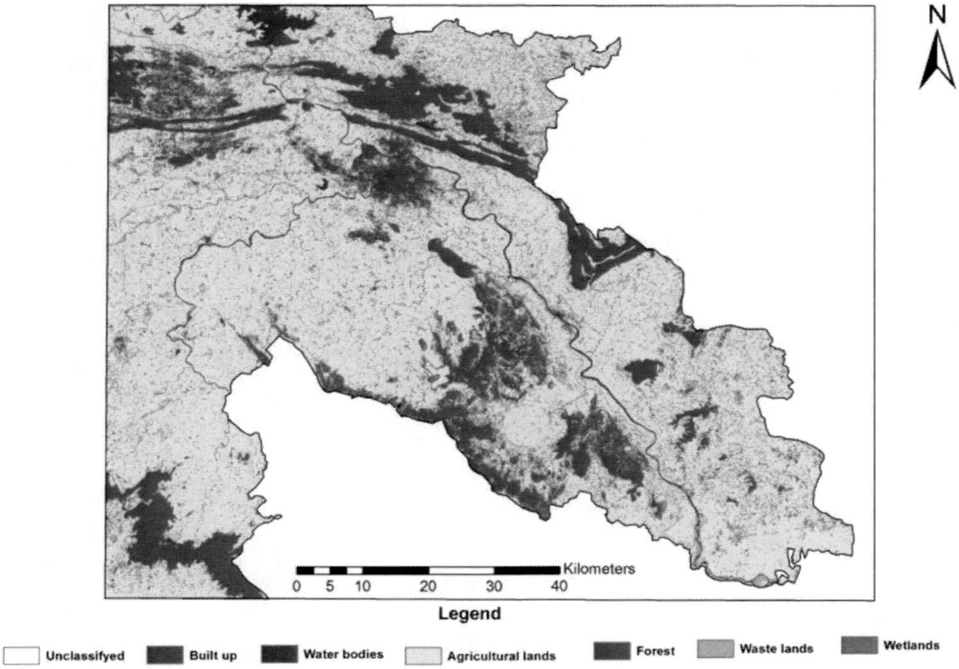

Figure 9. Land Use and Land Cover map in 2014.

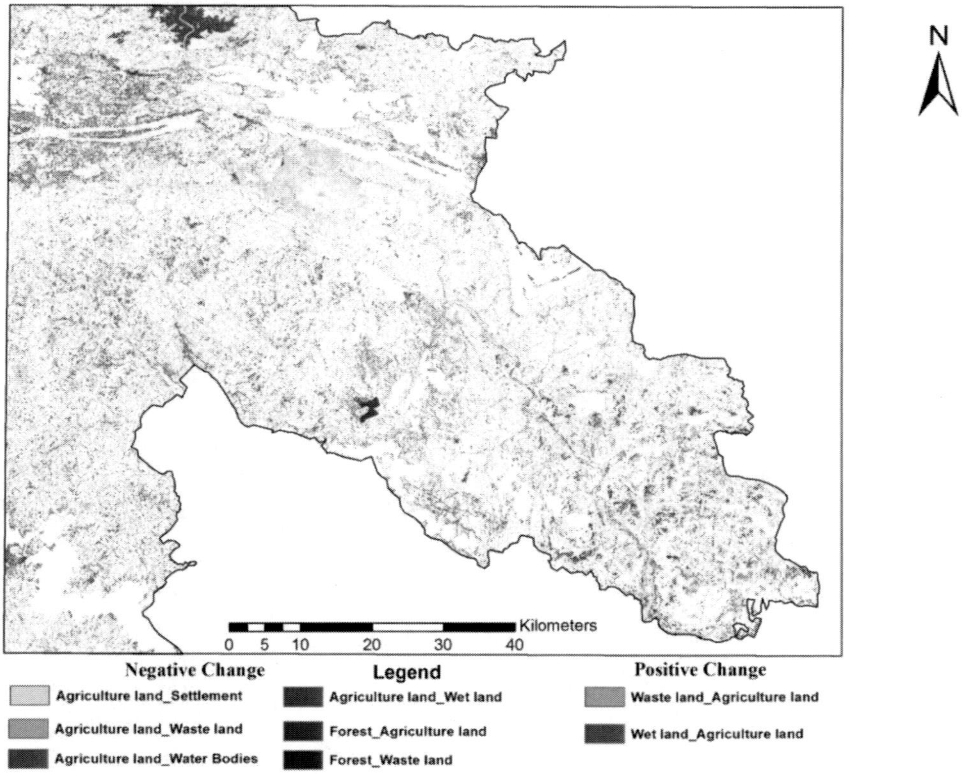

Figure 10. Distribution of negative (red) and positive (green) change categories 1989 to 2014.

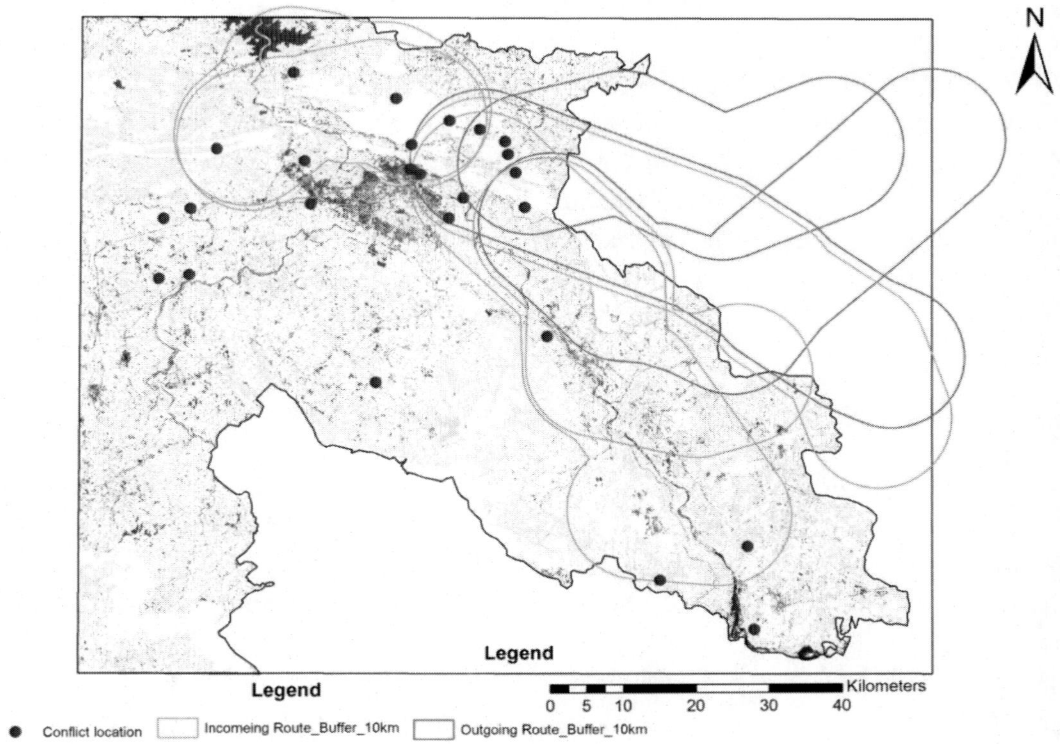

Figure 11. Potential conflict area.

CONCLUSION

This chapter has created maps of potential suitable areas for elephant presence and potential conflicts that will be useful for planners and researchers. The combination of GIS, remote sensing and field survey data has contributed clarity and understanding to a complex and debated issue, which is vital for human development and conservation biology. Using the geospatial model, dividing the Dalma Wildlife Sanctuary and its surroundings into five levels of suitability will help to plan future corridor areas for wildlife movements between the forests. The present elephant population has a deep association with the forest area, water sources and the migration routes, which are currently under threat from increased areas of human settlement. Such a holistic model may be developed to avoid future conflict, including more details on the relations between people and elephants, to ensure better management plans. This is an important contribution to the fields of geomatics and conservation biology.

ACKNOWLEDGMENTS

We are heartily thankful to the Sri Kamlesh Pandey, I. F. S., D. F. O., Wildlife, Ranchi, Deputy Conservator of forest cum field director elephant project, and Dalma Range officer Mr. Mangal Kashyap for their support to complete this project. Thanks to Vice Chancellor, Registrar and Head of the Department of Remote Sensing, Birla Institute of Technology, Mesra and all staff members of Remote sensing department, Birla Institute of Technology, Mesra for their support.

REFERENCES

Areendran, G., Raj, K., Mazumdar, S., Munsi, M., Govil, H. & Sen, K. P. (2011). Geospatial modeling to assess elephant habitat suitability and corridors in northern Chhattisgarh, India. *Tropical Ecology 52*(3): 275-283.

Baskaran, N., Varma, S., Sar, C. K. & Sukuma, R. (2011). Current status of Asian Elephants in India. *Gajah 35*: 47-54.

Bist, S.S. (2006). Elephant conservation in India. An overview. *Gajah 25*: 27-37.

Calabrese, A., Calabrese, J. M., Songer, M., Wegmann, M., Hedges, S., Rose, R. & Leimgruber, P. (2017). Conservation status of Asian elephants: the influence of habitat and governance. *Biodiversity and Conservation 26*(3): DOI: 10.1007/s10531-017-1345-5.

Chanda, S. (I. F. S) (1996 Batch). *Joint Director Sundarban Biosphere Reserve, West Bengal. Man - Elephant Conflict in South West Bengal.*

Chatterjee, N. D. (2016). Ecological biodiversity of Panchet Forest Division and Dalma Wildlife Sanctuary. In Chatterjee, N.D. (Ed.), *Man–Elephant Conflict*, 19 - 59. DOI 10.1007/978-3-319-31162-3.

Diallo, Y., Hu, G. & Wen, X. (2009). Applications of remote sensing in land use/land cover change detection in Puer and Simao Counties, Yunnan Province. *Journal of American Science*, 5(4):157-166.

Diemer, N. (2003). *Environmental Suitability Analysis for Asian Elephants*. Enschede, The Netherlands: International Institute for Geo-Information Science and Earth Observation.

Ellis, E. & Pontius, R. (2007). Land-use and land-cover change. In Cleveland, C.J. (Ed.), *Encyclopedia of Earth*. Washington, D.C.: Environmental Information Coalition, National Council for Science and the Environment.

Government of India (2012). *The Gazette of India: Extraordinary Part II, Section 3, Sub-section (ii) Government of India.* Ministry of Environment and Forests, Notification.

Retrieved from http://www.indiaenvironmentportal.org.in/files/file/Dalma%20Wildlife%20 Sanctuary.pdf.

Guha, M. (2017). Human-Elephant Conflict in South West Bengal–I: Study on Fodder Plants of the Elephants of Dalma Herd. *Journal of Economy, Environment and Society* *2*(1): 22-27.

Hazarika, R. & Saikia, A. (2013). The pachyderm and the pixel: an assessment of elephant habitat suitability in Sonitpur, India. *International Journal of Remote Sensing 34*(15): 5317-5330.

Imam, E. & Kushwaha, S. P. S. (2013). Habitat suitability modelling for Gaur (Bos gaurus) using multiple logistic regression, remote sensing and GIS. *Journal of Applied Animal Research 41*(2): 189-199.

Jharkhand Observer (2011). Retrieved from http://jharkhandobserver.com/wildlife.php.

Jharkhand Provisional Result-Census 2011-Data Sheet (2011). Retrieved from http://censusindia.gov.in/2011-prov-results/data_files/jharkhand/Jharkhand%20Provisional%20Result%20Data%20sheet%20for%20release.pdf.

Khanna, V., Ravichandran, M. S. & Kushwaha, S. P. S. (2001). Corridor Analysis in Rajaji-Corbett Elephant Reserve- A Remote Sensing and GIS Approach. *Journal of the Indian Society of Remote Sensing 29*: 24-38.

Kushwaha, S. P. S. & Roy, P. S. (2002). Geospatial technology for wildlife habitat evaluation. *Tropical Ecology 43*: 137-150.

Kushwaha, S. P. S., Munkhtuya, S. & Roy, P. S. (2000). Geospatial modelling for Goral habitat evaluation. *Journal of the Indian Society of Remote Sensing 28*: 293-303.

MOEFCC (2017). *Synchronised Elephant Population Estimation India.* Report prepared by Ministry of Environment and Forest and Climate Change, Goverment of India.

Mongkolsawat, C. & Chanket, U. (2007). *Predicting Asian Elephant Habitat Suitability with Satellite Data and GIS.* Geo-informatics Centre for Development of Northeast Thailand, Khon Kaen University, Khon Kaen.

Prasad, G., Shiny, R., Reghunath, R. & Prasannakumar, V. (2011). A GIS-based spatial prediction model for human–elephant conflicts (HEC). *Wildlife Biology in Practice* 7(2):30 - 40.

Rani, S., Mahato, S. K. & Mishra, A. T. (2014). Habitat Study of Sloth Bear in Dalma Wildlife Sanctuary, Jamshedpur. *The Indian Forester 140*(10): 960 - 964.

Ranjan, A., Anand, A., Kumar Singh, R. & Sivathanu, V. (2016). LU/LC change detection and forest degradation analysis in Dalma Wildlife Sanctuary using 3S Technology: A case study in Jamshedpur-India. *AIMS Geosciences 2:* 273-285. 10.3934/geosci.2016.4.273.

Reza, M. I. H., Abdullah, S. A., Nor, S. B. M. & Ismail, M. H. (2013). Integrating GIS and expert judgment in a multi-criteria analysis to map and develop a habitat suitability index: A case study of large mammals on the Malayan Peninsula. *Ecological Indicators 34*:149-158.

Roever, C. L., Van Aarde, R. J., Chase, M. J. (2013). Incorporating mortality into habitat selection to identify secure and risky habitats for savannah elephants. *Biological Conservation 164*: 98-106.

Sambath, S. (2014). Taxonomic Studies of Lepidoptera (Insecta) of Dalma Wildlife Sanctuary, Jharkhand (India). *Records of the Zoological Survey of India,* Occasional Paper No. 359.

Sitompul, A. F., Griffin, C. R., Rayl, N. D., & Fuller, T. K. (2013). Spatial and temporal habitat use of an Asian Elephant in Sumatra. *Animals 3*(3): 670-679.

The Pioneer (2015). *Elephants Trample Man to Death.* Retrieved from https://www.dailypioneer.com/2015/state-editions/elephants-trample-man-to-death.html

Tikhile, P., Gavade, V. V., Patil, R. R. & Palkar, J. M. (2013). Elephant Habitat Suitability in Southern Part of Kolhapur District with Geoinformatics Approach. *Academic Journal of Plant Sciences 6*: 82-88.

Zhao, C., Li, Y., Li, D., Guan, J., Xiong, Y., & Hu, J. (2014). Habitat suitability assessment of Sichuan sika deer in Tiebu Nature Reserve during periods of green and dry grass. *Acta Ecologica Sinica 34*(3): 135-140.

In: Geomatics and Conservation Biology
Editor: Michael O'Neal Campbell
ISBN: 978-1-53614-468-0

Chapter 4

REMOTE SENSING AND CONSERVATION BIOLOGY

Michael O'Neal Campbell
Camosun College, Victoria, Canada

ABSTRACT

Remote sensing is a vital aspect of geomatics, as images and recorded visual data are even more relevant than numerical counted data for the identification of environmental features. This does not negate the importance of numerical or statistical data but complements such sources. Originally based on human perceptions and artistic depictions, remote sensing has evolved from primitive photography to more advanced 'spy' space-based technology, which experienced massive boosts in power through computerization. Recent developments include LIDAR and a plethora of geomatics software, all of which are applicable to the spatial issues of conservation biology. This chapter examines the key issues of the application of remote sensing techniques to conservation biology, examining its history and current developments. Key points concern the ways in which remote sensing techniques have evolved or adapted to changes in the requirements of conservation biology research and assessment. The main relevant remote sensing technologies are those that enable more detailed identification of life forms (smaller species, at greater distances), perception of animal behavior (for example, movements, breeding, mating and feeding patterns), landcover change (including temperature, moisture and species change) and integration of such variables for the documentation of ecosystem variability. It is concluded that the applicability of remote sensing technology to keep abreast with the requirements of conservation biology may determine the future of the latter discipline.

INTRODUCTION

Remote sensing, an earth-observing (EO) technique, has been defined simply as "the acquiring of data about an object without touching it" (Jensen, 2007). Jensen (2007, 4) argues that this definition is simple, but all inclusive, including "virtually all remote sensing devices, including cameras, optical-mechanical scanners, linear and area arrays, lasers, radar systems, sonar, seismographs, gravimeters, magnetometers, and scintillation counters." A more focused definition is suggested: "Remote sensing is the noncontact recording of information from the ultraviolet, visible, infrared and microwave regions of the electromagnetic spectrum by means of instruments such as cameras, scanners, lasers, linear arrays, and/or area arrays located on platforms such as aircraft or spacecraft, and the analysis of acquired information by means of visual an digital image processing" (Jensen, 2007, 4). The next suggestion is that remote sensing is both a science and an art; the former because it involves the use of complex sensors for measuring electromagnetic energy leaving a geographical entity or area and using complex mathematical methods to extract the relevant information for the measurement of the object being scanned. It may also be an art, as the appraisal of the result may be based on the mind of the appraiser, involving personal and intuitive skills, understanding, experience and abilities in syntheses (Campbell, 2002; Bhatta, 2008).

Another way to define remote sensing is to describe its main operating tools and targets. In this sense, the instrument or sensor, and the object are the key components (Campbell, 2002; Bhatta, 2008) (Figure 1). The sensor generally emits electromagnetic radiation (EMR) that travels from the source to the target object. As noted by Livingston (2015, 12).

> Electromagnetic radiation can be described as a stream of photons moving at the speed of light in a wave-like pattern. Visible light falls between infrared and ultra violet and constitutes only a tiny fraction of energy. Radio waves, microwaves, infrared waves, ultraviolet waves X-rays and gamma rays make up the rest of the electromagnetic spectrum.

In water, sonar or acoustic waves may also be used (Jensen, 2007). Another informative description is to examine the remote sensing process (Figure 2). The full spectrum, from longest waves to the shortest, comprises radio waves which transmit radio and television signals(length <1 cm to 100 m +), microwaves (1 mm to 30 cm) infrared (from the visible region to about 1 mm), visible bands, 400 and 700 billionths of a meter (400 to 700 nanometers, nm), ultraviolet radiation (length 400 -10 nm), X-rays (10 nm – 10 picometers or 10 trillionths of a meter) and finally, gamma rays (<10 pm) (National Aeronautical and Space Administration, NASA 2018a). These numbers are shown in standard format in Figure 3.

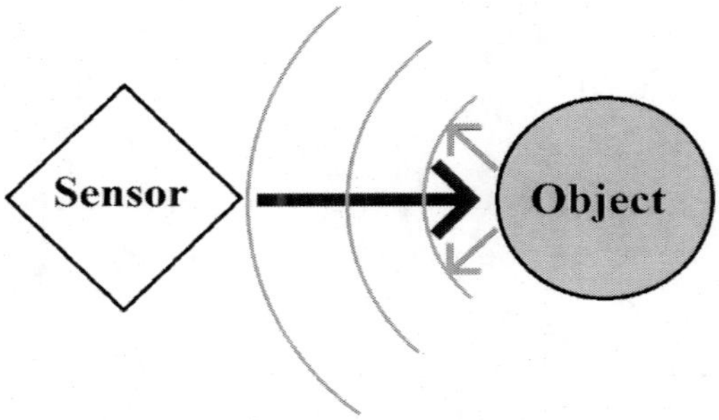

Figure 1. The emission of electromagnetic force from the sensor to the target object. Reading of the result eventually creates the remotely sensed image.

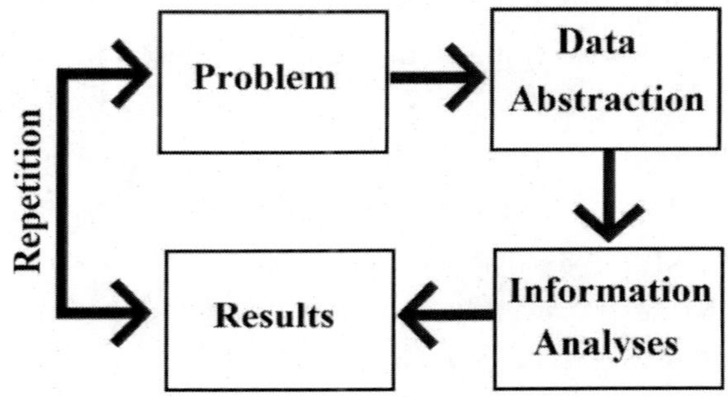

Figure 2. The 4 stages of remote sensing (Campbell, 2002; Jensen, 2007). The initial problem (1) include hypotheses and model selection, while the later stages involve decisions on data type and sensing methods (2) processing methods (3) and result presentation (4). Where vast areas are covered, or corrections are needed, the process may be repeated often (5). Comparing the problem (initial hypotheses) with the results may lead to the acceptance of rejection of the initial hypotheses, or their reframing.

Name	**Wavelength (m)**
Radio	$> 1 \times 10^{-1}$
Microwave	1×10^{-3} - $1 \times {}^{10-1}$
Infrared	$7 \times {}^{10-7}$ - $1 \times {}^{10-3}$
Optical	$4 \times {}^{10-7}$ - $7 \times {}^{10-7}$
Ultraviolet UV	$1 \times {}^{10-8}$ - $4 \times {}^{10-7}$
X-ray	$1 \times {}^{10-11}$ - $1 \times {}^{10-8}$
Gamma-ray	$< 1 \times {}^{10-11}$

Source: Adapted from NASA (2018).

Figure 3. Wavelengths on the electromagnetic Spectrum in meters.

The optical range refers to the visible bands of the spectrum, from red to violet in rainbow colors.

Jensen (2007) lists some advantages and limitations. It is systematic, which reduces the possibility of sampling bias which frequents other collection methods. It is also remote, and hence does not disturb the target object. Additionally, unlike geographical information systems (GIS) or cartography (the science and/or art of map making), remotely sensed data is directly acquired, rather than reliant on the analysis of data acquired by others. It is also capable of covering far greater areas than those covered by surveying, taking in both natural and cultural landcover and processes. Advanced techniques such as light detection and ranging (LIDAR) and radio detection and ranging (RADAR) or interferometric synthetic aperture radar (IFSAR) can produced more sophisticated products such as digital elevation models (Jensen, 2007).

The history of remote sensing has been divided into three main periods: (1) the early age (1839 - 1907); (2) the middle age (1908 - 1945); and (3) the modern age (1946 +) (Bhatta, 2008). During the early age, the main issue was the invention of the camera and flight (Jensen, 2007; Bhatta, 2008). The invention of the telescope may also have played a role, as although telescopes do not capture images, telescopic techniques assist in the detection of distant objects (Herbert, 2011). For the telescope, there is a clear timeline. Around 1608, two individuals named Jacob Metius and Hans Lippershey promoted the first telescope in the Netherlands, and later the Italian mathematician Galileo Galilei made an improved telescope that enabled space observation. Later, in 1872, the English scientist Isaac Newton created a more advanced telescope through the construction of a telescope that used mirrors, termed a reflector telescope, which was presented to the Royal Society. From there, the telescope advanced to the 1970s Hubble Space telescope, after the American astrologist Edwin Hubble. In 1990, this telescope was placed in an orbiting position around the earth, allowing the observation of the earth's surface and space without distortion (Herbert, 2011).

The camera followed a similar history from a rudimentary model to the current digital forms. The French inventor Joseph Nicéphore Niépce is commonly viewed as the inventor of photography, through heliography, in 1825 (Baatz, 1997). His associate, Louis Daguerre later invented the daguerreotype process, which greatly decreased the time for the exposure required for the development of photographs, from hours using bitumen, and the more, light sensitive resin, to minutes using a polished silver-plated surface, which was made light sensitive through treatment with fumes and after exposure treat the result with mercury vapor and chemical treatment (Coe, 1976). Others, such as the Brazilian inventor Antoine Hercule Romuald Florence developed rudimentary photography around the same period (Kossoy, 2018). Later developments in the late 19th and early twentieth centuries were film, color photography and more detailed photographs (Frizot, 1998).

The next important development was that of flight, starting with the hot-air balloon, followed by the hydrogen balloon. The first records of hot air balloons are from China as far back 220 to 280, during the Three Kingdoms era, when unmanned balloons were used for signaling (Deng, 2005). For Europe, the first development appears to be that of the brothers Jacques-Étienne and Joseph-Michel Montgolfier in Annonay, Ardeche, France, who demonstrated an unmanned flight on September 19, 1783 (Rolt 1966; Ege, 1974). Balloons designed by the brothers made more flights. Jean-Francois Pilatre de Rozier was a passenger in a tethered balloon on October 15, 1783 in the Faubourg Saint-Antoine (Crouch, 2008). A short while later, on November 21, 1783, aeronauts, de Rozier and the Marquis d'Arlandes, made an untethered ascent from the center of Paris to the suburbs, about 9 kilometers in about 25 minutes. On January 19, 1784, seven passengers flew 914 meters over Lyons, France (Macmillan, 1964; Kirschner, 1985). Balloons were used for observation in battle for the first time in 1794, when the French balloon named l'Entreprenant was used for observation during the Battle of Fleurus in 1794 (Paniccia, 2014).

The problem of a cooling air descent and possible fire in air-filled balloons led to the development of the hydrogen balloon, one successful test being that by Jacques Alexandre César Charles, a member of the French Academy of Science, on August 27, 1783, (Kirschner, 1985). The hydrogen balloon used silk materials strengthened with elastic gum but was vulnerable to fire from the highly flammable hydrogen. This pioneering balloon ascended from the Place des Victories in Paris to an altitude of around 914 meters and landed about 24 kilometers) away. The next flight, balloon manned La Charlière ascended from the Tuileries in Paris on December 1, 1783, carrying Charles and M. N. Robert, covering 43 kilometers to Nesle. The balloon featured advanced features such as a net covering that connected to the manned casket below (for even weight distribution), stones for ballast, top valve, and a barometer and thermometer for pressure and temperature measurement. Another flight, by Jean-Pierre Francois Blanchard and an American doctor John Jeffries, crossed the English Channel on January 7, 1785 (Kirschner, 1985).

These early developments were the foundation of the rudiments of remote sensing. The next developments, during the second period (1908 – 1945, including the First and Second World Wars, and the interwar period - 1919 - 1939) concerned advancements in airplane design, aerial photography, geographical coordinates such as longitudes and latitudes (allowing the synthesis of photograph mosaics), and the invention of satellites and satellite images, LIDAR and radar (Jensen, 2007; Bhatta, 2008). The First World War served as one of the catalysts for aerial reconnaissance, partly because the dominant trench warfare and the demise of the cavalry scout required a new form of intelligence gathering. As argued by Wakefield (2014, 1):

> The trench deadlock that descended on the Western Front at the end of 1914 acted as a catalyst for the development of aerial reconnaissance. Traditionally, cavalry had served as the eyes of an army, scouting behind enemy lines and reporting to commanders about troop concentrations and movements. In trench warfare, such freedom of movement was impossible, leading armies to devise new ways of looking behind the opposing front line.

The French had aerial reconnaissance, based on aerial photographic sections within the armed forces, with highly trained specialists, advanced cameras and aerial photographs analyses. The British Royal Flying Corps tried cameras, with the photography of the German lines at the Battle of Neuve Chapelle in March 1915. Later developments included the mosaiced, overlapping of photographic prints for the coverage of larger sections of the front lines; combinations of aerial photographs and ground survey techniques were used to create detailed maps. Large numbers of photographs were taken later; nearly 20,000 photographs and 430,000 prints the Battle of the Somme (1916) and 2.5 million aerial photographs in 1918 (Wakefield, 2014). The films included a combination of panchromatic film and a yellow (minus blue) filter to penetrate atmospheric haze (Howard, 1970; Moore, 1979).

This led to the development of advanced aerial photography for commercial purposes after the war in 1919 (Popular Mechanics, 1935; Hearst Magazines (1936). As noted by Moore (1979, 479) "by 1940, hundreds of papers had been published on photo interpretation in the fields of archaeology, ecology, geology, pedology, forestry, engineering, and geography (see also American Society of Photogrammetry, 1960, p. 7). Moore (1979, 480) further notes that "another large stimulus to photo interpretation was provided by military efforts in World War II and Korea. The US Army Air Force alone took 171 million negatives in World War II" (see also Infield, 1970).

During the third period (1946 +) after the Second World War, systematic aerial photography during the Cold War between the United States of America and the Soviet Union developed in sophistication, as modified combat aircraft were used for aerial surveillance. These aircraft included the North American Aviation P-51 Mustang (a fighter or fighter-bomber), the Lockheed P-38 Lightning (a fighter), the Douglas B-66 Destroyer (a light bomber) and the McDonnell Douglas F-4 Phantom II (a supersonic jet interceptor and fighter-bomber) (Moore,1979). Other airplanes were designed more specifically for the role of photographic espionage and reconnaissance. These included the Lockheed U-2, which is designed as an ultra-high-altitude reconnaissance airplane. Another important airplane was the Lockheed SR-71 "Blackbird" a long-range, strategic reconnaissance airplane. There was also the North American A-5 Vigilante a carrier-based supersonic bomber and the Grumman OV-1 Mohawk. This last airplane was equipped with radar imaging capability, namely Side-Looking Airborne Radar (SLAR) which could penetrate foliage and map terrain, and the images could show the both fixed terrain features and moving targets (Moore, 1979; Neubeck, 2007). For aerial photography, airplanes flew overlapping ranges to ensure continuous coverage, as shown in Figure 4 (a,b) (Jensen, 2007; Bhatta, 2008).

Satellite development was a more recent factor for the advancement of remote sensing (Falkner & Morgan, 2002; Bhatta, 2008). Satellite images have the advantage over aerial photographs in that they are taken from much higher altitudes and hence have much larger coverage. However, this also means that the level of detail may be coarse of resolution (Jensen, 2007). Here resolution refers to the area on the ground represented by one pixel or picture element in the image. For example, one pixel could represent one to 10 meters on the ground; where one pixel represents one meter, the image has a finer resolution with more detail than when a pixel represents 10 meters (Campbell, 2002).

The United States launched V-2 flight on October 24, 1946 took an image every 1.5 seconds, from a maximum altitude of 105 km. The most important development was Landsat program, which began on July 23, 1972 with the launch of the Earth Resources Technology Satellite, which later named the Landsat, and was succeeded by eight other satellites, the latest being Landsat 8, launched on February 11, 2013 (National Aeronautics and Space Administration NASA, 2018b). Other satellite-based imaging systems included the IKONOS (1999 – 2015), Nimbus (seven satellites, launched August 28, 1964 to October 24, 1978, last out of service 1994), RADARSAT (RADARSAT-1-1995–2013, RADARSAT-2 – 2007, of the Canadian Space Agency, CSA), Indian Space Research Organization IRS-1C (December 28, 1995 – 21 September, 2007), Upper Atmosphere Research Satellite (UARS) (15 September 1991 - 14 December 2005), EarlyBird-1 for Earth Watch Inc. (December 24, 1997), SPOT 1 – SPOT 7 and Earth Remote Observation System A (EROS-A) (Ramsay et al., (1998; Flett et al., 2001; Lindsey, 2005; National Aeronautics and Space Administration NASA, 2005, 2018c).

Some of these research groups within and outside Silicon Valley, including NASA Ames Research Center and Electromagnetic Systems Laboratory Inc. also developed image enhancement techniques (Ormsby, 1987; Showstack, 2004). IKONOS, started by Lockheed Corporation as the Commercial Remote Sensing System (CRSS) satellite and operated by DigitalGlobe, was the first commercial Earth observation satellite that collected publicly high-resolution imagery (the resolution was 1 meter for panchromatic images and 4 meters for multispectral images) (Livingston, 2015). Panchromatic refers to an emulsion produces a scene like its visible appearance. Digital panchromatic imagery is produced by some satellites, such as IKONOS, QuickBird and Cartosat, and is of a higher (spatial) resolution than multispectral imagery. QuickBird (2001 – 2015) panchromatic imagery has a pixel equivalent to an area 0.6 m × 0.6 m, but the multispectral pixels represent an area of 2.4 m × 2.4 m (Livingston, 2015).

Multispectral images refer to images produced in bands that represent a wider range of the electromagnetic spectrum, i.e., the ultraviolet, visible and infrared bands shown in Figure 3. Each image usually has three to ten bands, where bands refer to each image taken by the sensor. For example, Landsat 5, for example, produces 7 band images, between wavelengths of 450 and 1250 nm. Recently, synthetic aperture radar (SAR), transmits microwave energy to the observed surface and then collects the reflected energy, rather than the reflected solar energy from the earth's surface. This is more "active" than the passive collection of solar energy by an optical sensor satellite, because it removes the effect of day or night, or atmospheric interference such as cloud cover, smoke, and haze (Ramsey et al., 1998; Flett et al., 2001; Livingston, 2015, 16). Figure 5 shows the resolutions, dates and origins of some of the imagery developed in the past 20 years. Figure 6 shows the components of multispectral images, which have several bands.

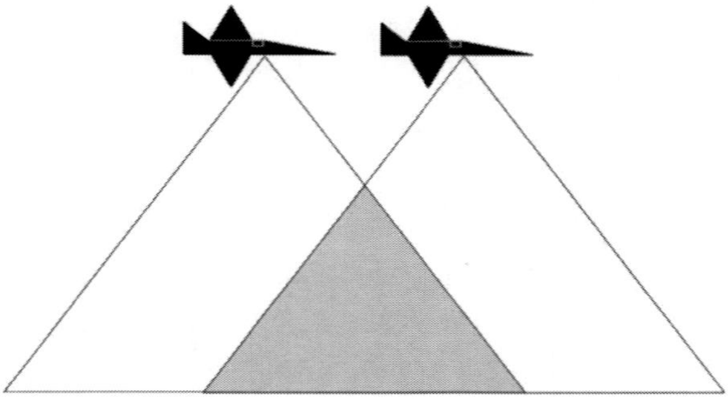

Figure 4a. Endlap or overlap between two consecutive photos in a flight line. The average is about 60 percent (Falkner, E. & Morgan, D. (2002).

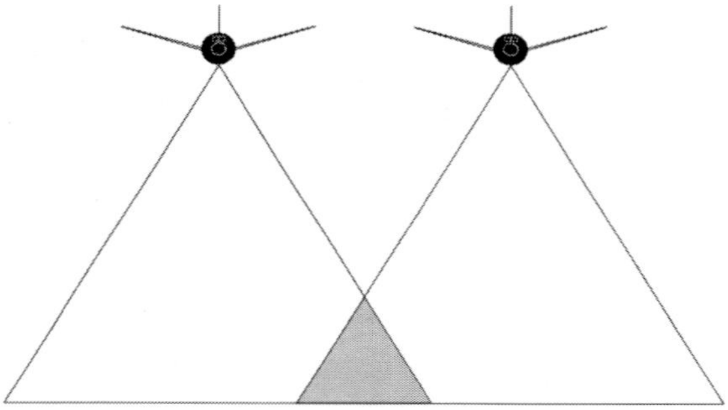

Figure 4b. Sidelap or overlap between two consecutive photos in a flight line. The average is about 30 percent (Falkner, E. & Morgan, D. (2002).

Name	**Nation**	**Active Dates**	**Spatial Resolution (m)**
EarlyBird-1	USA	Dec. 24, 1997 - Dec. 28, 1997	3 m PAN, 15 m MS
Ikonos	USA	Sept. 24, 1999 - Mar. 31, 2015	82 cm PAN 3.2 m MS
QuickBird	USA	Oct. 18, 2001 – Dec.17, 2014	60 cm PAN 2.4 m MS
OrbView-3	USA	Jun. 2003 - March 13, 2011	1 m PAN 4 m MS
WorldView-1	USA	Sept. 18, 2007 - 2018	0.5 m PAN
GeoEye-1	USA	Sep. 6, 2008 - 2018	0.41-46 m PAN, 1.65 m MS
WorldView-2	USA	Oct. 8, 2009 - 2018	46 cm PAN 1.84 m MS
WorldView-3	USA	Aug. 13, 2014 - 2018	31cm PAN, 1.24m MS, 3.7m SWIR, 30m CAVIS

Figure 5a. Recent satellite launched by the United States. Sources: Livingston (2015, 15-16), Indian Space Research Organization (2018); NASA (2018d). NB: PAN refers to panchromatic, MS refers to multispectral and SAR to synthetic aperture radar. CAVIS refers to Clouds, Aerosols, Vapors, Ice, and Snow. SW refers to short wave, IR to infrared, NIR to near infrared and SWIR to shortwave infrared imagery.

Name	Nation	Dates (Launch to Retirement)	Spatial Resolution (m)
RADARSAT-1	Canada	Nov. 4, 1995 - Mar.28, 2013	10 – 100 SAR
IRS-1C	India	Dec. 28, 1995 - Sept.21, 2007	6 m PAN
EROS-A	Israel	Dec.5, 2000 - May 2016	1.8 meter
Spot 5	France	May 4, 2002 -Mar. 31, 2015.	5 m PAN 10 m MS
Carosat-1	India	May 5, 2005 - 2018	2.5 m PAN
EROS-B	Israel	Apr. 25, 2006 - 2018	70 cm
Cartosat-2	India	Jan. 10, 2007 -2018	<1 m
EROS-B	Israel	Apr. 25, 2006 - 2018	70 cm
Cartosat-2	India	Jan. 10, 2007 - 2018	<1 m
Cosmo SkyMed-1	Italy	Jun. 8, 2007- 2018	SAR
TerraSar-X	Germany	Jun. 15, 2007	Variable, to 1 m SAR
SAR-Lupe 1 -5	Germany	Dec. 19, 2006 - Jul. 22, 2008 +	<1 m, SAR
Cosmo SkyMed-2	Italy	Dec. 9, 2007 - 2018	SAR
RADARSAT-2	Canada	Dec. 14, 2007 - 2018	3-100, <1 m SAR
Cartosat-2A	India	April 28, 2008	<1 m PAN
Cosmo SkyMed-3	Italy	Oct. 25, 2008 - 2018	SAR
TanDEM-X	Germany	Jun. 21, 2010 - 2018	2 – 4m
Cartosat-2B	India	Jul. 12, 2010 - 2018	<1 m PAN
Cosmo SkyMed-4	Italy	Nov. 6, 2010 - 2018	SAR
NigComSat-1R	Nigeria	Dec. 19, 2011 - 2018	2.5 m PAN 5 m MS
Spot 6	France	Sept. 9, 2012 - 2018	PAN - 1.5 m, MS– 6 m, (BGR, NIR)
Spot 7	France	Jun. 30, 2014 – 2018	PAN 1.5m MS- 6 m (BGR, NIR)
Pleiades-1A	France	Dec. 16, 2011 – 2018	PAN 0.5 m, 2 m MS
Pleiades-1B	France	Dec. 2, 2012 – 2018	PAN 0.5 m 2-m MS

Figure 5b. Recent satellite launched by nations other than the United States. Sources: Livingston (2015, 15-16), Indian Space Research Organization (2018); NASA (2018d). NB: PAN refers to panchromatic, MS refers to multispectral and SAR to synthetic aperture radar. SW refers to short wave, IR to infrared and SWIR to shortwave infrared imagery. BGR refers to blue, green and red. Note for the Spot satellites, Spot 1 – 5 were launched 1986 – 2002, and deactivated 1990 – 2015.

This technology is strongly capable of the type of applications required for conservation biology. Aerial photography may capture all features that may be captured by field photographs, but the coarse resolution of the former may require supporting field research, sometimes termed ground-truthing, to verify smaller features (Campbell, 2002). Stereographic pairs of aerial photographs are used in the production of topographic maps and for modelling terrestrial habitat features. Although conventional radar may be mostly used for aerial traffic control and some meteorological applications, and synthetic aperture radar is applied to digital elevation modelling of large scale terrain, they may also be used for habitat assessment in the case of conservation biology. Sonar may be used to detect shoals of fish or whales. Light detection and ranging (LIDAR) can detect atmospheric chemicals and ground features even more accurately than can radar technology. In fact, vegetation change is a key application of LIDAR. Resultant maps and models depict landcover, land use, and vegetation changes such as deforestation and desertification, and prediction of such changes. These are vital for the work of conservation biologists (Campbell, 2017).

Applications of Remote Sensing Techniques to Conservation Biology

There are several recent studies that illustrate the importance of remote sensing to conservation biology. As argued by McDermid et al., (2010, 193) "remote sensing has established a key role in modern wildlife ecology. For decades, remote sensing has been acknowledged as a critical data source of environmental information, theoretically capable of supporting a broad range of ecological applications." The important issues concern the spectral, spatial, radiometric, and temporal resolutions of the remote sensing tools. The impacts on conservation biology studies are strong, because they allow "an impressive suite of wildlife ecology and habitat attributes to be modeled, predicted, and monitored through time, including land cover physiognomy, vegetation structure and condition, forage characteristics, specific nutrient concentrations, overall productivity, and biomass" (McDermid et al., 2010, 193).

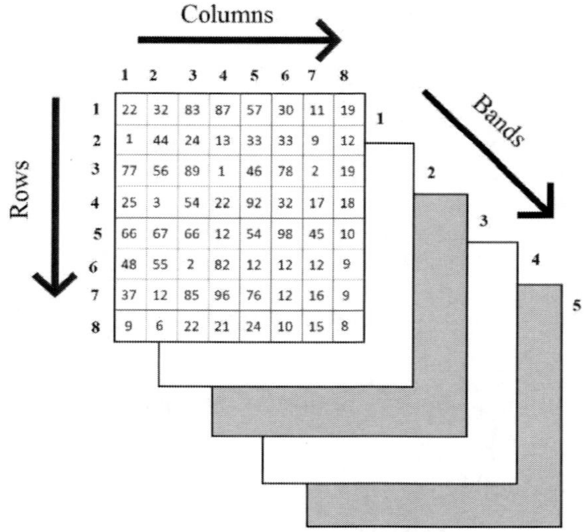

Figure 6. Pixels in a multispectral image. The numbers represent digital numbers of the pixels, which illustrates the intensity of the feature the pixel represents. Higher numbers represent higher intensity for brightness. This example represents a five-band image. A multispectral IKONOS image has four bands: Blue, Green, Red and Near Infrared. A Landsat TM multispectral image has seven bands: blue, green, red, near-IR bands, two SWIR bands, and a thermal IR band.

The most basic application is the mapping and analysis of the correlation between wildlife and landcover (Campbell, 2017). However, some have argued that remote sensing methods do not supplant traditional field methods, but only complement and sometimes spatially extend the field techniques (McDermid et al., 2010). Spatial resolution is of interest in this regard; this most often linked to the pixel size of a digital image. As noted in Figures 5a and 5b, this is variable, from less than one meter to over five meters in more recent imaging technologies (especially after the mid- 1990s) to 30 m x 30 m in the older Landsat Thematic Mapper images. In many cases, the images with smaller pixel sizes (e.g., representing one square meter) represent smaller areas, while those with larger pixels (e.g., representing 30m x 30m) cover larger areas of the earth's surface (Jensen, 2007; Bhatta, 2008; McDermid et al., 2010).

The spectral resolution is defined by the electromagnetic wavelengths where the measuring instrument acquires measurements. The multispectral sensors have varying numbers of bands, up to 20. Hyperspectral systems usually have more than 20, which are nearly contiguous. The quality is determined by the width of the bands, as the narrow bands give more detailed reflectance characteristics, and minimalize the generalization of the wider bands (Wulder et al., 2004; McDermid et al., 2010). This is particularly important, as the themes of conservation biology frequently involve detailed assessments of landcover, the better to discern changes and possible relations between the vegetation variation and wildlife (Campbell, 2017).

Temporal resolution, the frequency at which images are recorded for an area, is also important. The more frequent the survey, the higher the temporal resolution. The time sequence of the images determines the applicability to ecological change, including slower changes such as gradual deforestation and savanna expansion. Higher resolutions may be required to support shorter term studies on rapid urbanization impacts, natural hazards, or even forest and bush fires (Campbell, 1998, 2017). There is also radiometric resolution, which is the precision of the radiance measurement in a data type. This is shown by the range of digital numbers in each pixel. Natural Resources Canada (2016) notes that "the radiometric resolution of an imaging system describes its ability to discriminate very slight differences in energy … The finer the radiometric resolution of a sensor, the more sensitive it is to detecting small differences in reflected or emitted energy." This energy refers to electromagnetic energy. These different resolutions are important, but there are tradeoffs, as there may be different levels of these resolution types in each type of image. For example, datasets with coarse spatial resolution may also have more frequent temporal resolution (Campbell, 2002). Natural Resources Canada (2018) further notes:

> For high spatial resolution, the sensor has to have a small IFOV (Instantaneous Field of View).… this reduces the amount of energy that can be detected as the area of the

ground resolution cell within the IFOV becomes smaller. This leads to reduced radiometric resolution - the ability to detect fine energy differences.

Furthermore, if the radiometric resolution is to be increased, there must be an increase in the amount of energy detected. If this is done without reducing the spatial resolution, it would be necessary to broaden the wavelength range detected for the required channel or band, which would reduce the spectral resolution of the sensor. Coarser spatial resolutions also enable better radiometric and/or spectral resolution (Natural Resources Canada, 2018). As argued by (McDermid et al., 2010), there is no perfect remote sensing data, as the data may be better than others on some types of resolution and less detailed in other types. Datasets with coarse spatial resolution may also have more frequent temporal resolution (Campbell, 2002).

Boyle et al., (2014) argue that these resolution issues may be slowly resolved, as technology advances. These developments may however take time, as the recent developments in this field are expensive and rare, and hence may have limited application. These authors tested the hypothesis that imagery of high spatial resolution (≤5 m) was different in performance and applicability from lower-resolution imagery (≥30 m), with a case study from the Interior Atlantic Forest in Paraguay, a highly heterogenous and recently modified and fragmented landcover. The images were 4 m multispectral IKONOS and 30 m multispectral Landsat imagery.

The findings were that higher-resolution imagery gave a more accurate delineation of cover classes, and more accurately identified small patches and the shapes of these patches. Although these findings identified the stronger merits of the higher resolution images, a survey of three leading conservation journals (Biological Conservation, Biotropica, Conservation Biology) found that high-resolution imagery was limited in use (26.8% of land cover studies used analysis of satellite imagery, and 10.4% of these used imagery ≤5 m resolution). The authors concluded that high-resolution imagery was neglected in conservation research, but its clear utility in monitoring and evaluation of forest requires an increase in its use. The suggestion was for greater access to low-cost, multiband, high-resolution satellite imagery to enhance conservation management and decision-making.

These varied types of resolutions are relevant across the entire field of conservation biology. As argued by McDermid et al., (2010, 202), "the diversity of remote sensing systems and their varying spectral, spatial, temporal, and radiometric resolutions allows for an impressive suite of wildlife ecology and habitat attributes to be modeled, predicted, and monitored through time" The key parameters for such analyses include land cover physiognomy, vegetation structure and condition, forage characteristics, specific nutrient concentrations, overall productivity, and biomass (ibid.). For applications to wildlife studies, especially the determination of habitat suitability and human impacts, a major issue is the prediction of habitat suitability for some species or ecosystem types. This may involve image classification, where pixels in images are categorized to represent land cover classes, representing similarities in spectral and/or spatial properties.

McDermid et al., (2010) cite the instructive case of Cannon et al., (1982) who used Landsat MultiSpectral Scanner (MSS), with a spatial resolution of 80 m dated 1978 to classify shinnery oak (*Quercus havardii*, Rydberg) dominated landcover in Western Oklahoma, US. The findings documented a positive correlation between percentage of grassland habitat and the presence of male lesser prairie chickens (*Tympanuchus pallidicinctus*, Ridgway, 1873). Another study was that of Palmeirim (1988), which used a map derived from Landsat Thematic Mapper (TM) imagery, with 30m spatial resolution to estimate the number of avian species in northeastern Kansas. Another study (Luoto et al., 2004) used similar data to predict the presence of bird species in five classes of landcover in the boreal forests of Finland. An important finding was that geomatics-based methods increased the explanatory power of the topography-moisture models. Similarly, Miller and Conroy (1990) used SPOT High Resolution Visible (HRV) imagery (20 m spatial resolution), which has a finer resolution than Landsat TM, to create a predictive model for the potential habitat for the Kirtland's warbler (*Setophaga kirtlandii* Baird, 1852).

Another study by Rogan et al., (2003), which used multitemporal Landsat TM data, aerial photographs and field studies to analyze land-cover change in San Diego County (1990–1996), USA. The central issue was the degree of change, basically "three hierarchical levels of change classification of increasing detail: change vs. no-change; four classes representing broad increase and decrease classes; and nine classes distinguishing increases or decreases in tree canopy cover, shrub cover, and urban change" (Rogan et al., (2003, 793). Variables considered included variables representing brightness, greenness, wetness, and elevation, fire history, and slope. The study noted the necessity of including non-remote sensing sources (digital ancillary data) for greater and more accurate feature identification.

These applications may be relatively simple, as they consider land-class/species correlations. Analyses based on climate (water and energy) impacts on plant biomass and animal presence are more complex (Hawkins et al., 2003; Francis & Currie, 2003; McDermid et al., 2010). Plant primary production may be detected by the spectral reflectance of near-infrared and visible red bands of the electromagnetic spectrum, in multiple band images. Detection of high photosynthetic activity in vegetation is largely measured by the contrasting reflectance and absorption between the near-infrared region of the spectrum (high scattering and resultant reflectance signal) and the red region of the spectrum (high absorption of the red wavelengths by leaf pigments, and resultant low reflectance).

The Normalized Difference Vegetation Index (NDVI) is an important model based on this type of comparative analysis, measuring or estimating the vegetation attributes such as living including biomass, leaf area index, productivity, photosynthetic activity, and even the leaf nitrogen content (Bailey et al., 2004). These measurements may assess the link between plant biomass and related animal species presence (McDermid et al., 2010). Older studies based on coarse resolution data include that of Wallin et al., (1992), who used 1-km NDVI data from the NOAA Advanced Very High Resolution Radiometer (AVHRR) sensor, and successfully related vegetation landcover variation to predict breeding habitat of the Red-billed quelea (*Quelea quelea*, Linnaeus, 1758). The authors concluded that.

> These results demonstrate that coarse resolution satellite data can be used to monitor potential breeding habitat for the red-billed quelea in East Africa. Although this study was based on the comparison of data from occupied and unoccupied breeding sites, the results provide strong evidence that these satellite data provide an objective index to breeding site suitability (Wallin et al., 1992, 96).

The authors however point to the coarse scale of the imagery, in that "data derived from the Advanced Very High Resolution Radiometer (AVHRR) on board the NOAA series of operational, polar orbiting, meteorological satellites have previously been shown to be quite useful for monitoring vegetation dynamics at scales ranging from regional (10^4 km^2) to global" (ibid.87). Such studies may be replicated at more detailed scales using more sophisticated, recent imagery (Campbell, 2017). Other such studies using coarse, 1-km NDVI imagery include those of Jørgensen and Nøhr (1996) on correlations between avian species presence and vegetative biomass in Senegal and more recently, Skidmore et al., (2003), Bonn et al., (2004), (Amstrup et al., 2004), Stow et al., (2004), Nilsen et al., (2005), Rodriguez et al., (2005) and Van Bommel et al., (2006) on animal species presence and vegetation greenness.

More complicated than these are measurements of energy relations, which can take the form of vegetation productivity and biomass linked to estimates of land surface temperature (Campbell, 2002; McDermid et al., 2010). The Normalized Difference Vegetation Index increases over grassland, hence there is a linear decrease in surface temperature, possibly due to "an increase in latent heat flux away from the surface due to transpiration by grassland plant" and "can be exploited to describe the moisture availability of the landscape" (McDermid et al., 2010, 215). Liu et al., (2017, 4182) support this finding, writing that "the normalized difference vegetation index (NDVI) is a vegetation index derived from reflectance in red and near-infrared wavebands, and has been demonstrated to be strongly correlated with grassland AGB (aboveground biomass) in numerous studies," important as AGB "partially represents primary production" and in grassland ecosystems "acts as a significant indicator of vegetation activity."

Another complicated application of remotely sensed methods is the detection of vegetation characteristics using underlying pigment and chemical composition. These studies require very fine spectral resolution, such as those recorded by the Airborne

Visible/Infrared Imaging Spectrometer (AVIRIS), Compact Airborne Spectrographic Imager (CASI) and the Hyperion sensor on the Earth Observer (EO1) satellite platform. These have been used for recording both photosynthetic and non-photosynthetic (e.g., dead plant materials) landcover. As early as 1992, Roberts et al., (1992) used a hyperspectral AVIRIS airborne image to map non-photosynthetic vegetation (NPV), shade, and soil to predict fire possibilities wildlife habitat. More recently, Jia et al., (2006) used airborne AVIRIS imagery to study montane coniferous in Colorado, US, with substantial measurement of both NPV and bare soil related to fire and other landcover factors (McDermid et al., 2010).

The National Aeronautics and Space Administration (NASA) (2012) defines an active sensor as "a radar instrument used for measuring signals transmitted by the sensor that were reflected, refracted or scattered by the Earth's surface or its atmosphere." By contrast, the passive sensor is "a microwave instrument designed to receive and to measure natural emissions produced by constituents of the Earth's surface and its atmosphere." Therefore, the main distinction is whether the senor transmits its own emissions (active) or uses those produced by the Earth (passive).

McDermid et al., (2010) distinguish between the techniques that use passive reflected data from satellites and aerial sources, based on visible and infrared parts of the spectrum) and active remote sensing methods, which emit energy, and record the energy returned to the emitting instrument. These active sensors have a wider range of operation as they do not require solar illumination, defined as the light and associated energies from the sun (Hamelin et al., 2014). The active sensors are better for sensing where there may be obstructions such as cloud cover (Jensen, 2007). An example of an active sensor is synthetic aperture radar (SAR) which uses microwave wavelengths. For example, Imhoff et al., (1997) used aerial photography and field derived data with P-, L- and C-band SAR data from NASA's airborne AIRSAR system to analyze correlations between landscape variation and avian ecology at Kakadu National Park in Australia. They noted that SAR data documented detailed vegetation structures, which complemented the data derived from multispectral sensors, creating three-dimensional mapping of habitat dynamics (Imhoff et al., 1997).

A more advanced application concerns the documentation of animal movements, within variable landcover (Campbell, 2017). As argued by Neuman, et al., (2015, 8), "animal movement patterns in space and time are a central aspect of animal ecology... Remotely-sensed environmental indices can play a key role in understanding movement patterns by providing contiguous, relatively fine-scale data that link animal movements to their environment." This latter study reviewed published accounts of remote sensing applications to animal movement patterns in terrestrial landcover between 2002 and 2013. This study compared more advanced recent remotely-sensed products with older studies using images with coarser resolution and found merits for both types of studies. However, the study concludes that "greater documentation of the type and resolution of remotely-sensed products in ecological movement studies would enhance their usefulness" and "animal movement studies incorporate remotely-sensed products that provide time series of environmental response variables. This would facilitate wildlife management and conservation efforts, as well as the predictive ability of movement analyses." The solution would be a closer communication between ecologists and remote sensing experts, to enable more correlation between field and image-based research (Neuman, et al., (2015, 8). Particularly important issues concern recent sensor technology that can "link physiological data such as body temperature and heart rate to movement data, thereby directly linking animals' physiological conditions and movement behavior to landscape features" (Neuman, et al., 2015, 8; see also Signer et al., 2010, 75).

Recent developments include movement path analyses (such as state-space models or step selection functions) that can track animal movements based on resource selection over time (Patterson et al., 2010; Thurfjell et al., 2014; Campbell, 2017). Knowledge of animal ecology and related vegetation at different scales is important (Johnson et al., 2002; Leblond et al., 2010). Fine scale data on temporal movements may reveal smaller features such as migratory corridors and stop-over sites, which are important for understanding animal ecology (Takekawa et al., 2010; Morales et al., 2010; Buler & Moore, 2011; Börger & Fryxell, 2012; Cushman et al., 2011; Zeller et al., 2012).

One recent application that helps in the analysis of such complex patterns is LIDAR, or Light Detection and Ranging (National Ocean Service, 2018). This advanced technique uses light pulses that generate detailed three-dimensional information about features on the Earth's surface. Usually based on an aircraft platform, LIDAR apparatus is usually composed of a laser, a scanner, and a global positioning system (GPS) receiver. Ultraviolet, visible, or near infrared light to image several types of target, such as earth materials, precipitation and clouds, aerosols and chemical compounds. Resolutions can be as sharp as 30-centimeters. This precision is the main issue, as LIDAR can generate more accurate models and maps than older sensing methods. Early applications included the measurement of cloud formations by the National Center for Atmospheric Research and the mapping of the moon's surface by Apollo 15 mission in 1971 (Goyer & Watson, 1963; Jensen, 2007). LIDAR products are used for applications in many disciplines, such as geology archaeology, geography, seismology, forestry, atmospheric physics, and some narrower applications such in airborne laser swath-mapping (ALSM) and laser altimetry.

Applications include forest plot inventory, which are of sufficiently sharp resolution to record tree heights, crown widths and diameter, and canopy volume (Wulder et al., 2008, 2012). LIDAR may also be used to create terrain models and record three-dimensional information for the multiple resource inventory. These capabilities are based on the measurement of the three-dimensional (3D) structure of terrain, vegetation, and infrastructure. LIDAR and to a slightly lesser extent interferometric synthetic aperture radar (IFSAR or InSAR) can also distinguish between the vegetation and terrain at the 3D level, and analyze forest structure data, not the least of which is forest inventory and monitoring (Næsett et al., 2004). Development of LIDAR technology includes high-resolution (1–5 m) bare-ground digital elevation models (DEMs), canopy height models (CHM), canopy cover maps, LIDAR intensity images and all returns data set. The last is an archive of all the LIDAR returns including the "pulse number, return number, east coordinate, north coordinate, elevation, and return intensity for each LIDAR return and metadata documenting the LIDAR mission flight parameters, sensor type and settings, GPS control, horizontal and vertical datum, coordinate units and projection, and date and time of mission" (Reutebuch et al., 2005, 291).

CONCLUSION

This chapter has examined the development origins of remote sensing, including historical technologies and recent advanced technological developments such as LIDAR and the applications of these to conservation biology. The key issue concerns the extent to which remote sensing tools and technologies may adapt to the changing needs of conservation biology, especially concerned the requirement for greater detail and precision. Important issues concern the development of flexibility and more precise measurement of ecological patterns, this enabling the detection of increasingly elusive species, complex patterns of distribution, smaller scales and species and longer-range observation possibilities. This will involve the detection of smaller species at greater distances, in smaller ecosystems, more accurate recording of animal behaviors such as feeding, foraging and migratory movements, breeding, mating and the relation of these to landcover change (including decomposition, temperature, moisture, vegetation pattern and species change), and the integration and syntheses of these variables to the documentation of ecosystem variation. The future of conservation biology may be at least partially determined by the strong and flexible application of remote sensing techniques to these current and topical issues.

REFERENCES

American Society of Photogrammetry (1960). *Manual of Photographic Interpretation.* Falls Church, Virginia: ASP.

Bhatta, B. (2008). *Remote Sensing and GIS.* Oxford: Oxford University Press.

Börger, L. & Fryxell, J. (2012). Quantifying individual differences in dispersal using net squared displacement. Dispersal and Spatial Evolutionary Ecology. In Clobert, J., Baguette, M., Benton, T. G., Bullock, J. M. (Eds.) *Dispersal Ecology and Evolution,*

222 - 230. Oxford: Oxford University Press. 10.1093/acprof:oso/9780199608898.003.0017.

Boyle SA, Kennedy CM, Torres J, Colman K, Pérez-Estigarribia PE, de la Sancha NU (2014). High-Resolution Satellite Imagery is an Important yet Underutilized Resource in Conservation Biology. *PLoS ONE 9*(1): e86908. https://doi.org/10.1371/journal.pone.0086908.

Buler J. J. & Moore FR (2011). Migrant-habitat relationships during stopover along an ecological barrier: extrinsic constraints and conservation implications. *Journal of Ornithology 152*(Suppl1): S101–12.

Cannon, R. W., Knopf, F. L. & Pettinger, L. R. 1982. Landsat data to evaluate lesser prairie chicken habitat in west Oklahama. *Journal of Wildlife Management 46*: 916-922.

Coe, B. (1976). *The Birth of Photography: The Story of the Formative Years, 1800–1900*. London: Ash & Grant.

Crouch, T. D. (2008). *Lighter Than Air*. Baltimore: Johns Hopkins University Press.

Cushman SA, Raphael, M. G., Ruggiero, L. F., Shirk, A. S., Wasserman, T. N. & O'Doherty E. C. (2011). Limiting factors and landscape connectivity: the American marten in the Rocky Mountains. *Landscape Ecology 26*:1137– 1149.

Deng, Y. (2005). *Ancient Chinese Inventions*. Beijing: China Intercontinental Press.

Ege, L. A. T. (1974). *Balloons and Airships*. New York: Macmillan.

Falkner, E. & Morgan, D. (2002). *Aerial Mapping*. Boca Raton: Lewis Publishers.

Flett, D. G., Manore, M. J., Ramsay, B. R. & Falkingham, J. C. (2001). *Preparing for operational use of RADARSAT-2 data at the Canadian Ice Service*. 1. 493 - 495 vol. 1. 10.1109/IGARSS.2001.976200.

Frizot, M. (Ed.) (1998). *A New History of Photography*. Köln: Könemann.

Goyer, G. G., Watson, R. (1963). The laser and its application to meteorology. *Bulletin of the American Meteorological Society 44* (9): 564–575.

Hamelin, K. M., Kelley, D. E., Taggart, C. T. & James. M. C. (2014). Water mass characteristics and solar illumination influence leatherback turtle dive patterns at high latitudes. *Ecosphere 5*(2):19. http://dx.doi.org/10.1890/ES13-00158.1.

Hearst Magazines (1936). *Nine Lens Aerial Camera Films 600 Square Miles*. Popular Science. Hearst Magazines, 571.

Herbert, S. (2011). *The Timeline: Telescopes*. Retrieved from https://www.independent.co.uk/news/science/the-timeline-telescopes-2365084.html .

Imhoff ML, Sisk TD, Milne A, Morgan G, Orr T (1997) Remotely sensed indicators of habitat heterogeneity: use of synthetic aperture radar in mapping vegetation structure and bird habitat. *Remote Sensing of Environment 60*: 217-227.

Indian Space Research Organization (2018). *Indian Spacecraft*. Retrieved from https://www.isro.gov.in/spacecraft#block-md-megamenu-1.

Infield, G. B. (1970). *Unarmed and Unafraid*. New York: Macmillan.

Johnson, C. J., Parker, K. L., Heard, D. C. & Gillengham, M. P. (2002). A multiscale behavioral approach to understanding the movements of woodland caribou. *Ecological Application 12*:1840–60.

Jørgensen, A. F. & Nøhr, H. (1996). The use of satellite images for mapping of landscape and biological diversity in the Sahel. International. *Journal of Remote Sensing 17*: 91–109.

Kirschner, E. J. (1985). *Aerospace Balloons – From Montgolfiere to Space*. Fallbrook, CA: Aero Publishers, Inc.

Kossoy, B. (2018). *The Pioneering Photographic Work of Hercule Florence*. New York/London: Routledge.

Leblond, M., Dussault, C. & Quellet, J. P. (2010). What drives fine-scale movements of large herbivores? A case study using moose. *Ecography 33*:1102–12.

Lindsey, R. (2005). *Nimbus: 40th Anniversary*. Retrieved from https://earthobservatory.nasa.gov/Features/Nimbus/.

Liu, Shiliang & Cheng, Fangyan & Dong, Shikui & Zhao, Haidi & Hou, Xiaoyun & Wu, Xue. (2017). Spatiotemporal dynamics of grassland aboveground biomass on the Qinghai-Tibet Plateau based on validated MODIS NDVI. *Scientific Reports 7*. 10.1038/s41598-017-04038-4.

Livingston, S. (2015). Commercial remote sensing satellites and the regulation of violence in areas of limited statehood. *CGCS Occasional Paper Series on ICTs, Statebuilding, and Peacebuilding in Africa No.5*. Philadelphia, PA: University of Pennsylvania.

Luoto, M., Fronzek, S. & Zuidhoff, F.S. (2004). Spatial modeling of palsa mires in relation to climate in Northern Europe. *Earth Surface Processes and Landforms*. 29(11): 1373–1387, doi:10.1002/esp.1099.

Macmillan, N. (1964). *Great Flights and Air Adventures, From Balloons to Spacecraft*. London: G. Bell.

McDermid, G. J., Coops, N. C., Wulder, M. A., Franklin, S. E. & Seitz, N. E. (2010). Critical remote sensing contributions to spatial wildlife ecological knowledge and management. In Cushman, S. A. & Huettmann, F. (Eds.), *Spatial Complexity, Informatics and Wildlife Conservation*, 193-221. Tokyo: Springer.

Mochizuki, S & Murakami, Takuhiko. (2014). Application of remote sensing techniques in wildlife management. *Japanese Journal of Ecology 64*: 253-264.

Models of individual animal movement. *Trends in Ecology & Evolution 23*: 87–94.

Moore, G. K. (1979). What is a picture worth? A history of remote sensing. *Hydrological Sciences Bulletin 24*(4): 477-485, DOI: 10.1080/02626667909491887.

Morales, J. M., Moorcroft, P. R., Matthiopoulos, J., Frair, J. L., Kie, J. G., Powell, R. A., et al., Building the bridge between animal movement and population dynamics. *Philosophical Transactions of the Royal Society B: Biological Sciences 365*: 228–2301.

Næsset, E., Gobakken, T., Holmgren, J., Hyyppa, H., Hyyppa, J., Maltamo, M., Nilsson, M., Olsson, H., Persson, A. & Soderman, U. (2004). Laser scanning of forest resources: The Nordic experience. *Scandinavian Journal of Forest Research 19(*6): 482–499.

National Aeronautical and Space Adminstration (2012). *What are Passive and Active Sensors?* Retrieved from https://www.nasa.gov/directorates/heo/scan/communications/outreach/funfacts/txt_passive_active.html.

National Aeronautics and Space Administration NASA (2005). *Final Update: NASA's UARS Re-enters Earth's Atmosphere*. Retrieved from https://www.nasa.gov/mission_pages/uars/index.html.

National Aeronautics and Space Administration NASA (2018a). *Regions of the Electromagnetic Spectrum*. Retrieved from https://imagine.gsfc.nasa.gov/science/toolbox/spectrum_chart.html.

National Aeronautics and Space Administration NASA (2018b). *Landsat Science*. Retrieved from https://landsat.gsfc.nasa.gov/.

National Aeronautics and Space Administration NASA (2018c). *Early Bird 1*. Retrieved from https://nssdc.gsfc.nasa.gov/nmc/spacecraftDisplay.do?id=1997-085A.

National Aeronautics and Space Administration NASA (2018d). *SAR Lupe 5*. Retrieved from https://nssdc.gsfc.nasa.gov/nmc/spacecraftDisplay.do?id=2008-036A.

National Ocean Service (2018). *What is LIDAR?* Retrieved from https://oceanservice.noaa.gov/facts/lidar.html.

Natural Resources Canada (2018). *Radiometric Resolution*. Retrieved from http://www.nrcan.gc.ca/node/9379.

Neubeck, K. (2007). *OV-1 Mohawk – Walk Around No. 49.* Carrollton, TX: Squadron/Signal Publications, Inc.

Neumann, W., Martinuzzi, S., Estes, A. B., Pidgeon, A. M., Dettki, H., Ericsson, G., & Radeloff, V. C. (2015). Opportunities for the application of advanced remotely-sensed data in ecological studies of terrestrial animal movement. *Movement Ecology 3*(1), 8. http://doi.org/10.1186/s40462-015-0036-7.

Ormsby, J. P. (1987). Whitetail deer food availability maps from thematic mapper data. *Photogrammetric Engineering and Remote Sensing 53*(8), 1081-1085.

Palmeirim, J. (1988). Automatic Mapping of Avian Species Habitat Using Satellite Imagery. *Oikos 52*(1): 59-68. doi: 10.2307/3565982.

Paniccia, A. (2014). *Reshaping the Future: Handbook for a New Strategy: Volume 1 of Strategic Studies.* Venice: Mazzanti Libri - Me Publisher.

Patterson, T. A., Thomas, L., Wilcox, C., Ovaskainen, O., & Matthiopoulos J. (2008). State-space models of individual animal movement. *Trends in Ecology & Evolution 23*(2):87-94. doi: 10.1016/j.tree.2007.10.009.

Ramsay, B., Manore, M., Weir, L., Wilson, K. & Bradley, D. (1998). Use of RADARSAT data in the Canadian Ice Service. *Canadian Journal of Remote Sensing 24* (1): 36–42.

Reutebuch, S. E., Andersen, H. E. & McGaughey, R. J. (2005). Light detection and ranging (LIDAR): an emerging tool for multiple resource inventory. *Journal of Forestry 103* (6): 286-292.

Rogan, J., J. Miller, D. Stow, J. Franklin, L. Levien, and C. Fischer, 2003. Land Cover Change Monitoring in Southern California Using Multitemporal Landsat TM and Ancillary Data. *Photogrammetric Engineering & Remote Sensing 69*: 793-804. DOI: https://doi.org/10.14358/PERS.69.7.793.

Rolt, L. T. C. (1966). *The Aeronauts: A History of Ballooning* - 1783-1903. New York: Walker and Company.

Showstack, R. (2004). New exploration focus will not diminish Earth science agenda, NASA says. Eos. *American Geophysical Union. 85* (5): 46. doi: 10.1029/ 2004EO050003.

Signer, C., Ruf, T., Schober, F., Fluch, G., Paumann, T., Arnolds W. (2010). A versatile telemetry system for continuous measurement of heart rate, body temperature and locomotor activity in free-ranging ruminants. *Methods in Ecology and Evolution 1*: 75–85.

Thurfjell, H., Ciuti, S. & Boyce, M. S. (2014). Applications of step-selection functions in ecology and conservation. *Movement Ecology* 2: 4.

Wakefield, A. (2014). *A Bird's-Eye View of the Battlefield: Aerial Photography.* Retrieved from https://www.telegraph.co.uk/history/world-war-one/inside-first-world-war/part-eight/10742060/aerial-photography-world-war-one.html.

Wallin, D. O., Elliott, C. C. H., Shugart, H. H., Tucker, C. J., Wilhelmi, F. (1992). Satellite remote sensing of breeding habitat for an African weaver-bird. *Landscape Ecology 7*: 87–99.

Wulder, M. A., White, J. C., Nelson, R. F., Næsset, E., Ørka, H. O., Coops, N. C., et al., (2012). Lidar sampling for large-area forest characterization: A review. *Remote Sensing of Environment 121*: 196–209.

Wulder, M. A., Hall, R. J., Coops, N. C. & Franklin, S. E. (2004). High spatial resolution remotely sensed data for ecosystem characterization. *Bioscience 54*: 511-521.

Wulder, Michael A; Bater, Christopher W; Coops, Nicholas C; Hilker, Thomas; White, Joanne C (2008). The role of LiDAR in sustainable forest management. *The Forestry Chronicle 84* (6): 807–826.

Zeller, K. A., McGarigal, K. & Whiteley, A. R. (2012). Estimating landscape resistance to movement: a review. *Landscape Ecology 27*: 777–797.

In: Geomatics and Conservation Biology
Editor: Michael O'Neal Campbell
ISBN: 978-1-53614-468-0

Chapter 5

GIS AND CONSERVATION BIOLOGY

Michael O'Neal Campbell
Camosun College, Victoria, Canada

ABSTRACT

This chapter looks at the discipline of geographical information systems or geographical information science, both commonly referred to as GIS. GIS is defined as a system for the processing and production of information, usually about the earth's surface, aiming at increasing knowledge of the context under study. As such, it has very wide applicability to all the environmental sciences. The chapter examines some of the definitions of GIS and the applications of this science to conservation biology. Connections with the science of remote sensing are also explored. A case study is presented of the application of GIS methods to the uncovering of multidirectional environmental change. Multidirectional change in this sense refers to changes in landcover that do not follow the standardized assumption of linear progression (for example, deforestation, from a forested to open landcover type). In some cases, it is evident that there is more patch dynamic change, where multiple trajectories of change are occurring in different parts of the studied context, with for example deforestation, reforestation, species increases and decreases, etc. occurring. GIS techniques are particularly suited to such analyses, as they incorporate spatial analyses and image processing, and links to remote sensing and statistics, and offer more definitive conclusions on the trajectories of landcover change.

INTRODUCTION

GIS has a broad range of definitions, which include some common themes running. Bartelme (2012, 145) notes that there are many definitions that may be used, and distinguishes between the newer and the preferred older definitions, because the latter "pertain to the still-existing traditional GIS (in contrast to newer paradigms such as geospatial web services)." This author offers several definitions by others: "It serves for capturing, storing, analysis, and visualization of data that describe a part of the Earth's surface, the technical and administrative entities, as well as findings of geoscience, economics, and ecological applications"; "it is an information system with a database of observables of spatially distributed objects, activities, or events, which can be described by points, lines, or surfaces" (Dueker, 1979); "It is a comprehensive collection of tools for capturing, storing, retrieval, transformation, and visualization of spatial data of the real world for special applications (Burrough, 1979); "it is an information system containing all spatial data of the atmosphere, the Earth's surface, and the lithosphere, allowing the systematic capture, update, manipulation, and analysis of such data, based on a standardized reference frame" (Göpfert, 1991); and finally, "it is a system for decision support which integrates spatial data in a problem-solving environment (Cowen, 1988).

The GIS systems may include data capture, administration, analysis and presentation systems. Based on these and other definitions and descriptions, a GIS comprises several components, including:

> A collection of spatial data plus the corresponding functions for storage and retrieval, a collection of algorithmic and functional tools, a set of hardware and software components needed for the handling of geospatial data, a special type of information technology, a gold mine for answers to geospatial questions, and a model of spatial relationships and spatial reconnaissance (Bartelme, 2012: 148).

The history of GIS is usually linked to several key scientists who developed strands that would later combine to create the modern science and technology termed GIS. One important development occurred in 1960, when the federal Department of Forestry and Rural Development in Ottawa, Canada developed possibly the world's first GIS. This system was designed and developed by Roger Tomlinson, and named the Canada Geographic Information System (CGIS) (Tomlinson, 2003). The system stored and analyzed data for the Canada Land Inventory, which determined the land capability of rural Canada, including spatial information soils, agriculture, recreation, wildlife, waterfowl, forestry and land use. This CGIS development was rated as superior to previous attempts at computer mapping, due to the more advanced techniques for overlays, measurements, digitizing and scanning, and inclusion of a national coordinate system, and attribute and locational information. Other later developments were that of the Laboratory for Computer Graphics and Spatial Analysis at the Harvard Graduate School of Design (LCGSA 1965–1991), by Howard T. Fisher, which designed concepts of spatial data software code and systems, including SYMAP, GRID, and ODYSSEY that enabled later technological developments. These and other developments led to the 1986 Mapping Display and Analysis System (MIDAS), later renamed MapInfo for

Windows. Eventually, GIS were available on the internet and could be used or downloaded through the World Wide Web (Fu & Sun, 2010; Goodchild, 2010; Maliene et al., 2011).

GIS AND CONSERVATION BIOLOGY

The field of conservation biology is especially amenable to applications from GIS systems, due to the breadth of the two disciplines and the complex problems that require application and solution (Campbell, 2017). The objective focus of conservation biology was noted by Soulé and Wilcox (1980), who define conservation biology as "a mission driven discipline comprising both pure and applied science" and one that has "techniques to solve problems." Meine (2010: 7) also describes the "main thread" of conservation biology as "the description, explanation, appreciation, protection, and perpetuation of biological diversity" all these requiring spatial analytical models for advanced understanding (Meine, 2004, Campbell 2015).

The field of conservation is the base of conservation biology. Two definitions of the broad field of conservation are: (1) The "planned management of a natural resource or the total environment of a particular ecosystem to prevent exploitation, pollution, destruction or neglect and to ensure the future use of the resource" (Encyclopedia Britannica-Micropedia, 1998, 3, 533); and (2) "The concerns and strategies surrounding the protection of natural resources from overuse or degradation" (Encyclopedia Americana 1998, 17, 618). Beattie (1995) gives three related reasons for conserving biodiversity: function or acknowledgment of the role of biodiversity in the ecosystem; beauty; and profit. This reveals the complex nature of conservation mindsets, which have dominated the perception of conservation. These include those of: scientists focused on the role of experts in environmental protection; policy makers, deriving their legitimacy from scientific positions, emphasised in protection schemes; and some social scientists who argue for local participation in conservation schemes as the most likely positive approach (Campbell, 2017).

The establishment and merit of protected areas is internationally significant, but many efforts at conservation are insufficient and hence unsuccessful, despite the work of organizations such as the World Conservation Union, and subunits such as the Species Survival Commission (Rabb & Sullivan, 1995). One problem is the insufficient application of technological bases for assessment and analysis of the issues at hand (Campbell, 2017). Curio (1995) also argues that some failures result from limited funding for such projects, the result of the low scientific and political priority that such projects obtain in policy, donor and science circles.

Even where funding is adequate, the very broad reach and requirements of conservation projects may require the application of GIS technology to widely varying topics, both social and environmental (Campbell, 2002, 2017). For example, Burbridge and Wallace (1995), Subak (2000) and Zumeta (2001) point out that for biodiversity conservation to succeed, there must be an integration of the daily social activities of various actors (individuals, communities, NGOs, business groups and governments) with the aims and methods of conservation policy and management, combined with inventories of environmental variables (Ludwig, 1995; McNeely, 1995; Prance 1995; Cole, 1995; Lajeunesse et al., 1995; Nilsson & Grelsson, 1995; Akcakaya, 2000; Trakolis et al., 2000). The competing land-uses of the areas near conserved areas must also be analyzed (as argued by Burroughs and Clark, 1995, in a study of Yellowstone and Georges Bank National Parks) and the human activities must be evaluated as part of the larger socio-environmental system (Salwasser et al., 1995; Quon et al., 2001).

The development of information bases and associated methodologies for the effective monitoring and evaluation of such areas is vital. Optimal strategies for landscape assessment are debatable, because the analysis of land-use is necessarily an interdisciplinary endeavor, being an assessment of diverse human activities in relation to complex spatial and temporal variations in the landscape (Campbell, 2017). An important issue concerns the integration of the interpretative and analytical methods of the social sciences with the arguably more rigorous methods of the natural sciences, within a GIS model. For example, Harberl et al., (2001) describe the importance of comparative studies and the specification of human impacts, where interfacial studies of landscape ecology and social history are envisaged. Within the narrower field of protected area management, new difficulties arise, concerned primarily with short term and small-scale assessments of impacts, and flexibility of methods, in addition to the interfacial issues of the natural and social sciences. The goal is an adequate information base for regular, possibly daily, management decisions (Campbell, 2017).

The principal contribution of GIS to the conservation sciences is the recording of more accurate measurements of organism behavior and locations, and temporal and spatial variations of land-use change and relations between contextual variables, such as landcover, habitat, topography and land-use (Campbell, 2017). For example, the examination of the landscape variation changes in visible representations of plant structures (crown height, position, perimeter canopy dynamics and open spots, etc. assessed by the parameters of tone, shape, texture, size and pattern of the features) over time reveal phytogeographical change. Also, the size and shape of urban features also give evidence of the rate and character of urbanization (Jensen, 2007). Finally, these may be analysed in relation to animals as habitat variables (Campbell, 2017).

For landcover assessment, remotely sensed images may be the starting point for GIS processes. Aerial photographs generally have a finer resolution than satellite images and this allows more detailed feature identification (Campbell, 2002). SPOT and especially Landsat TM images have a lower spatial resolution. This creates difficulties for the measurement of the parameters of tone, texture, shape, pattern and size, especially without supporting information, such as field data or aerial photographs (Turner, 1990;

Innes & Koch, 1998). Nevertheless, problems may emerge even with high-resolution aerial photographs, especially where the coverage is of areas of high species and stand density (Campbell, 2002). This may reveal image interpretation as "an art of probability, based on the assessment and judgement of evidence" (Rubben, 1960, 109).

For the effective assessment of environmental relations and subsequent competent conservation policy and management, there must be reliable information bases and analytical methods. The accurate documentation of environmental change became a very important issue in the 1990s (Fairhead & Leach, 1995; Leach & Mearns, 1996; Tiffen et al., 1994) and has further developed during the 21st century with light detection and ranging (LIDAR), radio detection and ranging (RADAR) or interferometric synthetic aperture radar (IFSAR) which can produce more sophisticated products such as digital elevation models, LIDAR-based images and other advanced results (Klemas, 2001; Wing & Johnson, 2001, Jensen, 2007). Frequently, in the past there were inaccurate assessments of socioenvironmental relations, based on less reliable data and inadequate syntheses. These have created dominant paradigms and mindsets that have influenced information bases, factored for the development of GIS and contributed to policy actions.

Many of the applications of such landcover studies have been based on the African continent. This is generally because the ecological changes on that continent are perceived to be severe, hence requiring a strong research effort for the documentation of processes and promotion of solutions (Campbell, 2002). One major area of research is on the dynamics of forest/savanna boundary change. Past attempts to document such landscape variation, based on field measurements and existing literature with an inadequate spatial and historical base, have created disputed narratives of socioenvironmental relations and environmental change.

Several recent studies have used integrated research methods with GIS (aerial photographs and/or satellite imagery, ground surveys, social research and historical data, with tabulated information) to examine the forest/savanna mosaics of Africa. The results frequently challenge the currently popular narrative of linear deforestation under the influence of human mismanagement (Elliot & Campbell, 2002). In some cases, critical analyses rather indicate a scenario of multi-directional environmental change. Environmental, economic and socio-cultural dynamics create multiple effects on the environment, from degradation to enrichment. In one famous study, conducted in the forest margin zone of Guinea, Fairhead and Leach (1995, 1023) used aerial photographs, the testimony of local respondents, colonial records and oral history, and thus derived a "a vegetation history that sharply contradicts the deforestation analysis…" and including "incontrovertible evidence" of increased or stable forest areas between 1952 and 1994 (*ibid.*). This contrasted with the established narrative of linear deforestation. In a similar study, of the Machakos Reserve of Kenya, Tiffen et al. (1994) found that land degradation, called the "Machakos problem" (*ibid.* 3) which was believed to be caused by the five-fold population growth and agricultural intensification, was less pronounced than expected. Photographic evidence from 1937 and 1991, showed that in many areas badlands had been replaced by farmlands, "erosion scars" (*ibid.*) were reduced, tree numbers increased, and grazed fields mostly remained unchanged.

Other studies have supported the existing narrative of deforestation. For example, in the Guinea savanna of the Upper West region of Ghana, Gyabaah (1992) used similar research tools to Fairhead and Leach (Land Satellite Multi-Spectral Scanner (MSS) images from 1972 and 1989, local ground surveys, and semi-structured interviews). The results showed deforestation as the main form of change. This was like a study by Lieberman (1979) on the coastal savanna of Ghana, which used aerial photographs, quadrats, social surveys and archival data, and found deforestation (Lieberman, 1979; Campbell, 1998). In a broader study, using similar research tools, and covering eight countries (Ghana, Congo, Cote de Ivoire, Ethiopia, Mali, Senegal, South Africa and Zaire), Millington et al. (1992) found that deforestation was evident in several areas, with cases where drier savanna woodland was replaced by farm/shrub/grass mosaics, with dynamic forest boundaries. There has also been an increase in the application of GIS methods to such research. This is largely due to the current availability of such technology and the higher awareness of the greater scientific precision and possibilities for quantification, statistical analysis and synthesis of multiple sourced data (Campbell, 2017). However, despite the high analytical possibilities of GIS, the problems inherent in the images as described above exist. These include the reliability of statistical analyses, imprecise rectification and registration of images, insufficient synchronization of times series images and field studies, and consequently misleading results (Campbell, 2017).

In the North American context, an important issue for GIS application is the monitoring of landscape change (largely forest fragmentation and urban encroachment). For example, Wing and Johnson (2001) demonstrated in a study of the McDonald forest in Western Oregon that forest managers could use GIS methods to analyze recreation density information. Wickham et al. 2000) also described a GIS-based analysis of the relationship between forest fragmentation and economic development. As the loss of temperate forests is argued to be a serious concern (as also argued by Wilcove et al., 1986; Opdam, 1991; and Wickham et al., 2000) their study investigates the hypothesis that forest fragmentation can be quantitatively related to urbanization.

In the case study of the southside economic region of Virginia, USA, the results showed that there was a moderate correlation between forest fragmentation and the degree of urbanization pressure. It was argued that this finding is like the results of other studies (LaGro & DeGloria, 1992; Turner et al., 1996) which show that the extent of landcover change declines with increasing distance from urban areas. An important note is however made that many studies consider a situation with a single urban centre. More complex is a situation where there are several urban centres each exerting pressure on the forest ranges from several points, an issue that Wickham et al. (2000) argue must be more fully investigated.

These research tools have also been applied to zoogeographical research. The common view is that habitat destruction leads to animal species decimation and even local extinction, as is well described by Tilman et al. (1994). Some studies have integrated the methods of the field sciences and GIS to investigate this hypothesis. A particularly interesting study of this type was that of Mace et al. (1999). The objective of this research was the development of "a method for evaluating the cumulative effects of human activity on grizzly bear (*Ursus arctos*, Linnaeus, 1758) habitat in Montana" (ibid. 367). TM satellite imagery, radio collaring of bears and categorization of human activities by roads, trails and movements were analyzed with geomatics software, revealing land-use/mammal presence relationships important for local ecology.

An important development relevant to GIS and conservation biology is the advancement of the science of geostatistics. Geostatistics is defined as the "study of phenomena that vary in space and/or time" (Deutsch, 2002). Olea (2012, 1) further defines geostatistics "as a collection of numerical techniques that deal with the characterization of spatial attributes, employing primarily random models in a manner similar to the way in which time series analysis characterizes temporal data." This is echoed by Isaaks and Srivastava (1989): "Geostatistics offers a way of describing the spatial continuity of natural phenomena and provides adaptations of classical regression techniques to take advantage of this continuity." The Oxford American Dictionary (2018) defines geostatistics as dealing "with spatially autocorrelated data." Zhang (2009, 1) distinguishes between statistics and geostatistics in that: "classic statistics is generally devoted to the analysis and interpretation of uncertainties caused by limited sampling of a property under study… geostatistics however deviates from classic statistics in that geostatistics is not tied to a population distribution model that assumes, for example, all samples of a population are normally distributed and independent from one another."

Geostatistics aims to predict a property's possible spatial distribution, usually through maps (when the prediction is through estimation, an estimate map is produced; when it is through simulation, several possible maps may be produced, the differences between these multiple products displaying the quantification of the uncertainty) (Zhang (2009, 1). Hengl et al. (2009, 492) note that as "geostatistics evolved in the mining industry, it for a long time meant statistics applied to geology…geostatistical techniques have been successfully applied to applications in numerous fields, ranging from soil mapping, meteorology, ecology, oceanography, geochemistry, epidemiology, and human geography, to geomorphometry." It now refers to "a branch of statistics that specializes in the analysis and interpretation of any spatially (and temporally) referenced data."

Bellier et al. (2010) demonstrate the application of geostatistics techniques to animal counts. They describe a hierarchical model based on geostatistics to analyze spatial structures such as trends, patchiness and random fluctuations. This model is intended to cope with the heterogeneity of wildlife population habitats, which may compromise the analyses offered by other methods. What is required is a hierarchical model that assumes that the count data follow a Poisson distribution (a discrete frequency distribution that gives the probability of a number of independent events occurring in a fixed time). The results of their analysis, using the example of fin whale (*Balaenoptera physalus*, Linnaeus, 1758) data, illustrates the utility of this method for the mapping of animal density in a context assumed to be non-stationary. This demonstrated the usefulness of geostatistics for studies related to conservation biology (Bellier et al. (2010).

Case Studies Using GIS to Map Multi-Directional Change

Two case studies are presented based on publications by Elliott and Campbell (2002) and Campbell (2013), covering the environs of Accra, Ghana and the greenbelt of Ottawa, Canada, respectively (Figure 1 and 2). Both use a multidirectional methodology, which uses matrices that depict multi-directional landcover change. Table 1 shows how the results of GIS analysis (image stitching, digitizing, map creation, crosstabulation analysis, tabulation of the results) can be tabulated into a matrix that depicts multidirectional change, using a fictional example. In multidirectional change, some areas change from one landcover type to another in both directions, while some other areas retain their landcover. Linear percentage change may be recorded, but it is multidirectional change that explains the environmental dynamics of the area under study.

In this example, the variables in the column lose area to the variables in the rows, and where there is common overlap, the figure represents no change. In this example, the forest cover declined from 50% to 30% between 1990 and 2005. However, this simple linear statement is not the whole story. The column shows that in the studied area (the 100% in the bottom right hand corner) forest cover in 1990 overlapped with the forest cover in 2005 by in 20% (second row, second column) of the studied area. As shown the first column, the forested cover lost area to shrubs (15% of the study area), grass (10% of the study area) and bare soil (5% of the study area). The forest cover also gained area (as shown in the second row), from shrub cover (7% of the study area) and grass cover (3% of the study area). It did not gain against bare soil covered areas.

Table 1. A Fictional Example of Multidirectional Change (1990 – 2005)

Landcover	Forest	Shrub	Grass	Bare Soil	2005
Forest	20	7	3	0	30
Shrub	15	10	8	7	40
Grass	10	2	0	3	15
Bare Soil	5	1	7	2	15
1990	50	20	18	12	100

Numbers represent percentages. The total area is the 100% in the bottom right corner.

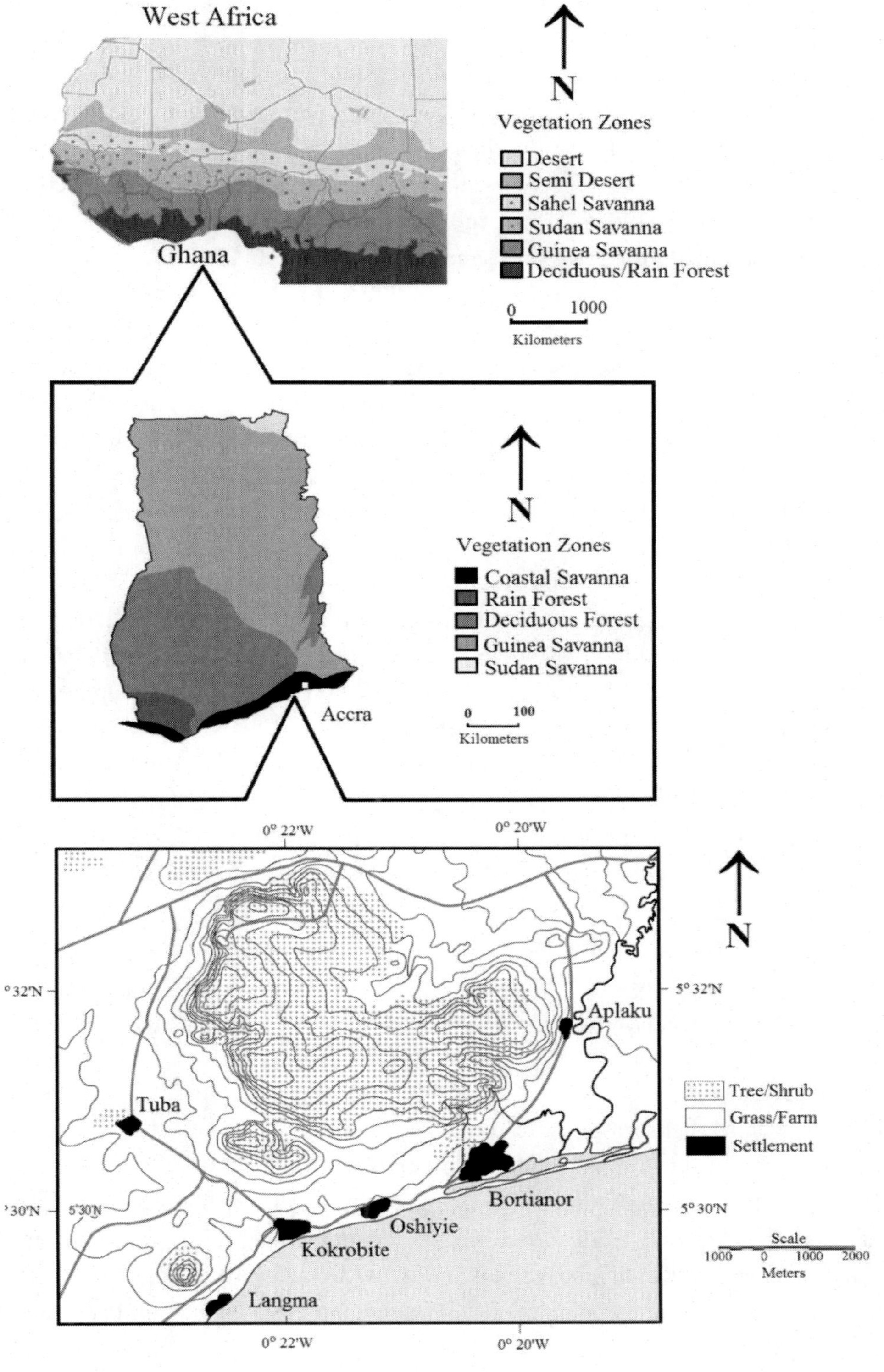

Figure 1. The Tuba-Aplaku Plains of the Coastal Savanna of Ghana.

Shrub covered areas increased from 20% (1990) to 40% of the studied area (2005, third row and third column). However, this was not a linear increase, as the column shows that shrub covered landcover lost to forest (7% of the studied area), grass (2% of the studied area) and sand (1% of the studied area). Hence, shrub cover losses amounted to 10% of the studied area. Only 10% of the study area that was shrub cover in 1990, was still so in 2005. Conversely, shrub cover gained against forest (15% of the studied area), grass cover (8% of the studied area) and sand (7% of the studied area). Similar assessments may be made for grass and sand covered areas. In the case of grass cover, none of the area covered by grass in 1990 was still covered by grass in 2005, indicating that the grass cover in 2005 represents new growth. Overall, this example exhibits patch dynamic change, with a general trend towards deforestation. This type of matrix may be reversed, so that the variables in the rows lose against those in the column, which require the switching of the dates in the bottom row (left) and the right column (top). It may also use actual areas, rather than the percentages of the total studied area as this example does. It is also possible to state the percentage of the landcover types lost or gained (i.e., the percentage of forest or grass lost or gained) rather than expressing the gains or losses as percentages of the total area, as in this example.

In both case studies (Accra and Ottawa), similar matrices were used, and the main issue was forest change, within a larger environment that included settlements and more open vegetation cover. The Accra case study was based on the Tuba/Aplaku plains of the coastal savanna of Ghana (Figure 1). located about 20 km southwest of the capital, Accra, containing the villages of Bortainor, Oshiyie, Kokrobite, Tuba, Aplaku and Langma, with populations increasing from 2000 persons in 1948 to over 7000 by 1984 and over 10,000 by 1996. The vegetation type was composed of a patch mosaic of trees, shrubs and grasses, mostly *Azadirachta indica*, *Mangifera indica, Andropogon guyanus* and *Panicum maximum* (Campbell, 1998).

In the case of Accra, the study materials were black and white aerial photographs at the scale of 1:10000, flown in September 1960 and a SPOT satellite image dated September 1998 (Elliot & Campbell, 2002). The information derived from the images, using the IDRISI GIS system was supported by ground truthing and social surveys, which included semi-structured interviews with over 500 respondents in the Tuba-Aplaku plains of the coastal savanna that surrounds the city of Accra.

Despite the existing narrative of linear deforestation in the study area, the GIS based analysis found substantial multidirectional change, as shown for two areas in Table 2 and 3 below. Forest included all tall, dense vegetation, including primary and secondary forest. This was necessarily subjective, as it is difficult to define the difference between a tree and a shrub. As argued by Menaut (1983) there is no clear distinction between trees and shrubs. However, this study went by the definition that trees are plants with a distinct bole or trunk exceeding seven to eight meters in height, while shrubs may be shorter and low branched with short, "contorted" stems, although such distinctions may be "unrealistic", as under varying conditions, tree/shrub definitions may overlap (ibid). A limitation for this definition is that firewood cutting may create low stemmed and branched trees that pass for shrubs: a good example being the 'stunted' neem trees (*Azadirachta indica*) that dominate certain 'thicket' contexts of the coastal savanna of Ghana (Campbell, 1998).

In these examples, there was significant deforestation, with the main replacement being shrubland and farmed areas. In the first study area, the Aplaku/Densu, the shrub dominated areas expanded, but in the second study area, the Kokrobite/Oshiyie area, the same landcover type declined. In this case study, factors for deforestation were short farming fallows, plot expansion, firewood cutting, bush-burning for hunting, cattle herding and grazing, road building and urban development (housing and road construction). These activities varied between the two selected locations within the case study, hence the different levels of forest stand change (Campbell, 1998; Elliott & Campbell, 2002).

Table 2. Matrix of landcover change in the Aplaku/Densu area of the Coastal Savanna of Ghana

Landcover	Forest	Shrub	Farm	Urban	Other	1998
Forest	6	1	<1	<1	0	7
Shrub	31	24	2	<1	1	59
Farm	7	6	<1	<1	<1	14
Urban	8	5	<1	1	2	17
Other	<1	<1	0	0	1	2
1960	52	36	4	4	4	100%

The variable in the column lost area to the variable in the row between the dates, leaving only the overlap (e.g., where the forest column meets the forest row. The numbers refer to the percentage of landcover out of the total study area.

Table 3. Matrix of landcover change in the Kokrobite/ Oshiyie area of the Coastal Savanna of Ghana

Landcover	Forest	Shrub	Farm	Urban	Other	1998
Forest	31	11	3	<1	<1	45
Shrub	11	5	<1	<1	<1	15
Farm	7	6	1	<1	<1	14
Urban	16	2	<1	<1	<1	21
Other	1	1	0	<1	<1	5
1960	66	24	5	3	2	100%

This area shows less deforestation than the Aplaku/Densu area, but also less shrub expansion, but similar farm and urban expansion.

The gains by forest against other landcover types is particularly important, as it illustrates the possibilities of reforestation in scattered areas, in zones considered to be undergoing linear deforestation. The main factors for reforestation appeared to be longer farm fallow periods (less intense cultivation), cessation of farming (due to factors such as farmer bankruptcy, emigration, or change in livelihood activity), vegetation protection for religious purposes and/or changes in farming patterns. While the shift in cultivation practice in the previous decades from shifting cultivation (3 – 5-year fallow periods) to permanent plots contributed to deforestation, the later shift from permanent plots to other non-farming activities (such as trading and urban jobs) contributed to some reforestation, balanced by increased urban land use that removed some forest stands. In the field survey, new forest stands were composed of both the species in older and previous stands, or new species entirely, creating a mixed secondary forest variably related to the local primary forest stands. Therefore, in some cases the vegetation after human modification was richer, pointing to human activities as factors for both landcover degradation and enrichment. The result was a patch dynamic landcover that may only be fully evaluated by complex GIS modelling (Campbell, 1998; Elliott & Campbell, 2002).

In the Ottawa study (Figure 2), landscape change in the Western Greenbelt of Ottawa was analyzed using IDRISI, PCIWorks, ArcView and Corel Draw and air photos and satellite images dated between 1934 and 1999 for some areas, and 1955 and 1999 for others. This study was based on the Greenbelt, which is part of the National Capital Region (NCR) of Ottawa. The Greenbelt covers 20,000 hectares in eastern Ontario, within the Saint-Laurent Ecoregion (Campbell, 2002, 2013). The GIS model was supported by information from five sources. Satellite images (Landsat TM, dated 1994 and 1999), with a resolution of 30 meters per pixel aerial photographs (dated 1934, 1955, 1964, 1970, 1977, 1987, 1994, 1999), maps (urban, vegetation and farming) dated 1879 and 1991, field measurements (field photographs, vegetation observations), and historical literature and demographic data for landcover identification purposes.

For the construction of the matrices, there were six attribute classes for the landcover. Urbanized areas comprised buildings, roads, pavements other concrete infrastructure, including vegetation (tree lined avenues, parks, grassed edges and suburban gardens). Forests included all tall, dense vegetation, such as coniferous, deciduous and swampy forest. Shrubland included light vegetation; grasses, shrubs, denuded areas in non-farmed areas, fallows and all cultivated areas, with some mixed fallow grasses. The category classified as "other" included open water such as ponds and bogs, and quarry areas with large expanses of bare soil and rock surface.

Table 4 shows figures covering the whole study area for the period 1955 to 1999. Table 5 shows the matrixed results of the crosstabulation of images dated in pairs (1955 and 1999). Table 4 shows there has been very dynamic landscape change between 1955 and 1999. Comparison of similar classes for the two dates shows that in no case was more than 50 percent of the area under any category in 1955 still under the same category in 1999. The largest change was urban expansion into shrubland (farmland, grassed meadows, and light mixed vegetation). Forest expanded into areas previously under lighter vegetation.

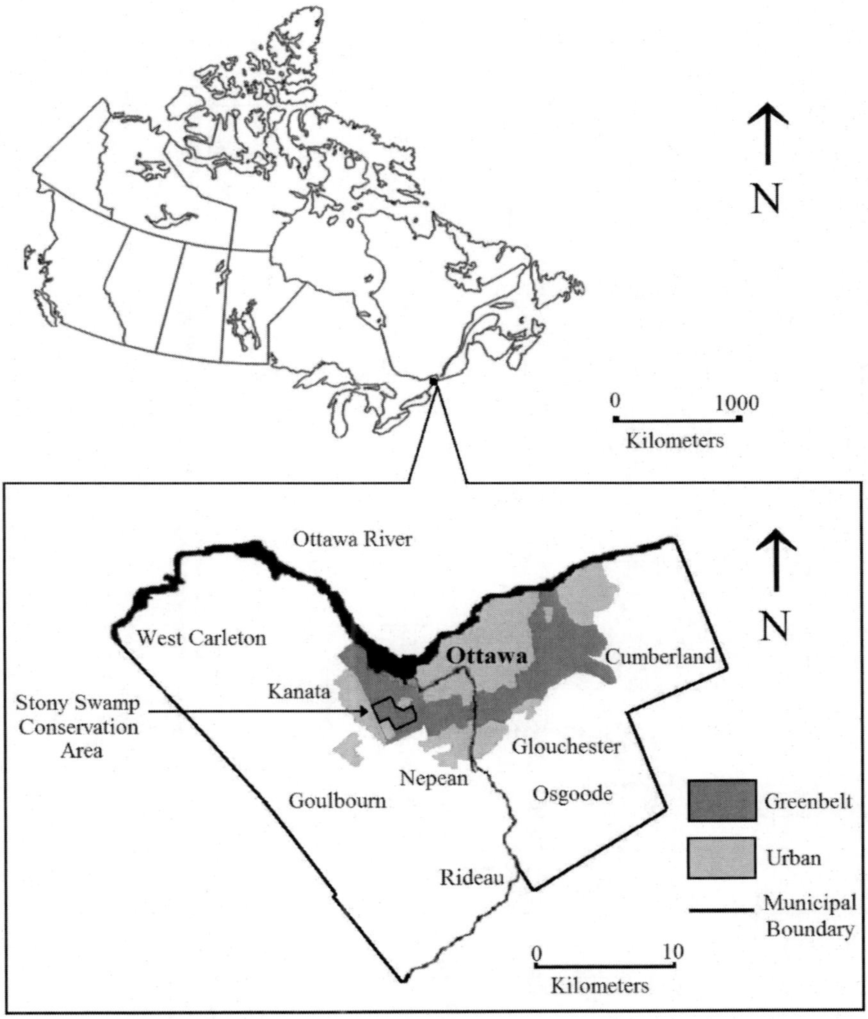

Figure 2. The Stony Swamp Conservation Area of the Western Greenbelt of Ottawa, Canada.

Table 4. Landscape change in the study area by percentage (1955-1999)

Classes	1955	1964	1970	1977	1987	1994	1999
Forest	17	21	22	26	26	26	28
Shrub	80	71	66	59	52	49	43
Urban	3	8	11	12	19	21	25
Other	0	0	1	3	3	4	4
Total	100%	100%	100%	100%	100%	100%	100%

Despite the increased forest cover, only about half of the area under forest in 1955 was still forest in 1999, indicating new forest growth. Shrubs lost to both forest and urban, while forest lost only minimally.

Table 5. Matrix of landscape change in the study area by percentage (1955-1999)

Classes	Forest	Shrub	Urban	Other	Total 1999
Forest	9	18	<1	0	28
Shrub	4	40	<1	0	43
Urban	3	19	3	0	25
Other	1	3	0	0	4
Total 1955	17	80	3	0	100%

Landscape change in the study area by percentage (1955-1999), shows the cross-tabulation of the results for 1955 (columns) against 1999 (rows) based on the percentage of total area (approximated to the first decimal point: changes less than 0.1 are taken to be insignificant).

The result in Table 4 may be summarized as: a large decline of farmland and related cover (80 to 43 percent); an increase in urban areas (3 to 25 percent); and an increase in forested areas (17 to 28 percent). Table 5 shows these figures in matrix format.

Table 6. Landscape Change in Stony Swamp by Percentage (1955 to 1999)

Classes	1955	1964	1970	1977	1987	1994	1999
Urban	1.5	1.5	1.9	4.1	6.9	7.2	7.6
Light Veg	68.7	63	62.1	43.8	37.6	34.6	29.8
Dense Veg	29.6	35.4	35.7	50.2	53.8	56.1	60.4
Other	0	0.1	0.1	1.5	1.7	1.9	0.8
Total	100	100	100	100	100	100	100

Table 7. Landscape Change in Stony Swamp (1955-1999)

Classes	Urban	Light Veg	DenseVeg	Quarry	Total 1999
Urban	1.2	5.9	0.5	0	7.6
Light Veg	0.1	23.5	6.2	0	29.8
Dense Veg	0.1	38.2	22.3	0	60.4
Other	0	0.4	0.4	0	0.8
Total 1955	1.5	68.7	29.6	0	100

Table 8. Collisions/Road Kill of Wild Animals (1994 – 2000)

Year	1994	1995	1996	1997	1998	1999	2000	Total
Total Kill	283	266	390	493	520	551	685	3188
Wild Kill	268	248	363	476	493	536	671	3055
Domestic	15	18	26	17	27	15	14	132
Deer Collision	261	241	355	469	479	502	628	2935
Human Injury	4	8	8	16	18	17	17	88
Car Damage	257	233	347	453	461	485	611	2847

Source: Adapted from Brown (2000).

Within the western Greenbelt area under study, is the Stony Swamp Conservation Area. This area was under stricter conservation protection, so it is instructive to make a comparison with the wider greenbelt area. Table 6 shows the changes of features in the Stony Swamp Conservation Area from 1955 to 1999, and Table 7 shows the same changes in matrix format. These show an increase in urban features within Stony Swamp. There has also been a high exchange of light vegetation and dense vegetation (6.2 percent from light to dense and 38 percent from dense to light). Dense vegetation also lost some area to urbanization. This result, evaluated in conjunction with the overall percentage of the two vegetation classes (dense vegetation, 29.6 to 60.4 percent, 1955 to 1999; and light vegetation 68.7 to 29.8 percent, 1955 to 1999) shows a largely linear expansion of forest in this area. Both light vegetation (5.9 percent) and dense vegetation (0.5 percent) lost some area to urban expansion. The increased road kill (Table 8) could be due to the narrowing of wildlife corridors, road widening, increased human proximity and/or traffic increase (Table 9). Reforestation may have increased wild mammal populations. Most of the road kills where small mammals such as deer, porcupines and skunks (Campbell, 2002).

Table 9. Vehicles Visiting Stony Swamp Trails (December 1999)

Day	Number of vehicles	Percentage of total
Monday	531	12
Tuesday	534	13
Wednesday	585	14
Thursday	565	13
Friday	519	12
Saturday	719	17
Sunday	817	19
Total	4269	100

Source: National Capital Commission NCC (1999).

CONCLUSION

This chapter has demonstrated key points in the application of GIS to conservation biology topics. The field to be covered is vast, hence only the selected topics were cited. The main point concerns the analytical capabilities of GIS and the vast possibilities for application in the emerging field of conservation biology. Both GIS and conservation biology are emerging and changing disciplines. The future of their interactions is positive, with the key issues being both the possibilities for analysis of large areas and the variable landcover of such areas, and the detection of environmental variables at ever more detailed scales of resolution. Other points concern the capabilities of GIS applications to analyze spatial correlations, areas that traditionally were covered by conventional statistics. These must also be linked to remote sensing tools that produce the required images. The syntheses of these strands will determine the future of GIS applications to conservation biology.

The key idea of this chapter is that despite numerous problems, GIS methods provide more precision and hence reliability to socioenvironmental analysis, as a basis for the knowledge bases for planning. The evidence presented in the case studies supports this contention. However, some suggestions must be made for the more effective application of GIS methods. The technical aspects of the work (software development, accessing scanning, registration, rectification, image mosaics, digitizing, data tabulation and analysis must be done critically, and any inaccuracies noted. This will minimalize the creation of inaccurate results, which may form the basis of unreliable conclusions. There must be adequate secondary, supporting, data, such as social and field studies. For this to be possible, the fieldwork must be comprehensive, and well synchronized with the software-based analysis. The fieldwork methodology must be participatory, and multiple-sourced (archival, government documented, field science and even public media), the better to reduce the likelihood of the construction of inaccurate or ill-informed results. In terms of the resolution and scale of the time series images, the scale must be large enough for details (such as vegetation identification, habitat assessments and micro-impacts of the livelihood activities). For the analysis of features, there must be variable, numerous and reliable indicators (for example vegetation, soil, settlement). These allow more accurate identification of features in similar or different situations and images.

GIS methods in application to conservation biology topics may therefore contribute the quantification of habitat analyses, wildlife counts, linear land cover change and multi-directional, patch dynamic and contextual changes that result from continually transitional socio-environmental relationships. The development such reliable appraisal methods, especially in areas of conservation requirements may enable more informed decision making and multiple actor collaboration, and better ecological understandings, especially with continual technical and social developments.

REFERENCES

Akcakaya, A. R. (2000) Conservation and Management for Multiple Species: Integrating Field Research and Modelling into Management Decisions. *Environmental Management*, *26*, S75-S83.

Bartelme, N. (2012). Geographic Information Systems. In Kresse, W. & Danko, D. M. (Eds.) *Springer Handbook of Geographic Information*, 145 – 174. New York: Springer.

Bellier, E. & Monestiez, P. & Guinet, C. (2010). Geostatistical modelling of wildlife populations: a non-stationary hierarchical model for count data. In Atkinson, P. M. &

Lloyd, C. D. (Eds.) *GeoENV VII – Geostatistics for Environmental Applications*, 1-12. 10.1007/978-90-481-2322-3_1.

Brown, C. (2000). *Deer Collisions*. Ottawa: City of Ottawa.

Burbridge, A. A. & Wallace, K. J. (1995). Practical Methods for Conserving Biodiversity. Conserving Biodiversity. In Bradstock, R. A., Auld, T. O., Keith, D. A., Kingsford, R. T., Lunney, D. & Sivertsen, D. P. (Eds.). *Conserving Biodiversity: Threats and Solutions*, 11 - 26. Chipping Norton: Surrey Beatty and Sons.

Burrough, P. A. (1979). *Principles of Geographical Information Systems for Land Resources Assessment*. Wotton-under-Edge: Clarendon.

Burrough, P. A. & Mcdonnell, R. (1998). *Principles of Geographical Information Systems*. Oxford: Oxford University Press.

Burroughs, B. H. & Clark, T. W. (1995). Ecosystem management: a comparison of Greater Yellowstone and Georges Bank. *Environmental Management, 1* (5), 649 - 663.

Campbell, M. (2015). *Vultures: their Evolution, Ecology and Conservation*. Boca Raton: CRC Press.

Campbell, M. (2013). *Biogeography, Environmental History and Sustainability in Coastal Ghana*. New York: Nova Science Publishers.

Campbell, M. (2002). *The Assessment of Biogeographical Dynamics in the Western Greenbelt of Ottawa using Geomatics*. MA. Thesis, Carleton University, Ottawa.

Campbell, M. (1998). *Interactions Between Biogeography and Rural Livelihoods in the Coastal Savanna of Ghana*. PhD. Thesis, University of London (Imperial College).

Cowen, D. J. (1988). GIS versus CAD versus DBMS: What are the differences? *Photogrammetric Engineering & Remote Sensing*, *54*, 1551–1555.

Curio, E. (1995) Conservation needs ethology. *Trends in Ecology and Evolution*, *11*(6), 260-263.

Deutsch, C. V. (2002). *Geostatistical Reservoir Modeling*. Oxford: Oxford Univeristy Press.

Dueker, K. J. (1979). Land resource information systems: A review of fifteen years of experience, *Geo Processing*, *1*, 105–128.

Elliott, J. A. & Campbell, M. (2002). The environmental imprints and complexes of social dynamics in rural Africa: cases from Zimbabwe and Ghana. *Geoforum*, *33*(2), 221-237.

Fairhead, J. & Leach, M. (1998) *Reforming Deforestation: Global Analysis and Local Realities - Studies in West Africa*. London: Routledge.

Fu, P. & Sun, J. (2010). *Web GIS: Principles and Applications*. Redlands: ESRI Press.

Goodchild, M. F. (2010). Twenty years of progress: GIScience in 2010. *Journal of Spatial Information Science, (1)*. doi: 10.5311/JOSIS.2010.1.2.

Göpfert, W. (1991). *Raumbezogene Informationssysteme*. [Spatial information systems.] Heidelberg: Wichmann.

Haberl, H., Erb, K. H. & Krausmann, F. (2001). How to calculate and interpret ecological footprints for long periods of time: the case of Austria 1926 - 1995. *Ecological Economics*, *38*(1), 25 - 45.

Hengl, T., Minasny, B. & Gould, M. (2009). A geostatistical analysis of geostatistics. *Scientometrics*, *80*(2), 491–514.

Innes, J. L. & Koch, B. (1998). Forest biodiversity and its assessment by remote sensing. *Global Ecology & Biogeography Letters*, *7*, 397-419.

Isaaks, E. H. & Srivastava, R. M. (1989). *An Introduction to Applied Geostatistics*. New York: Oxford University Press.

Jensen, J. R. (2007). *Remote Sensing of the Environment: An Earth Resource Perspective, 2nd Edition*. Upper Saddle River: Pearson Education Inc.

Klemas, V. V. (2001). Remote sensing of landscape-level coastal environmental indicators. *Environmental Management*, *27* (1), 47–57.

LaGro, J. A. & DeGloria, S. D. (1992). Land-use Dynamics within an Urbanising NonMetropolitan County in New York State (USA). *Landscape Ecology*, *1*(4), 27

Lajeunesse, D., Domon, G., Drapeau, P.D., Cogliastro, A. & Bouchard, A. (1995). Development and application of an ecosystem management approach for Protected Natural Areas. *Environmental Management*, *19*(4), 481- 495.

Leach, M. & Mearns, R. (1996). *The Lie of the Land: Challenging Received Wisdom on the African Environment*. London: James Curry.

Lieberman, D. D. (1979). *Dynamics of Forest and Thicket Vegetation on the Accra Plains, Ghana*. Ph.D. Thesis, University of Ghana, Legon, Accra.

Ludwig, D. R. (1995). Assessment and management of wildlife diversity in an urban setting. *Natural Areas Journal*, *15*(4), 35.

Mace, R. D., Waller, J. S., Manley, T. L., Ake, K. & Wittinger, W. T. (1999). Landscape evaluation of grizzly bear habitat in Western Montana. *Conservation Biology*, *13*(2), 367 -377.

Maliene, V., Grigonis, V., Palevičius, V. & Griffiths, S. (2011). Geographic information system: Old principles with new capabilities. *Urban Design International* 1–6. doi:10.1057/ udi.2010.25.

McNeely, J. A. (1995). Keep the all the pieces: Systematics 2000 and world conservation. *Biodiversity and Conservation*, *4*(5), 51.

Meine, C. (2004). *Correction Lines: Essays on Land, Leopold, and Conservation*. Washington, D.C: Island Press.

Meine, C. (2010). Conservation Biology: Past and Present. In Sodhi, N. S. & Ehrlich, P. S. (Eds.), *Conservation Biology for All*, 7 – 26. Oxford: Oxford University Press.

Menaut, J. C. (1983). The vegetation of African savannas. In Boulrliere, F. (Ed.), *Tropical Savannas*. Amsterdam: Elsevier Scientific Publishing Company.

Millington, A. C, Styles, P. J. & Critchley, R. W. (1992). Mapping forests and savannas in Sub-Saharan Africa from advanced very high resolution (avhrr) imagery. In

Furley, P. A., Proctor, J. & Ratter, J.A. (Eds.), *Nature and Dynamics of Forest-Savanna Boundaries*, 37-62. London: Chapman and Hall.

National Capital Commision (1999). *Project of Research on Gatineau and the Greenbelt: Study on Traffic in Stony Swamp-Summer Season*. Ottawa: NCC.

Nsiah Gyabaah, K. (1992). *Environmental Degradation and the Threat of Drought and Desertification in the Upper West Region of Ghana: A Search for Sustainable Development Solutions.* Ph.D. Thesis University of London, Wye College.

Opdam, P. (1991). Metapopulation theory and habitat fragmentation: a review of Holarctic breeding bird studies. *Landscape Ecology*, *5*(2), 9.

Olea, R. A. (2012). *Geostatistics for Engineers and Earth Scientists.* Berlin: Springer Science & Business Media.

Oxford American Dictionary (2018). *Geotatistics.* Oxford: Oxford University Press.

Quon, S. P. Martin, L. R. G. & Murphy, W. (2001). Effective planning and implementation of ecological rehabilitation projects: A case study of the Regional Municipality of Waterloo (Ontario, Canada). *Environmental Management*, *27*(3), 421-433.

Rabb, G. B. & Sullivan, T. A. (1995). Coordinating conservation: global networking for species survival. *Biodiversity and Conservation*, *4*, 536-543.

Salwasser, H., Doyle, H. T. & Kessler, W. B. (1995). Biodiversity generates new views on old issues. *Forum for Applied Research and Public Policy*, *10*(2), 121 - 123.

Soulé, M. E. & Wilcox, B. A. (Eds.) (1980). *Conservation Biology: An Evolutionary Ecological Perspective.* Sunderland: Sinauer Associates.

Subak, S. (2000). Forest protection and reforestation in Costa Rica: Evaluation of a Clean Development Mechanism Prototype. *Environmental Management*, *26*(3), 283-297.

Tiffen, M., Mortimore, M. & Gichuki, F. (1994) *More People, Less Erosion: Environmental Recovery in Kenya.* Chiche: John Wiley and Sons.

Tilman, D., May, R. M., Lehman, C. L. & Nowak, M. A. (1994). Habitat destruction and the extinction debt. *Nature*, *371*, 65–66.

Tomlinson, R. F. (2003). *Thinking about GIS: Geographic Information System Planning for Managers.* Redlands: ESRI Press.

Trakolis, D., Platis, P. & Meliadis, I. (2000) Biodiversity and Conservation Actions on Mount Voras, Greece. *Environmental Management*, *26*(2), 145 - 151.

Turner, M. G. (1990). Spatial and temporal analysis of landscape patterns. *Landscape Ecology*, *4*, 21–30.

Turner, M. G., Wear, D. N. & Flamm, R. O. (1996). Land ownership and land-cover change in the South Appalacian Highlands and the Olympic Peninsula. *Ecological Applications*, *6*(4), 1150-1170.

Wickham, J. D. O'Neill, R. V. & Bruce-Jones, K. (2000). Forest fragmentation as an economic indicator. *Landscape Ecology*, *15*, 171-179.

Wilcove, D. S., McLellan, C. H. & Dobson, A. P. (1986). In Soule, M. (Ed.), *Habitat Fragmentation in the Temperate Zone. The Science of Scarcity and Diversity*, 237-256. Sunderland: Sinaur Associates.

Wing, M. G. & Johnson, R. (2001). Quantifying forest visibility with spatial data. *Environmental Management*, *27*(3), 411– 420.

Zhang, Y. (2009). *Introduction to Geostatistics — Course Notes*. Retrieved from ftp://ftp.ciat.cgiar.org/DAPA/projects/Cursos_Talleres/Curso_R/DOCUMENTOS/Geosta1.pdf.

Zumeta, D. C. & Ellefeson, P. V. (2000). Conserving the biological diversity of forests: Program and organisational experiences of state governments in the United States. *Environmental Management*, *26*(4), 393-402.

In: Geomatics and Conservation Biology
Editor: Michael O'Neal Campbell
ISBN: 978-1-53614-468-0

Chapter 6

GPS, Animal Tracking and Conservation Biology

Michael O'Neal Campbell
Camosun College, Victoria, Canada

Abstract

This chapter examines the development of Global Positioning Systems (GPS) for animal tracking and their applications to the key issues of conservation biology. It also presents case studies of wildlife conservation and enumerates parameters that would be the focus of advanced GPS and associated techniques. GPS are the basis for advanced techniques in animal tracking and compare favorably with other tracking methods. Cases are cited from the relevant literature, and hypothetical cases are also examined, where ecological problems may be solved using tracking devices to record measurable parameters. The object would be to record the movement of animals, as individuals or as part of migratory patterns, to determine their ecological relations with their encountered environments and the spatial dynamics that may affect their presence and movements. This involves the examination of the parameters that would be used for the applications of geomatics techniques, especially recent, advanced techniques for tracking and locational analyses of target animals. It is concluded the GPS systems are vital for current studies of conservation biology, but such studies and methods must be supported by related geomatics methods, parameter identification and social studies.

Introduction

Global Positioning System tracking devices are a recent development in animal tracking (Campbell, 2017). The history of wildlife tracking is an interesting trajectory of technological development, priorities in ecological management and environmental change. One development is the emergence of the field of biotelemetry, which is defined as the "remote detection and measurement of a human or animal function, activity, or condition" (Merriam-Webster, 2018). Slater (1965, 81) defines biotelemetry as "the instrumental technique for gaining and transmitting information from a living organism and its environment to a remote observer." As argued by Cooke et al. (2004: 334):

> Remote measurement of the physiology, behavior and energetic status of free-living animals is made possible by a variety of techniques that we refer to collectively as 'biotelemetry'. This set of tools ranges from transmitters that send their signals to receivers up to a few kilometers away to those that send data to orbiting satellites and, more frequently, to devices that log data.

These authors note that "the last major reviews on biotelemetry" (by this they are referring to studies by Burggren (1987), and Butler (1989)) "concluded that, throughout the 1990s, we could anticipate major advances in our understanding of how animals interact with each other and their environment owing to these technologies."

Prior to the development of GPS and biotelemetry, there were more primitive methods available to record animal movements. For hundreds of years, bird ringing or banding was developed, which allowed the tracking of birds over long distances (Wood, 1945). Later, in the 1940s, radar was used to observe flocks of migrating birds (Russell & Gauthreaux, 1998). There was also acoustic telemetry, developed in the 1950s on the principles of sonar (acronym for sound navigation and ranging), which was developed in the early 20th century to detect submarines, with later methods able to detect animal movements in water (Cronkite et al., 2007; Hastie, 2012).

Farve (2018) lists the main tracking methods currently available as "VHF (very high frequency) radio-tracking, GPS tracking, satellite tracking using the Argos system and a combination of the advantages of GPS and the Argos satellites into a GPS/satellite telemetry method." VHF telemetry, may involve using a hand-held antenna to receive VHF transmissions from a VHF transmitter (possibly from a collar attached to an animal). The transmitter location (and that of the animal) can be determined by triangulating the transmissions from several receivers (Farve, 2018). GPS tracking telemetry uses a GPS receiver to connect with signals from some of the GPS satellites orbiting the earth. Signals from three satellites locate a receiver in two dimensions (e.g., latitude and longitude), while signals from four satellites locate a receiver in three dimensions.

These methods usually involve the attachment of a GPS receiver (collar or harness) to an animal, the receipt of signals being the indication of the animal's location over time. The location and time of the animal's movements are stored in the receiver, this data can later be retrieved by capturing the animal or remotely to a receiver, or remotely to the Argos Satellite system (termed GPS/ satellite telemetry), which was established in 1978

(Tomkiewicz et al., 2010). The last system mostly uses the two polar-orbiting Argos satellites, which are operated jointly by the United States (National Oceanic and Atmospheric Administration and National Aeronautics and Space Administration) and the French government (French Space Agency) and data processing centers in Landover, Maryland and Toulouse, France.

Animal detection is reasonably accurate, usually between 100 meters (m) and 4 kilometers (km). Hence, a combination of GPS and satellite technology allows more technical accuracy. The data stored in the GPS/satellite located on the tracked animal can be transmitted to the Argos satellite, enabling the distribution of higher resolutions of movement data (Farve, 2018). This finer resolution is particularly important, as ecosystem studies may require tracking of animals of varying sizes and movement ranges, which may be difficult to detect using the satellite technology without the GPS, and such applications are fundamental to some geomatics-based applications in conservation biology (Campbell, 2017).

Recent Advanced GPS Methods in Conservation Biology

Several new methods related to GPS have been developed, in adaptation to the changing demands of conservation biology. These largely involve recording the life activities of organisms within their complex ecologies. Some of these developments have been undertaken by private companies undertaking contract, commercial or consultancy work. For example, Wildlife ACT (2018) documents the use of life saving tracking collars, intended for the tracking of endangered species like cheetah, African wild dog, rhino and leopard. This method employs various types of tracking collars, such as radio, GPS and satellite collars. Wildlife ACT's monitors conduct daily tracking of the target animals, enabling research on injuries, illegal trapping, movement patterns, population demographics and inter-species interactions (Wildlife ACT, 2018).

Crucially, Wildlife ACT is also developing anti-poaching collars. These are designed with reinforced plates and rivets, ostensibly to prevent animals from choking in poachers' snares. The collars can send out signals when the target animal is stationary for short period, allowing the monitors to track the animal and release it if it is ensnared. There are also Rhino Ankle Collars (replacing the less suitable and generally unsuccessful neck collars and horn implants), designed by Wildlife ACT and the radio tracking company, Animal Trackem. The rhino ankle collars allow for larger devices and batteries, GSM (cellphone) and Satellite units.

There is also anti-poaching transmitter technology, which is fitted on animals (e.g., on the horn of a rhinoceros) and transmits information about the animal's location and movements and can transmit an alarm when the animal is stressed, and even if (in the case of a rhino) the horn is being hacked off. This technology is vital, enabling the capture of the poachers in the act of the removal of the animal from snares. There is also remote camera trapping, which is based on remote-activated camera traps. Movements trigger the camera to take snapshots through infrared or motion sensors, and record animal presence. This is especially effective in the monitoring of more elusive species, the scarcity of which may hamper conservation efforts (Wildlife ACT (2018). The evolution of animal monitoring devices in response to such changing demands for conservation biology has been noted, especially in the 21st century. Clark et al. (2006, 334) note that

> An improved global positioning system (GPS)–based animal tracking system is needed to meet quickly evolving demands of ecological research, range livestock production, and natural resource management. Commercially available tracking systems lack the data storage capacity needed to frequently collect animal location data (e.g., 15-minute intervals or less) over long-term deployment periods (e.g., 1 year or more). Some commercial systems have remote data–download capabilities, reducing the need to recapture tagged animals for data retrieval, but these systems download data via satellite (Argos), global system for mobile communications (GSM), cellular telephone, or telemetry radio frequencies.

Satellite systems are argued to be very expensive, and the coverage of the GSM cellular system within the United States is limited. Additionally, the radio-based systems are limited by the narrow-band, very-high or ultra-high frequencies, which require the user to adjust for narrow frequency selections. "None of these existing systems were designed to provide continual, real-time data access" (Clark et al., 2006, 334). They therefore tout the Clark GPS Animal Tracking System (Clark ATS), designed for various stakeholders, including ecologists, animal ethologists, livestock producers and natural resource managers.

The Clark ATS system is state of the art in that it uses memory-card technology, which allows storage of up to 8 gigabytes and data stored can be remotely downloaded by spectrum radio transceivers, without the requirement of narrow-band radio frequency allocations. Remote data Program uploading is enabled by the same method. The radios transmit a GPS-location beacon, using a user-defined time interval, to Clark ATS base stations up to 24 km distant. Crucially, and like some of the Wildlife ACT models, the Clark ATS advanced applications allow the detailed evaluation of target animal behavior over extended periods. This allows the monitoring, not only of animal distribution patterns, but also activities and responses to short-term disturbances, and the simplification of relocation efforts for tagged animals (Clark et al., 2006, 334).

Numerous criticisms have been leveled against the recent, highly technical methods of tracking, such as harm to animals, high cost, lack of trained personnel and neglect of human understanding of nature from a more hands on approach (Tomkiewicz et al., 2010; Meikle, 2011; Matthews et al., 2013). Andrews (2013) compares "traditional tracking" and "electronic surveillance" and poses the question: "has wildlife research become too high-tech?" The older form of tracking, listed as observation of paw prints, scats and chews on wood, are increasingly popular, especially among the amateur wildlife enthusiasts. Matthews et al. (2013: 65) point out that "common problems encountered were associated with collar design, the GPS, VHF and timed-release components, and unforeseen costs in retrieving and refurbishing collars." Examples of problems include the testing of "shock collars" on wolves, these giving the animal shocks when the animals ventured into fences of sensors under satellite control; the devaluation of tourist experiences due to sightings of collars on most observed animals; discomfort of collars on some animals (e.g., on howler monkeys, some attempting removal due to grooming needs); overenthusiasm among the scientific community for highly technical applications; and the reduction of knowledge about primal nature with no technical interference.

Andrews (2013) therefore notes the concerns of some critics, who argue that "old-fashioned wildlife tracking… not only results in usable data but heightens awareness of sounds, smells, visual clues and surroundings…surveillance technology tends to erode a sense of wildness and a feeling of mystery around that "other world," out there. This could hinder conservation efforts rather than promote them. These issues may be balanced with the arguments for more highly technical applications, supporting arguments for critical, balanced and highly multidisciplinary methods for wildlife surveillance. Applied social research has been linked to the development of technology, this encompassing the coevolution of human values (such as interest and support for conservation and ecotourism,), the organization of the frameworks of research and policy, technological advancements and applications, and the environment. This may be explained or modelled using the basic model of Norgaard's (1994) Coevolution Framework, depicted in Figure 1.

Wilmers et al. (2015) document the rise of ever more sophisticated and widely used technology and point to the application of such developments to conservation biology. They note that:

> Great leaps forward in scientific understanding are often spurred by innovations in technology. The explosion of miniature sensors that are driving the boom in consumer electronics, such as smart phones, gaming platforms, and wearable fitness devices, are now becoming available to ecologists for remotely monitoring the activities of wild animals. While half a century ago researchers were attaching balloons to the backs of seals to measure their movement, today ecologists have access to an arsenal of sensors that can continuously measure most aspects of an animal's state (e.g., location, behavior, caloric expenditure, interactions with other animals) and external environment (e.g., temperature, salinity, depth) (Wilmers et al., 2015: 1741).

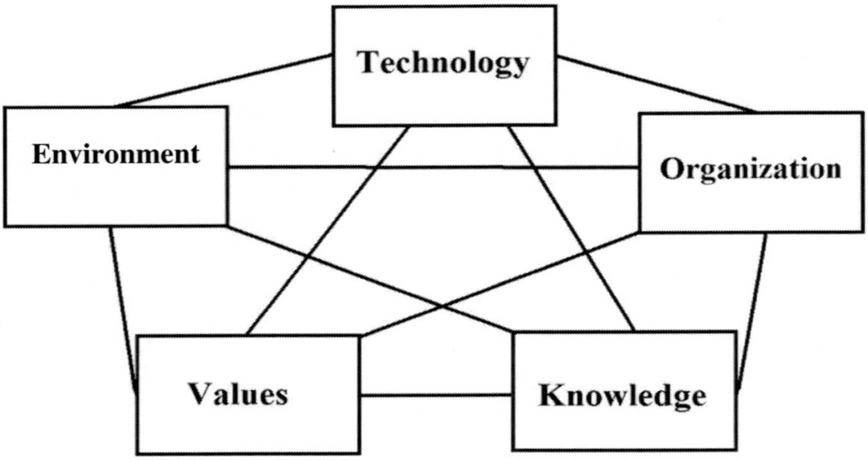

Figure 1. The coevolutionary relationship of technology and its social and environmental context. Derived from Norgaard (1994). Here technology centers around the development of GPS technology and the social and environmental issues of its application.

Specifically, the contributions to conservation biology and ecology include the answering of questions concerning the animal's behavior, physiology, ecology, elusiveness, and developing and analyzing the veracity of theoretical hypotheses. These analyses include the collection of ecological data, as well as "the physiological performance, energetics, foraging, migration, habitat selection, and sociality of wild animals" (Wilmers et al., 2015: 1741). Exploring the roots of these applications, the authors examine the recent proliferation of technological advancements from which the applications to ecology emerge. These include smart phones and miniature sensors such as accelerometers, magnetometers, GPS and digital cameras.

The most important development is the ability of these advanced machines to measure the more detailed behaviors and abilities of target animals, including its movements and physical, livelihood activities. Wilmers et al. (2015) list examples, including tri-axial accelerometers, which can detect a fish opening and closing its mouth (Viviant et al., 2014), the number of calories that a cougar burns in stalking and attacking prey (Williams et al., 2014), and the detection of fungi on insects such as cockroaches (Wilson et al., 2014). In some cases, cameras can even be designed to observe animals eating (Watanabe & Takahashi 2013). There are also tri-axial digital compasses, or magnetometers, which can record minute movements and even the gaze direction of animals (Mitani et al., 2010). Some sensors are able to measure the daily energy budgets, heart and stomach functions, and metabolic rates, swimming speeds, body temperatures, body and jaw acceleration, use of oxygen when diving and oxygen stores, heat flux, stomach pH and temperatures, hunger and satiation indicators, torpor and hibernation extensions, and reproductive states (Cooke et al., 2004, 2008, 2013; Costa & Sinervo,

2004; Ponganis, 2007; Rutz & Hays, 2009; Barron et al., 2010; White et al., 2013). Wilmers et al. (2015: 1741) further argue that "this technology is revolutionizing our ability to study animal ecology by allowing researchers to (1) answer questions about the physiology, behavior, and ecology of wild animals that would have previously been limited to tests on model organisms in highly controlled settings; (2) study cryptic or wide-ranging animals that have previously evaded investigation; and (3) develop and test."

One theoretical method for the application of these sophisticated methods is termed the ''laboratory-to-field'' method (Wilmers et al., 2015: 1741). Here, the sensors are calibrated to adapt to similar species or the same species to be observed in laboratory conditions. The example is cited of research by Williams et al. (2014) who calibrated a GPS–accelerometer collar on caged cougars, which walked on the ground and on a treadmill, to match the collar accelerometer signatures to the energetic costs of locomotion, focused on specific gaits, conditioned by the open-flow respirometry on the treadmill movements. These parameters were then tested on wild ranging cougars, the results indicating that the energy cost of kills by the cougars was positively correlated with the size of the prey killed. Increased levels of low activity, such as stalking, and ambush decreased the hunting energy cost. Evolving from these sophisticated methods, interest in the energy expenditure of animal transport, termed the "energy landscape" has increased research on environmental and social factors that may affect the energy of animals and the relations with range (Shepard et al., 2013). Another relevant study is that of Mosser et al. (2014), which also examined the energy landscape of woodland caribou, by using combined accelerometer–GPS collars on captive caribou in the first instance, then applying the method to wild caribou in northern Canada. This was a novel application, because it was capable of recording caribou calorie expenditure in normal activities such as foraging in different ecological settings, such as variable snow depths, temperature levels and vegetation densities. This represented an important source of information for the factors for the decline of caribou populations (Mosser et al., 2014; Wilmers et al., 2015).

Case Studies of Applications of GPS

The impact of GPS on conservation biology currently follows the trends noted above. As the issues of conservation biology become more complicated, generalization is increasingly difficult (Campbell, 2017). Therefore, GPS must adapt to these changes and include more detailed and flexible techniques. The most important requirement is to move beyond the simple documentation of animal movements and record the micro-activities of animal life, such as their foraging, breeding and predator avoidance on an hour by hour or even minute by minute basis. Such an application would utilize the applications described above (Clark et al., 2006; Wildlife ACT, 2018). The following case studies examine the problems and prospects of the applications of such technology to the current issues of conservation biology. Important points concern the elusive nature of some animals, their individualist and adaptive behavior, small scattered populations and recent interactions with human dominated landcover.

Tracking the Energy Expenditure of Wolves: A Case Study of Wolves

Augliere (2015) gives a case study of GPS tracking of wolves (*Canis lupus*, Linnaeus, 1758), documenting their everyday activities; arguing that "conventional collars show scientists where in the forest a wolf is, but new technology also tells you what the animal is doing hour by hour." An important justification concerns failure of older tracking systems that merely pointed to the location of the wolves and could not record the metabolic cost of the different activities of the wolves. Without this information, it would be difficult to assess the impact of environmental and human induced stresses on the wolves and their ecological relations. The case study concerns the work of scientists at the University of California at Santa Cruz, which analyzed the impacts of outfitting wolves with advanced, transmitting collars. The technology was based on the type of accelerometers used for smartphones and fitness trackers, which record the burning of calories during athletics or running. The parameters for calorie burning were set in laboratory work, to detect the correlation between calorie burn and behaviors with different levels of exertion, in other words devising an energetic "signature" for each measured activity. The resultant technology was tested on wolves and other animals, including polar bears, narwhals, and mountain lions.

The Metabolism and Activities of Cheetahs

The understanding of the ecological relations of cheetahs (*Acinonyx jubatus*, Schreber 1775) has so far been hampered by the lack of appropriate methodological tools (Wilmers, et al., 2015). Cheetahs are usually assessed as the fastest land mammals, but most studies of the mechanics of their speed and hunting acumen use only speed cameras and videos (Weise et al., 2017). Wilson et al. (2013) attempted to answer more detailed questions on cheetah hunting ability by using combined GPS, accelerometer, and gyroscope collars on wild cheetahs in Botswana, with the sampling data calibrated at up to 300 times a second. These measurements were based on measurements of the velocity, forward and lateral acceleration, and travel paths of 897 runs of five cheetahs over 17 months. Several important findings emerged from this novel study. The cheetah's maximum hunting speed was 25.9 m/s (58 mph). The mean top speed of 14.9 m/s generally lasted for only 1–2 seconds per chase. Rapid deceleration at the close of the chase was the most important factor for the success of the chase, rather than maximum speed, number of turns, or distance traveled. Foraging and chases occurred in both open grassland (48% of kills) and dense vegetation (24% of kills). This vegetation cover was insignificant as a factor for kill success.

This type of study contributes to studies that used older methods such as field observation and radio collars. For example, Mills et al. (2004, 178) used these older techniques to test several hypotheses:

> Our hypothesis is that cheetah hunting behavior varies as a function of habitat, and that the species is not exclusively a hunter on open plains. It is able to hunt adequately in savanna woodland habitats, which may even provide benefits not found on open plains. Testable predictions of this hypothesis are: (i) chase distances vary with habitat and are shorter in woodlands; (ii) hunting success rates (kills/hunting attempts) are comparable across habitats but will be higher in more open habitats; and (iii) cheetah kills are kleptoparasitised less in wooded habitats than in open habitats.

To test these hypotheses, seven adult cheetahs were radio-tracked during 1987-1990 in south eastern Kruger National Park, the data comprising: radio-location recordings of animal locations, surrounding habitat type and possible kill data; continuous short-term observations of cheetah movements for 2-15 hours; and continuous long-term observations over 14 days. The continuous tracking was carried out with vehicular transport and visual contact. Other data included parameters of prey encounters, such as: species, sex, approximate age of prey (using different tooth forms for juvenile and adult full permanent dentition), chase distance using the vehicle's odometer, time spent at the carcass and kleptoparasitism). Kills could be failures or successes, or cheetahs did not attempt to kills. This study used personal observation, with possible error. GPS type studies such as that by Wilson et al. (2013) seek to overcome personal error by using more advanced technology. Hopefully, the added detail of these more advanced studies will contribute to such studies (Wilmers et al., 2015).

Another study examined the possibilities of predator competition on cheetahs, in this case the impacts of invading lions (*Panthera leo*, Linnaeus, 1758) and hyenas (*Crocuta crocuta*, Erxleben, 1777) on feeding cheetahs (Scantlebury et al., 2014). Here, GPS collars were used to locate cheetahs for hunting observations, these methods supported by the collection of feces and double-labeled water techniques for assessment of energy expenditure. The findings were that theft levels of 25% by lions and hyenas influenced cheetahs to hunt 1.1 more hours each day. This increased the daily energy expenditure of the cheetahs by 12%. These findings illustrated the utility of animal borne sensors to both locate animals and record aspects of their life activities. Wilmers et al. (2015) nevertheless argue that the study could have gone further and used sensors to estimate energetics.

Foraging Ecology of Herbivores

The foraging ecology of herbivores is crucial for the herbivores, the vegetation they feed on, and the predators that depend on these species. Older techniques for analyzing foraging behavior involved simple visual observation. Newer methods may use animal-borne sensors to measure previously undetectable parameters. Wilmers et al. (2015) mention the use of tilt sensors, which are accelerometers designed to measure the orientation of the observed animal's head in terms of gravity). There are also devices that use sound to measure activities such as chewing or silence to detect nearby predators. Lynch et al. (2014: 77) describe a study using acoustic data collection, in which ten wild adult female mule deer (*Odocoileus hemionus*, Rafinesque, 1817) were studied using collar mounted acoustic recording devices (positioned against the animal's throats) to track audible behaviors and related sounds. Sounds detected included those of vocalizations, footsteps, foraging, rumination (regurgitation, mastication, and swallowing), and respiratory processes. The tests on wild, free ranging animals was preceded by experiments on captive deer. GPS radio collars also recorded the movements of the deer.

An important finding of the study was that pauses between feeding and chewing were linked in some cases to awareness of predators, and this was influenced by the surrounding landcover. Lynch et al. (2014, 79) conclude:

> Our results demonstrated that pauses during mastication bouts (chewing) were structured by landscape properties that are expected to influence perceived risk and impede visual vigilance. Consistent with predictions, we found that deer paused more extensively where concealment cover abounded and thus where visual vigilance was likely to be less effective. In addition, deer allocated a larger proportion of time to pausing during mastication bouts at night, implying that auditory vigilance is an important defense mechanism when darkness reduces the effectiveness of visual scanning.

These results were particularly important as such details on auditory vigilance were overlooked in older studies, the authors citing Lima (1987), Altendorf et al. (2001) and Lind (2010), a further merit for such an advanced, technical study. Other similar studies of herbivores include those of accelerometers located on the jaws of marine predators that record the movements of the predators' jaws in prey capture (Viviant et al., 2014) and other accelerometers on bird wings that record wing beat frequencies that may change with body mass change, which may be due to feeding. Therefore, the combination of accelerometers and GPS can measure and map the energy expenditure across the range of a species and determine the link between this variation and ecological factors (Wilmers et al., 2015).

Possible Applications to the Activities of Dangerous Animals

There are myriad possible applications of GPS, accelerometers and acoustic recording devices. One method for examining future applications is to look at cases where such sophisticated techniques were not used and identify the parameters that could be measured using such techniques in the future. Three examples are cited: the complexities of human-big cat relations in Central and South America; the urban behavior and spatial ecology of urban cougars in North America; and the spatial conflicts of vultures and airplanes at airports world-wide. In each of these cases all the types of technological application discussed above are relevant. Parameters to be measured include the spatial location of the animals, their population clusters or densities, their feeding habits and habitat preferences, and the locational character of their conflicts with people. These parameters, measured with advanced geomatics techniques, may be documented and then compared with human perceptions, the better to find solutions to problems and correct misconceptions based on human error and biases.

Jaguars and Cougars and Human Perceptions

The relationship between jaguars (*Panthera onca*, Linnaeus, 1758) and cougars (*Puma concolor*, Linnaeus, 1771) and people in the Americas is a complex issue that centers on their size and perceived power, and their ability to intrude into human spaces and access human assets, such as safety, livestock and even human lives. Although North American big cats generally have less violent relations with people than those in Africa and Asia, substantial conflict emerges (Campbell & Torres Alvarado, 2011; Campbell, 2017). The jaguar, the world's third largest cat and mostly larger than the cougar, has a wide range from the southern United States, southwards to Central and South America and on to northern Argentina (Sanderson et al., 2002; Morrison et al., 2007; Zeller, 2007; McCain & Childs, 2008). Jaguars are generally less studied than lions, tigers and leopards (Brodie, 2009) with most studies focused on their non-modified habitats (Sanderson et al., 2002; Sollmann et al., 2008; Tôrres et al., 2008; Cavalcanti et al., 2010; Rabinowitz & Zeller, 2011; Campbell & Torres Alvarado, 2011). Most of the problematic episodes between people and jaguars concern the reduction of the latter's habitat through deforestation and the conversion of mixed forest/grass landcover into cultivated and urban landcover. This results in the hunting of the big cats and the extirpation of their prey species through over-hunting and habitat landcover conversion (Conforti & Azevedo, 2003; Zimmermann et al., 2005; Caso et al., 2008; Campbell, 2017). Concerning the justification of human fear for the jaguar, the big cat was feared and worshiped in prehistoric and historic times by Native American tribes; through these times and up to the current era, jaguars were more feared than cougars (Benson, 1998; Saunders, 1998; Santos et al., 2008). The recorded attacks on people are however rare, less frequent than those by tigers, lions and leopards in Asia and Africa (Guggisberg, 1975; McDougal, 1987; Bailey, 1993).

The smaller, more widely distributed cougar is similar in size to the leopard (Lopez-Gonzalez & Gonzalez- Romero, 1998; Campbell & Torres Alvarado, 2011). Having a wider range than jaguar, the cougar occurs through the whole of the jaguar's range (Scognamillo et al., 2003). The cougar is considered more elusive, adaptable and less dangerous to people and possibly livestock (Campbell & Torres Alvarado, 2011; Campbell, 2017). The perception of the jaguar as more dangerous, is mostly due to its larger size and stronger build, and preference for larger sized prey, including larger livestock predation (Farrell et al., 2000; Polisar et al., 2003; Scognamillo et al., 2003; Palmeira et al., 2008; Rosas-Rosas et al., 2008; Rosas-Rosas & Bender, 2012).

One important application of geomatics, in this case GPS tracking satellite mapping of habitats, concerns the disputed range of this species in the southwestern United States. The jaguar past range included much of the southwestern United States, but now most records are in northwestern Mexico (Brown & López-González, 2001; Gutiérrez-González & López-González, 2017). This Mexican population has the lowest density across the entire range of this species in Central and South America (Gutiérrez-González et al., 2012). The survival rates are mostly unknown, but generally considered to be low, possibly due to the dry climate and open, human-modified, non-forested ecological system (Brown & López-González, 2001; Campbell, 2017). In this area, there are more records of cougars than jaguars, possibly due to their greater adaptability to both human proximity and open, human modified landcover (Logan & Sweanor, 2001; Sunquist & Sunquist, 2002; Campbell, 2017; Gutiérrez-González & López-González, 2017).

Another important application for geomatics techniques is the documentation of the relations between jaguars and cougars, especially concerning conflicts, competitive exclusion, differences in prey selection and reactions to human presence. This would require very detailed studies of location, stress, food conflicts and avoidance strategies, all of which would be applications of the sophisticated GPS techniques mentioned above. The relationship between the jaguar and cougar creates several ecological intersections such as the possible dominance of the smaller predator (the cougar) by the larger predator, the more adaptable species (the cougar) contrasted with the less adaptable, and possible differences and similarities in prey species between the predatory species. Parameters for the assessment of such relations include avoidance behavior by the smaller predator, through different foraging areas or times, numbers of outright conflicts and the outcomes of such conflicts, or the permanent exclusion of one species from the conflict zones (Donadio & Buskirk, 2006; Ramesh et al., 2012; Vanak et al., 2013). Another parameter concerns how such relations may change with human proximity or interference, which may involve habitat degradation and landcover change, or the killing of one or both predators, or prey species, or the introduction of vulnerable livestock. In the case of large competing predators, the presence of large herbivores is important for the coexistence of large carnivores (Odden et al., 2010; Mitchell & Hebblewhite, 2012; Carter et al., 2015; Gutiérrez-González & López-González, 2017).

Detailed studies of the parameters of such large carnivore conflict are largely absent (Campbell, 2017). Competition between jaguars is even disputed (Yáñez et al., 1986; Nowell & Jackson, 1996; Farrell et al., 2000). Competition avoidance has been argued by some, as the jaguar may take larger prey, and the cougar more wide-ranging prey (Jaksic et al., 1981; Rabinowitz & Nottingham, 1986; Iriarte et al., 1990; Emmons, 1991). Scognamillo et al., (2003; 269) argued that where jaguars and cougars coexist, "although several studies have focused on the interactions between these two predators, the ecological and behavioral factors that promote their coexistence remain unclear." This study examined the factors for the coexistence of jaguars and cougars in the Venezuelan llanos, finding both presence overlaps and some interspecies avoidance. Jaguars took larger and medium-sized prey (capybara *Hydrochoerus hydrochaeris,* Linnaeus, 1766), collared peccary *Pecari tajacu,* Linnaeus, 1758) and to a lesser extent spectacled caiman (*Caiman crocodilus,* Linnaeus, 1758) and white-tailed deer (*Odocoileus virginianus,* Zimmermann, 1780). Cougars took medium-sized prey such as the collared peccary and occasionally, caiman. Abundant medium-sized prey and habitat heterogeneity, favored by both species, possibly encouraged the cats' coexistence (Scognamillo et al., 2003). Nevertheless, Harmsen et al. (2009: 612) on the Cockscomb Basin of Belize noted that "apart from their overall spatial similarities, jaguars and pumas avoided using the same location at the same time. This interspecific segregation was detectable over and above the spatial and temporal segregation of individual jaguars" (Harmsen et al., 2009: 612). Evidently, the "physically smaller pumas avoid jaguars"; evidence from Mexico indicated that may even be eaten by jaguars.

Another important parameter concerns the predation by both species on livestock (Campbell, 2017). This is a key area that would benefit from geomatics and GPS based studies, as no definitive conclusions have been reached using other means. Therefore, the extent of predation and prey selection is disputed in the literature. As noted by Azevedo (2008, 494) who studied the "the last significant population of jaguars in southern Brazil… located inside a protected area, the Iguacu National Park" and point out that the "importance of livestock predation on jaguar and puma coexistence is poorly known." Poachers and other types of hunters killed off most of the natural prey of the big cats, these including the white-lipped peccary (*Tayassu pecari*, Link, 1795). The dominant prey for jaguars were collared-peccary, coati, and deer (66.7% of prey biomass). Large wild species (50%) and medium-sized wild species (49.5%) comprised the main prey biomass. The dominant prey for cougars were large species (44.0%) and medium sized species (55.4%) (Azevedo, 2008). Oliveira (1994, 2002) and Rosas-Rosas & Valdez

(2010) also found that jaguars preferred peccary species, such as the collared peccary and white-lipped peccary, and cougars preferred deer species (e.g., *Odocoileus* spp.). Livestock predation was one consequence of the decline of these species.

Cattle (*Bos taurus,* Linnaeus, 1758), mostly calves are predated by both jaguars and cougars. Azevedo's (2008) study found that jaguars were more likely than cougars to prey on livestock, but this was not related to the numerical presence of livestock, hence the conclusion was that the return of large wild ungulates might reduce wildlife predation (Azevedo, 2008: 494). Gutiérrez-González and López-González (2017) also argued that big cat presence was positively correlated with peccary presence, and jaguar presence was positively correlated with deer presence, with cougar absence. While it is accepted that both cat species may prey on livestock, the apportionment of blame for livestock killings uncertain, hence this an area where the focus of geomatics (Oliveira et al., 2002; Polisar et al., 2003; Palmeira et al., 2008; Rosas-Rosas & Bender, 2012).

There is insufficient evidence of interspecies differences in livestock predation and of inter-species differences in adaptation to human reactions to livestock predation (Saenz & Carillo 2002; Sanderson et al., 2002). In cases, the big cats may be blamed for livestock kills by other species, such as feral dogs (*Canis lupus familiaris*, Linnaeus, 1758), coyotes (*Canis latrans*, Say 1823) and even ocelots (*Leopardus pardalis*, Linnaeus, 1758) (Campbell & Torres Alvarado 2011). Some studies have used identification techniques that are arguably less effective than the current geomatics techniques, including scat and genetic (deoxyribonucleic acid - DNA) studies (these may use bile powder for DNA analyses) and animal remains (Höss et al., 1992; Paxinos et al.,1997; Wasser et al., 1997; Farrell et al., 2000). Cameras may also be used to identify individuals, using fur patterns (Sarmiento, 2004; Silver et al., 2004; Salom-Perez et al., 2007; Maffei et al., 2011; Rosa-Rosas & Bender, 2012). These techniques may have limitations, such as the failure to record individual behavioral differences, such as evasiveness and territorial movements, which may differ between males and females (Wallace et al., 2003; Salom-Perez et al., 2007). Surveys of human opinions may also be unreliable (Campbell & Torres Alvarado 2011).

Another important issue, possibly resolvable by detailed geomatics surveys, was that jaguars and cougars are more likely to share habitat than to avoid each other. There is no conclusive answer using current survey methods. This may be disputed by some as the jaguar's superior strength is perceived as a general factor for competitive exclusion (Donadio & Buskirk, 2006). Some authors point to the competitive exclusion of the cougar, especially in the forests were jaguars attain their largest size (Emmons, 1987; Scognamillo et al., 2003; Sollmann et al., 2012). Other authors argue against competitive exclusion of cougars. Romero-Muñoz et al., (2010) found jaguars were not strongly dominant over cougars in a Bolivian dry forest, possibly due to the latter's greater adaptation to the dry landcover. Gutiérrez-González and López-González (2017: 7) also determined that "jaguars were not dominant over pumas in our study area" of arid

Northern Mexico, unlike the findings of other studies (Novack et al., 2005; Harmsen et al., 2009; Romero-Muñoz et al., 2010). Gutiérrez-González and López-González's (2017) hypothesis was that the cougar's superiority in prey selection, and adaptation to environmental variability and human presence, compensates for the jaguar's superiority in strength and thick forest specialization (see also Iriarte et al., 1990; López-González & Miller, 2002; Polisar et al., 2003; Scognamillo et al., 2003; Ruth & Murphy, 2010; Tortato et al., 2015; Hayward et al., 2016).

Other studies argue that jaguars are less able to cope with human-induced landcover change, due to their generally larger size, preference for denser vegetation and larger prey preferences (deer *Capreolinae* spp., Brookes, 1828; tapirs (*Tapirus terrestris*, Linnaeus, 1758), peccaries, giant anteaters (*Myrmecophaga tridactyla,* Linnaeus, 1758), capybaras, caimans and turtles (*Podocnemis*, Wagler, 1830 and *Trachemys,* Agassiz, 1857 spp.), and livestock), and very importantly, the more negative public image of the jaguar (Campbell & Torres Alvarado, 2011). However, in the relative absence of detailed, advanced geomatics-based studies, it is difficult to determine the impact of human-induced landcover change on jaguars (Campbell, 2017). Boron et al., 2016, 1), note that "information on jaguar densities across unprotected landscapes is still scarce and crucially needed to assist management and range-wide conservation strategies." These authors further note that the commonest habitat for jaguars are "wetter and prey-rich habitats such as lowland tropical forests" (Boron et al., 2016: 7). Jaguars are usually rare where there is human hunting activity (Quiroga et al., 2013) and forest degradation (Paviolo et al., 2008), but commoner in areas of cattle rearing (Arispe et al., 2007; Soisalo et al., 2006; Tobler et al., 2013; Boron et al., 2016).

Cougars in Urban Land

Geomatics applications are particularly relevant to cougars in urban substrates, because of the elusive behavior of the urban. The elusive behavior of the urban cougar is extreme to the point of invisibility, and there is insufficient understanding of the cougar's foraging and even reproductive behavior in urban contexts (Campbell, 2012a). Urban cougars are much more invisible than urban bears in North America, and unlike the latter animals, do not target human waste and gardens. Important general issues concerning the urban cougar are: (1) the interactions between cougars and people; (2) the urban substrates that attract or repel cougars; and (3) cougar adaptation to urban landcover (Campbell, 2012b). Recent literature records increasing interactions between people and cougars in urban North America (Beier 1993; Torres et al., 1996; Weaver et al., 1996; Campbell 2012a,b). Some cougars use urban substrates for transit rather than habitation (Dickson & Beier 2002, Dickson et al., 2005). Beier et al. (2010) observe that "probably every adult mountain lion in the coterminous United States has seen humans, crossed paved roads, and encountered human settlements." Cougar penetration into urban substrates is especially common in the cities of Denver/Boulder, Missoula, Los Angeles and San Diego, which have seen conflicts with people since the 1970s than other cities with similar records of cougar presence (Shuey, 2008).

The topic of cougar presence in urban areas, and the need for more advanced geomatics-based studies on the urban behavior and location, is justified largely by social attitudes to this species. Wolch et al. (1997, 106) pointed to a dominant narrative in the American media, which described the cougar as "a disruption to urban life, a nuisance to society, or a threat to humans." Cougars were linked to the themes of "serial killers", premeditated criminal behavior" and "popular worries about rising crime and lawlessness" (ibid., 109). Notably, as argued by Perry (1994). "in any civilization, killers aren't allowed to run loose." Cougars might also be developing a taste for companion animals, because they were easier prey wild animals (Wolch et al., 1997, 110). Large carnivore conservation for social benefit has nevertheless been defended, if public support can be garnered (Linnell, 2001; Karanth & Chellam, 2009). This links to urban planning and design, which may require detailed ecological information at species and population levels (Fonturbel & Simonetti 2011; Campbell 2012a,b). Human attitudes to the big cats must be compared with the facts of aggressive behavior.

As cougars share urban North America with other large mammals, especially black bears, an important research area that would complement geomatics-based tracking of urban predators concerns comparisons of social attitudes to the different species (Campbell, 2012a,b, 2017). Several studies have found that cougars, compared with black bears, are perceived as more predatory, unpredictable and secretive, and more a threat to humans especially children (Beier, 1991; Davis et al., 2001; Campbell & Lancaster 2010; Campbell 2012a). Compared to brown or grizzly bears, the cougar is however less feared (Kellert et al., 1996: 5). This is because of the brown bear's larger size, reputed ferocity and greater visibility, although the threat of an attack from an elusive cougar is often rated at similar levels (Pelton et al., 1976; Kellert 1994; Campbell 2012a). Brown bears rarely penetrate urban substrates, so the comparison is usually confined to rural or conservation areas (Campbell, 2017).

An important parameter for geomatics studies would be to investigate if such conflicts emerge from human penetration of cougar habitats through urbanization and suburban developments, or it is the cougars that adapt to human presence and invade previously unsuitable land. Traditionally, the latter view was more popular; urban expansion was to blame (Spencer et al., 2001, Dickson & Beier, 2002; Thornton & Quinn, 2009). Studies have argued that all types of cougars, not only sub-adults and transients occur in exurban, suburban and residential areas (Shuey, 2008). Some studies veer from the traditionalist view and argue for cougar adaptation to human presence. For example, Keston et al. (2013) argue that in some cases where sub-adult cougars penetrate urban landscapes more frequently than adult cougars, it may be due to their attempt to establish home ranges near their previous territory (see also Beier 1995; Beier et al., 2010; Logan & Sweanor 2001; Maehr et al., 2002; Ruth,1991; Van Dyke et al., 1986). Kertson et al., (2013) further attribute cougar presence in urban areas to individualist rather than population level behavior, which makes prediction difficult. This unpredictable, elusive behavior of the cougar contrasts with that of the urban black bear, which predictably invades urban substrates for food; seasonal fruits, urban green spaces and organic garbage (Merkle et al., 2013). Black bears are much more frequently linked to property damage and general nuisance behavior (Decker et al., 1981; Herrero & Higgins, 1999; Freedman et al., 2003; Morzillo et al., 2007).

Another important, potentially measurable parameter concerns the correlations between cougar presence and variations in urban landcover (Campbell, 2012a,b). Put simply, do cougars prefer some parts of cities more than others? This is relevant to urban planning, if the urban areas preferred or avoided by cougars are planned for future development. Important externalities would be killings of prey species and companion animals and changing public attitudes to cougar presence in these different landcover types. Generalizable hypotheses and findings from the current literature cluster around three main points. Firstly, cougars appear to prefer urban green spaces to dense urban substrates, especially when connected along corridors with favorable cover, and conflicts may emerge as these green spaces are also valued for human recreation health. Secondly, as cougars usually do not scavenge (unlike bears), their motives are more like predatory, justifying concerns for companion animals and children in such recreational, residential and transit areas (Campbell & Lancaster, 2010; Campbell, 2012a). The third hypothesis or finding emerges from the first two, in that these issues may affect human quality of life, through human fear and location avoidance and fuel negative public attitudes to cougars. Geomatics-based surveys in this regard would provide the definitive evidence to either support or deny the veracity of these perspectives, but most studies have yet to use the advanced methods described in this chapter (Campbell, 2017).

Taking the case of urban green spaces, as likely the preferred cougar urban habitat, there are several relevant parameters. Urban green spaces are areas with variable proportions of trees, shrubs and grasses, surrounding urban areas (greenbelts), inside urban areas (greenhearts) and in scattered areas within an urban area (in the case of smaller, isolated green spaces) (Kuhn, 2003; Campbell, 2009). Urban green areas are mostly recorded as expanding, due to the increased rates of urbanization, the recreational and health interests of urban dwellers, and wildlife conservation (Bird et al., 1996; Belant, 1997; Jokimaki, 1999, 2003; Pirnat, 2000; Fernandez-Juricic, 2000; Savard et al., 2000; Chiari & Seeland, 2004; Cornelis & Hermy, 2004; Erickson, 2004).

An important parameter for tracked cougars would be the duration of the time they spend in the green spaces under study, after locating them and recording their movements. This could be covered by the latest GPS technology, as outlined by Augliere (2015), Wilmers et al. (2015) and Lynch et al. (2014). This is important, because cougars in urban North America may be classified as transients, occasional inhabitants or even permanent residents (Campbell 2012b). Cougars may use urban green spaces for hunting, transit, territorial marking, relaxing, family support and/or breeding, largely due to the presence of cover, and prey species (Burdett et al., 2010; White et al., 2011; Keston et al., 2011a, b, 2013). If time and location are important parameters, these must also be linked to green space size and vegetation configuration, two issues largely neglected in the literature (Campbell, 2012). Studies have noted that cougars prefer green spaces and tend to frequent areas with higher heterogeneity, including both green and built landcover (Campbell 2012).

These studies complicate the green space hypothesis. Shuey (2008) found no linear relation between cougar presence and green areas, as there were more cougar-human encounters in grassland and suburban land-cover than there were in the preferred cougar habitat of evergreen tree growth, even when such tree growth covered the more land than the suburban land-cover. In heterogeneous landcover, more cougar-human encounters occurred near human-modified land cover distant from natural cougar habitat. Therefore, urban centers with highly heterogeneous landcover might have more cougar-human conflicts in developed areas. Cougars may use residential landcover when green spaces and corridors are absent, rapidly moving between green spaces through urban substrates by night to reduce (Kertson et al., 2011b). This allows a "high potential for coexistence" between people and cougars (Kertson et al., 2013: 1). Kertson et al. (2013) further argue that "cougars and people appear to coexist better than previously perceived and use of areas close to people is likely to continue if these environments provide ample stalking and security cover, adequate prey resources, and limited human interference." This adaptive behavior is important, because it disproves the common assumption correlations between green spaces and urban cougar presence and may impact on human quality of life and human-cougar conflicts in non-green space areas (Campbell, 2012a).

Lions in Rural Land in West Africa

The topic of lions (*Panthera leo*, Linnaeus, 1758) in West Africa is like that of urban cougars in North America, because lions in West Africa are currently very rare and their locations and ranges are disputed, with no definitive or reliable methods of assessment. Hence, the West African lion (*Panthera leo senegalensis* or *Panthera leo leo*) has evolved from a common predator into an elusive, rare and locally extirpated species (Campbell, 2018). The factors for lion decline are declining populations of large prey herbivores such as antelopes and African buffalo, increased human populations and increased intensity of landuse, mostly farming, livestock grazing, hunting and urbanization, and the killing of lions in response to perceived or real threats to people (Bauer et al., 2015; Campbell, 2017, 2018). As lions generally avoid deciduous and rain forests, partly due to the general absence of larger herbivores, deforestation is not a problem (Campbell, 1998). As lions are wide ranging species, and likely to impact on human life activities such as cattle herding, farming, urban and suburban development and reactional activities, their precise location and behaviors must be critically recorded. This underscores the possible contribution of geomatics-based techniques, such as GPS and associated techniques (Campbell, 2017).

Table 1. Conservation Areas and Lions in West Africa

Western Africa					
Country	**Area**	**No. NP**	**N. Park Area**	**Conserved %**	**Lion Presence**
Ghana	238,534	7	36,019	15.1	No
Cameroon	475,442	12	51,823	10.9	Yes
Côte d'Ivoire	322,463	6	13,500	4	No
Mauritania	1,030,000	1	1600	1.7	Yes
Gambia	10,380	2	436	4.2	No
Liberia	111,369	1	2,784	2.5	Yes
Mali	1,240,192	1	104,176	8.4	No
Niger	1,267,000	1	222,992	17.6	Yes
Senegal	196,723	6	49,574	25.2	Yes
Sierra Leone	71,740	4	7,820	10.9	No
Guinea-Bissau	36,125	5	5,021	13.9	No

Sources: Campbell, 2018; Drewniak et al. (2012). Note the figure for Cameroon includes 9615 km^2 in the forest area outside normal lion habitat. International Union for the Conservation of Nature (1992), African Wildlife Foundation (2011), International Bank for Reconstruction and Development/The World Bank (2015), Partenariat Régional pour la Conservation de la zone côtière et Marine en Afrique de l'Ouest (2017). Note that some of these numbers vary, according to the definition of National Park, Conservation Area and Nature Reserve

Henschel et al. (2015, 4) estimate the West African lion population at just over 400 animals (<250 mature individuals) and in a declining situation. The lion was recorded as "native for Benin, Burkina Faso, Niger, Nigeria and Senegal; possibly extinct for Ghana and Guinea; and regionally extinct for Côte d'Ivoire, Gambia, Guinea-Bissau, Mali, Mauritania, Sierra Leone and Togo." The United States Fish and Wildlife Service (2017), citing (Bauer et al., 2015) gives a classified list of the West African lion status, distinguishing between "lions considered recently extirpated"; Côte d'Ivoire, Gabon, Gambia, Guinea-Bissau, Mali, Mauritania, Niger, Nigeria, Senegal, Sierra Leone; "lions considered possibly extirpated (Ghana, Guinea, Togo) and lions still occurring (Benin, Burkina Faso, Cameroon, Central African Republic, Chad, Congo, Democratic Republic of the Congo, Liberia, Niger, Nigeria, Senegal). Table 1 shows conservation areas of West Africa and Table 2 shows the status of prey species.

Table 2. Status of some lion prey species

English name	Scientific name	Status
Roan Antelope	*Hippotragus equinus* (É. Geoffroy Saint-Hilaire, 1803)	Least Concern, declining in West Africa.
African Buffalo	*Syncerus caffer* (Sparrman, 1779)	Least Concern, declining in West Africa.
Common Warthog	*Phacochoerus africanus* (Gmelin, 1788)	Least Concern, declining in West Africa.
WA Giraffe	*Giraffa camelopardalis peralta* Thomas, 1898	Threatened, isolated populations.
WA Hartebeest	*Alcelaphus buselaphus* (Pallas, 1766 A. b. major (Blyth, 1869)	Least concern, sharply declining, locally extinct.
WA Topi	*Damaliscus lunatus ssp.korrigum* (Ogilby, 1837)	Vulnerable to endangered.
Dama Gazelle	*Nanger dama* (Pallas, 1766)	Critically Endangered
Dorcas Gazelle	*Gazella Dorcas* (Linnaeus, 1758)	Vulnerable, declining
Red-fronted Gazelle	*Eudorcas rufifrons* (Gray, 1846)	Least concern, declining
Oribi	*Ourebia ourebi Zimmermann*, 1782	Least concern, declining.
Western Giant Eland	*Tragelaphus derbianus derbianus (Gray, 1847)*	Much rarer than Eastern Giant Eland, declining.
Kob	*Kobus kob* (Erxleben, 1777)	Least concern, vulnerable.

Sources: Campbell, 2018; International Union for the Conservation of Nature IUCN (2008a, b, c, d; 2016a, b, c, d); de Jong et al. (2016); Fennessy et al. (2016)

The current range of the West African lion is mostly composed of isolated patches with no effective corridors between the patches, with some areas which were believed to be occupied having no current records, e.g., in Ghana and Guinea (Bauer & Van Der Merwe, 2004; Angelici et al., 2015; Henschel et al., 2015). An important habitat complex is termed the WAP-Complex, comprising the W, Arli, and Pendjari National Parks in Burkina Faso, Benin, and Niger. Other records are in the Niokolo-Koba National Park in Senegal and the Kainji Lake National Park. Howard (2014: 1) noted "lions may soon disappear entirely from West Africa unless conservation efforts improve, a new study predicts" and argues that studies have "found an estimated total of only 250 adult lions occupying less than one percent of that historic range. The lions form four isolated populations: one in Senegal; two in Nigeria; and a fourth on the borders of Benin, Niger, and Burkina Faso. Only that last population has more than 50 lions." The habitat in the Sahel and Sudan savanna may still be suitable for lions, but lion populations have failed to recover.

Crucially, there is little agreement on the exact range and numbers of the West African lion (Campbell, 2018). For example, DeMarco (2015) argues that currently, the status of lions may be overestimated. The IUCN and Henschel et al. (2014) studied 17 protected areas, and found lion presence in three of these protected areas, in Senegal (Niokolo-Koba National Park, with fewer than 60 lions), Benin, Burkina Faso and Niger (the tri-national W-Arly-Pendjari Complex, with 246 to 466 lions) and in Nigeria (the Kainji Lake National Park, 23 to 63 lions) (Henschel et al., 2014). These records were derived from track identification. The research failed to find lions in some areas previously believed to be current habitats; these included the Mole National Park in Ghana and in Guinea's Haut Niger and Kankan Reserves, and several areas in Togo, Guinea-Bissau, Mali and Côte d'Ivoire. The findings indicated a population of between 269 and 583 lions in West Africa, with only about 121 to 375 mature animals, and "continuing calamitous declines in prey populations are almost certainly causing concomitant declines in lions" (Henschel et al., 2014).

Measurable parameters for the study of the populations and ranges of West African lions would include presence/absence data, migratory and foraging movements in classified landcover types, and monitoring of prey animal movements in different seasons, using techniques such as those described above (Wilson et al., 2013; Wilmers, et al., 2015; Wildlife ACT, 2018). However, lion studies have so far been rudimentary (Campbell, 2018). Generalized assumptions on lion foraging habits have been applied to the changing landcover, to derive hypotheses about the declining ranges of lions in some areas of West Africa. Important parameters may be derived from such hypotheses and descriptions (Campbell, 2018).

Urbanization contributes to the reduced natural landcover between urban centers, transport network development and increased human presence and landuse. It is a parameter that may be correlated with records of lion presence and pattern of movement. In the general absence of studies of the West African lion's range, ranges may be hypothesized or constructed by examining parameters derived from other African lions in other parts of Africa. For example, Loveridge et al. (2009) argue that lion ranges and lion social groups (prides) increased when food sources declined. This finding indicates that the declining numbers of prey species (due to urbanization, agriculture, hunting and transport), would require increased ranges for lions, which is nearly impossible as human activities and landuse are increasing in spatial dimensions. Intensive use of GPS monitoring would be required to document the details of lion movements, ranges and feeding habits, and those of people, prey animals and landcover change. Parameters for GPS and associated technology would include individual lions, individualized movements, individualized and population movements of prey animals, feeding episodes of lions and those of any competing predators. The derived data would then be modeled, and these compared with those of lions in other regions of Africa with denser lion, prey animal and possibly human populations, such as East Africa.

Another set of parameters would be the numbers or potential livestock (and possibly humans) killed by lions (based on distance between the animals). This is important, because livestock rearing, especially cattle herding, is a key issue in West Africa (Campbell, 1998; Schwartz, 2017). Evidence that lions kill livestock was recorded in the *Daily Telegraph* (2012) a top British newspaper, which noted "Lion 'on a rampage' kills two in Nigeria", which "killed two herders and 30 livestock in a settlement in northeastern Nigeria's Yobe State." Chardonnet et al. (2010: 12) noted that "in Mali, for example, lion predation on cattle is one of the main reasons for the drastic reduction in the number of lions." In the savannas of Central African Republic, lion population decimation is mostly due to shooting by pastoralists during the dry seasons. Stephens (2015: 14754) argues that "sparing ...extensive tracts is unlikely to be possible in many parts of the world, so we must usually choose between coexistence (sharing lands with large mammals) and the confinement of species in areas too small for them to play a meaningful ecological role." Löe and Röskaft (2004) have ranked lions as third behind tigers (*Panthera tigris,* Linnaeus, 1758) and leopards (*Panthera pardus,* Linnaeus, 1758) for killing people in the twentieth century. In East Africa, lions have killed many people; for example, in Tanzania., where 563 people were killed and 308 injured by lions between 1990 and 2006 (Packer et al., 2005, 2006). In Uganda, lions attacked 275 people between 1923 and 1994 (Ugandan Game Department archives, 1923–1994). In West Africa, the lower lion population may account for the lower number of kills (Chardonnet et al., 2010), but the East African figures could be modeled if lion conservation or population increases are examined.

The parameter of prey animal numbers and distributions are particularly important for West Africa (Campbell, 2018). Bauer et al. (2015: 14897) note that "lion trends are consistent with time series data on their main prey species: whereas herbivore population sizes increased by 24% in southern Africa, herbivore numbers declined by 52% in East Africa and 85% in West Central Africa between 1970 and 2005" see also Craigie et al., 2010). Henschel et al. (2014) note that the killing of the lion's prey animals is an even bigger problem than the killing of lions for livestock predation. Loomis (2014) suggests that in Senegal, the bush meat trade has decimated prey especially buffalo and roan antelope in Niokolo Koba National Park by 95 percent over the past 20 years. The parameter of lion shootings or other killings is currently poorly recorded. Chardonnet et al. (2010, 42) argue that in comparison with other African regions "in West Africa there is currently much less lion hunting, though pastoralists tend to shoot and poison lions to protect their stock. Little information is available on the effects of hunting in the region." As mentioned above, there is scanty evidence of people killing lions in retaliation for cattle killing, especially in Mali and the Central African Republic (Chardonnet et al., 2010).

Vultures in Urban Areas

Large birds are also frequent denizens of urban areas and as such have many conflicts with people (Campbell, 2015). Vultures are among the commoner species that interact with people in urban and built up areas. Parameters for assessing this situation, like those for large mammals, would involve the tracking of the vultures and deriving from this information the factors that attract or repel vultures. The key issues for human-vulture relations in urban areas are scavenging in urban substrates, invasions of human spaces, airplane collisions and the killing of vultures by pylons and wind turbines. As noted by Campbell (2009: 341), "a central, emerging research problem concerns the mutual co-evolution of the human–avian relationship, and how this is manifested in different land uses within the urban areas." Urban development has positive and negative effects on vultures; positive in terms of food provision – road kills, human discards, buildings and green spaces for nesting; and negative in terms of habitat loss, noise, pollution, shooting, lower biodiversity and in some cases absence of food. All these provide parameters, which can be measured and correlated with GPS tracking of vultures and technical monitoring of vulture behavior.

Visual observations have been enough to determine which species of vultures are attracted to urban developments. An important parameter would be the numbers of vultures in the urban areas. Another would be their roosting and foraging locations, and the proximity of these to feeding areas. A third relevant parameter would be the feeding habits of the vultures. All of these could be recorded by the advanced GPS accelerometers and tilt sensors, which are accelerometers designed to measure the orientation of the observed animal's head in terms of gravity (Wilmers et al., 2015). For example, among the *Cathartid* 'New World' vultures, the Black vultures (*Coragyps atratus*, Bechstein 1793), are very common in urban areas, common problems resulting from this are roosting pollution due to defecation, related property damage and aircraft collisions (Buckley, 1999; Lowney, 1999; Avery, 2004; Avery & Cummings 2004; Blackwell & Wright, 2006; Novaes & Cintra, 2013). Black vultures are attracted to rubbish dumps and urban markets (Campbell, 2015). An important impact of Black vulture presence in urban areas is the ingestion of artificial mostly plastic products. Iñigo Elias (1987) studied defecated pellets and discovered synthetic products, mostly plastic bag materials in 39.1% of the total pellets examined. Black vultures in Manaus, Brazil, scavenge in street markets, garbage dumps, open sewers and polluted streams (Novaes & Cintra, 2013).

The parameter of roosting locations may be measured by advanced techniques that would document movements of individual birds, rather than only flocks (Campbell, 2015). Using simple observation, Novaes and Cintra (2013, 1) found that "black vultures adjusted to the nearest possible roost to the food source to reduce the cost of movement." Roosting was in small to large green spaces and vultures ignored intensive human presence when close to refuse sites. Hill and Neto (1991, 173) found that "the black vulture has benefited immensely from human post-Columbian colonization of the New World... feeding on the garbage of dumps, fish killed in polluted waters, livestock dying in pastures and animals killed along highways, the Black Vulture's numbers have increased so dramatically that it is now considered the most abundant bird of prey in the Western Hemisphere."

Black vultures were found nesting in office buildings in downtown Sao Paulo, Brazil. Hill and Neto (1991; 175) have also argued that "skyscraper-nesting by the black vulture appears to be common in Brazil"; this mostly comprises window nesting and in some under roof-mounted water storage tanks. These authors attribute the urban nesting of the black vulture to their dependence on human waste, the increase of such waste with increased human populations, the decline in natural locations for traditional nesting sites and the fact that there are "similarities that large, unkempt window planters and roof recesses have to natural cliff ledges and caves" (Hill & Neto, 1991, 176). Other measurable parameters, which could be correlated with GPS surveys and tracking of urban black vultures include the reduction of food supplies, thorough garbage and sewage collection, replacing open garbage cans with closed garbage cans and stopping garbage disposal by local communities (Novaes & Cintra, 2013).

Turkey vultures also forage and roost in urban areas, but they are less communal in nature (Campbell, 2015). Some have argued that urbanization may reduce the Turkey vulture population (Garrett & Dunn 1981; Unitt 1984; Kiff, 2000), but Turkey vultures have been recorded in urban nests (Houston et al., 2007; Igl & Peterson, 2010). Urban parks are common nesting areas, but there is evidence that vultures from urban nests foraged outside the urban areas in more pristine habitat (Airola, 2011). These observations could be verified if geomatics applications were applied. Hence, an important parameter would be the recording of vultures that nested in urban areas and comparing those vultures that foraged in urban areas (if any) with those that foraged in rural landcover. Other Cathartid vultures are not urban scavengers (Campbell, 2015). These include the King vulture (*Sarcoramphus papa*, Linnaeus, 1758), which is forest dependent (Ellis et al., 1983; Stiles, 1985; Clinton-Eitniear, 1986; Reid, 1989; Whitacre et al., 1991, Berlanga & Wood, 1992). The Yellow-headed vulture (*Cathartes burrovianus*, Cassin, 1845) forages and nests in a wider range of environments but is not known as an urban scavenger (Ferguson-Lees & Christie, 2001; Braun et al., 2007; Zilio et al., 2013).

Some *Acciptrid* 'Old World vultures' are urban foragers, especially near areas with "human excrement, carcasses, offal, bones and fresh meat associated with slaughterhouses" (Ssemmanda, 2005: 10). For example, the Hooded Vulture (*Necrosyrtes monachus*, Temminck, 1823) is the commonest vulture species in many towns, even in areas were other vulture species are more common in the nearby rural landcover (Brown et al., 1982; Wacher et al., 2013). The Hooded vulture has been described as very adaptable to all types of surroundings, including urban, desert, savanna and forest landcover. In urban areas, favored areas are rubbish dumps and slaughterhouses, especially as there is little competition from other larger vultures that are largely absent from urban substrates (Anderson, 1999; Ogada & Buij, 2011). In some areas and countries, hooded vultures are commoner in urban than in rural areas (Richards, 1982; Walsh, 1987; Halleux, 1994; Thiollay, 2006; Rondeau et al., 2008; Ssemmanda & Plumptre, 2011). For example, in urban Uganda, urban declines are attributed to improved abattoir hygiene (Ssemmanda & Pomeroy, 2010). Ssemmanda and Plumptre (2011: 17), note that "across most of Uganda this is an urban dwelling species with the largest population in Kampala," possibly partly due to "new abattoirs in emerging towns that have attracted some of the birds." Competitors in Uganda included the palmnut vulture (*Gypohierax angolensis*, Gmelin, 1788) and the Egyptian vulture (*Neophron percnopterus*, Linnaeus, 1758) (Campbell, 2015). In other parts of Africa, there is little information on other vulture species, possibly due to the "lack of active resident ornithologists" (Ssemmanda & Plumptre, 2011).

In Asia, vultures are strongly present in urban areas, even though in some cases they roost in rural areas. Harris (2013) notes that the white-rumped vulture (Gyps *bengalensis*, Gmelin, 1788) forages in urban areas and nests in trees near human habitation. The Indian vulture (Gyps *indicus*, Scopoli, 1786) is also "found within cities, towns and villages associated with cultivated areas" (Harris, 2013). The slender-billed vulture (*Gyps tenuirostris*, Hodgson (in Gray), 1844) and red-headed vulture (*Sarcogyps calvus*, Scopoli, 1786) forage in open forests, especially near urbanized areas, mostly scavenging at rubbish dumps and slaughterhouses (Arun & Azeez, 2004; Campbell, 2017). All these studies give clear visual descriptions of vulture presence and would benefit from more detailed geomatics-based surveys, as the details of individual and small group foraging are missing (Campbell, 2017).

Another important relationship between people and vultures in urban areas concern the collisions of the birds with airplanes, generally near large airports in larger cities (Lowney, 1999; Avery & Cummings, 2004; Blackwell & Wright, 2006). These collisions occur mostly in the tropics, as most vulture species are found there (Campbell, 2015). For these events, parameters would be the number of vultures, their individual and clustered locations, common flying altitudes, reactions to approaching or passing airplanes, foraging behavior and favored landcover near airports. Other parameters would be species that collide with airplanes and reactions to human proximity.

Some species appear to collide with airplanes more often than others (Campbell, 2015). For example, Satheesan & Satheesan (2000) point out that between 1955 and 1999 more than 33 aircraft were destroyed by collisions with seven species of vultures in seven countries in Asia, Africa, Europe and North America. Some were at very high altitudes, such as that of 29 November 1973, when a Ruppell's griffon collided with an airplane at 11,280 m over Abidjan, Cote d'Ivoire (Laybourne, 1974). In the United States of America, such collisions with black vultures cost more than US$ 25 million to the US Air Force (USAF 2009). In Brazil, the Aeronautical Accidents Investigation and Prevention Center (CENIPA) recorded over 980 strikes involving vultures between 2000 and 2011. In Manaus, 65 vulture-aircraft strikes occurred between 2000 and 2012 (CENIPA 2012).

In India, the cost was $70 million annually from 1980 to 1994 (Satheesan, 1994, 1996, 1998a,b). In India, all crashes involved the white-rumped vulture, while the long-billed griffon and Egyptian vulture recorded minor accidents. In Kenya, the African white-backed vulture (*Gyps africanus*, Salvadori, 1865), white-headed vulture (*Trigonoceps occipitalis*, Burchell, 1824) and Ruppell's Griffon (*Gyps rueppelli*, Alfred Brehm, 1852) have caused crashes. In the United States and Spain, the losses are $10 million to $17 million per accident (Satheesan & Satheesan, 2000).

Vulture soaring altitudes are the main measurable parameter for collisions, when correlated with airplane altitude. Individual and communal thermal soaring and gliding between thermals, vulture size (compared with smaller birds) and slow reflexes of vultures all contribute to these accidents. Most collisions happen when the plane is cruising or during descent; in cases of known altitude, about half the collisions were below 200m. Military fighter jets were particularly vulnerable to collisions, at high speeds and low altitudes (Satheesan & Satheesan, 2000). DeVault et al. (2010) studied collisions between aircraft and black and turkey vultures in South Carolina, USA, and found that average altitudes for black vultures were 169 ± 115 (SD) m, and turkey vultures 163 ± 92 (SD) m. Influences were food availability, inter- and intra-specific interactions, and physiological demands, rather than weather and time variables. The conclusion was that collision avoidance strategies require attention to variable vulture flight behaviors across their ranges, rather than the effects of local conditions on flight behavior.

Thus, the parameters of bird flight behavior can be measured by some of the newer GPS techniques, some of which can measure flight speed, maneuvering and physiological processes (Mitani et al., 2010; Wilmers et al., 2015). Solutions outside of bird behavior include removal of vulture food sources within 100 km in radius of a civil airports and 200 km for military airfields. Shooting of vultures is also possible; between 1991 and 1997, some hired shooters at JFK International Airport, USA killed 52,235 gulls in 6369 person-hours of shooting (Dolbeer 1998). Shooting has however been criticized for conservation reasons (Satheesan & Satheesan, 2000). Other solutions include more effective sanitation near airports, to prevent the attraction of scavengers (Blokpoel, 1976, Ali & Grubh, 1984, Satheesan 1992, 1994, 1996).

Another issue in the vulture-human relationship concerns vulture collisions with power lines and wind turbines (Moccia & Arapogianni, 2011; Martin et al., 2011a,b). Vultures are vulnerable to collisions with such structures, despite their sharp vision (Land & Nilsson, 2002). This issue is current for international conservation, as a factor for the species decline (Drewitt & Langston, 2008; De Lucas et al., 2008; Carette et al., 2009, Jenkins et al. 2010). Wind turbines in some cases kill more raptors than power lines (Barrios & Rodríguez 2004; Telleria, 2009a,b; De Lucas et al., 2012a,b). An important parameter to avoid these impacts is the distance between vulture breeding and foraging sites and wind turbine locations. This knowledge is however scanty, with the example of

Spanish griffon vultures cited (*Gyps fulvus*, Hablizl, 1783) (Telleria, 2009a). In the absence of detailed, geomatics-based vulture location studies, turbines are generally located in highlands and mountain rides, which are frequented by foraging vultures (Telleria, 2009a,b). One estimate is of a death rate of eight griffon vultures per year in an area near the Salajones wind plant, in Navarre, Southern Spain (Lekuona, 2001). Griffon vultures comprised 63.1% of bird fatalities near the wind plants of Navarre (Lekuona & Ursula, 2007). In one study, griffon vulture mortality was reduced by about 50% from 2006 to 2009, when mitigation measures were introduced (De Lucas et al., (2012a).

Vultures also collide with electric power lines, resulting in death through injury or electrocution (Campbell, 2015). Collisions are commoner if power lines cut into vulture foraging or nesting habitats (Hunting, 2002). Important parameters for factors for vulture presence in such areas include perching availability, vegetation form and carcass presence. Power lines and towers may also be used for nesting or roosting, especially in non-tree landcover (e.g., deserts and open plains (APLIC, 2006). Vultures in Africa, especially those in open plains, such as the gregarious Cape and African white-backed vultures may crowd onto powerlines for lack of other perching sites. Electrocution is commoner for larger species, as they are more likely to touch two energized components of the pole structure, creating short circuits as electricity flows through the bird (Prinsen et al., 2011).

Other factors for electrocutions are raised topography, which encourages birds to perch to scan the landscape (Nelson & Nelson, 1976; Benson, 1981; APLIC, 1996), nesting periods (Harness & Wilson, 2001), rain wetting of feathers (Nelson, 1979, 1980) and wind direction diagonal or parallel to the cross arms as landings and takeoffs are more difficult (Nelson & Nelson, 1976; Benson, 1981). Some species, such as Egyptian vultures in Fuerteventura, the second largest Canary Island, roost on power lines (probably due to the lack of trees in the desert, as it is rarer were there are trees (Donazar et al., 1996; Sigismondi & Politano, 1996). Electrocutions are therefore likely (Janns, 2000). Six Egyptian vultures were found dead in one area near power lines (two dead by collision and four others by electrocution (Lorenzo, 1995). This affected species populations in the area (Ferrer, 1993; Ferrer & Janns 1999). Other affected species are cinereous vultures in Bulgaria (Prinsen et al., 2011); Eurasian black vultures, griffon vultures and Egyptian vultures in Spain (Guzmán & Castaño, 1998; Martínez, 2003; Palacios, 2003); cinereous vultures in Kazakhstan (Haas & Nipkow 2006; Lasch et al., 2010); and griffon vultures in Israel (Prinsen et al., 2011).

In Africa, vultures have been killed in Egypt, Sudan, Ethiopia and Kenya (Niklaus, 1984; Virani, 2006; Smallie & Virani, 2010; Haas, 2011), with estimates of over 5,000 dead (Angelov et al., 2011). Virani (2006) and Smallie & Virani (2010) list species at risk as Egyptian, white-headed, lappet-faced, African white-backed and Rüppell's vultures. In South Africa, Cape vultures and African white-backed vultures roost or perch communally on power lines, with multiple electrocutions (Lehman et al., 2007; Prinsen et al. 2011). Kruger et al. (2004) recorded the electrocution of 46 Lappet-faced vultures, 24 African white-backed vultures, 4 Cape griffons and 12 other, unidentified vultures in South Africa. Boshoff et al. (2011) reported 1,504 electrocuted cape griffons, white backed vultures, lappet faced vultures and other species between August 1996 to May 2011.

Collisions are a leading cause of vulture injuries. Surviving birds may be severely injured and require rehabilitation, especially in South Africa (Naidoo et al., 2011). Common vultures rehabilitated in conservation centers are Cape griffon and African white-backed vultures. Injuries are caused by the collisions with pylons, which cause both soft tissue and skeletal injuries. Urbanization and the attraction of vultures to food sources in urban areas have been blamed for some of the collisions (Naidoo et al., 2011). Vulture behavior was analyzed, and many electrocutions and collisions were blamed on flight and perching behavior, including the vultures' gregarious nature, the limit of alternative natural perching areas and the small sizes of pole-tops. Collisions occurred when vultures failed to see wires. Electrocution occurred when vultures landed on wires or pole tops and slipped, touching two live wires due to their large size (Kruger et al., 2004). Vultures, despite their sharp eyesight were found to be limited in their perception of wires. Martin et al. (2012: 5) found that "the visual fields of vultures contain a small binocular region and large blind areas above, below and behind the head." This enables good eyesight for foraging by sun protection and good lateral vision, but gives narrow, frontal binocular vision. Therefore, Martin et al. (2012: 5) found "by erecting structures such as wind turbines, which extend into open airspace, humans have provided a perceptual challenge that the vision of foraging vultures cannot overcome." Solutions would include attracting vultures away from powerline and wind turbine areas, to other areas of the landscape. This might require more knowledge of vultures, and the use of the recent developments in GPS and associated technologies, which are easily able to detect animal behaviors at detailed levels (Mitani et al., 2010; Wilmers et al., 2015).

CONCLUSION

This chapter has examined the development of Global Positioning Systems (GPS) for animal tracking and their applications to the key issues of conservation biology. Case studies of applications were taken, largely from studies of large wild mammals and their relations with people in urban or semi-urban contexts. The main issue was to discuss the recent developments in GPS and related technology and assess the capabilities of these new techniques in the assessment of animal ecology and conservation applications. The evidence points to a wide range of possible applications, with possibilities of achieving previously elusive solutions and understandings, due to the previous coarser resolutions and generalized methodologies of previous technologies. The chapter also examined the parameters that would be examined in the application of the technology, to explain animal behavior or to be correlated with measurements of animal behavior. These would provide the background understandings for the applications of the technology. It can be safely concluded that Global Positioning Systems (GPS) and associated technologies have a bright future in the science of animal tracking and for the status of this field within the larger sciences of geomatics and conservation biology. Numerous difficulties must be overcome, partly because knowledge of animal behavior is incomplete, and animal behavior is adapting to landcover and ecological issues that are changing under human influences. GPS applications must also be supported by related geomatics methods, parameter identification and social studies. Future studies may compare the effectiveness of the different GPS and associated techniques, include case studies of wider selections of animals and landcover types, and evaluate the contribution of these studies to conservation efforts.

REFERENCES

African Wildlife Foundation (2011). *Wildlife Conservation in Zimbabwe*. Nairobi: AWF Conservation Centre.

Airola, D. A. (2011). Dynamics of an urban Turkey Vulture roost in Sacramento, California. *Central Valley Bird Club Bulletin 14*:1-8.

Ali, S. & Grubh, R. B. (1984). *Ecological Study of Bird Hazards at Indian Aerodromes Phase II. First Annual Report (1982-83)*. Delhi: Bombay Hindan: Bombay Natural History Society.

Altendorf, K., Laundré, J., López González, C. & Brown J. (2001). Assessing effects of predation risk on foraging behavior of mule deer. *Journal of Mammalogy 82*:4 30–439.

Anderson, M. D. (1999). Africa's hooded vulture: a dichotomy of lifestyle. *Vulture News 41*: 3–5.

Andrews, C. G. (2013). *Traditional Tracking vs. Electronic Surveillance: Has Wildlife Research Become Too High-Tech?* Retrieved from http://goodnature.nathab.com/traditional-tracking-vs-electronic-surveillance-has-wildlife-research-become-too-high-tech/

Angelici, F. M. (Ed.) (2015). *Problematic Wildlife. A Cross-Disciplinary Approach*. New York: Springer.

Angelov, I., Hashim, I. & Oppel, S. (2011). *Persistent electrocution mortality of Egyptian Vultures Neophron percnopterus over 28 years in East Africa*. Unpublished Report.

APLIC (Avian Power Line Interaction Committee) (2006). *Suggested Practices for Avian Protection On Power Lines: The State of The Art in 2006*. Washington, D.C: Edison Electric Institute.

Arispe, R., Rumiz, D. & Venegas C. (2007). *Censo de jaguares (Panthera onca) y otros mamíferos con trampas cámara en la Concesión Forestal El Encanto*. Santa Cruz, Bolivia.

Arun, P. R. & Azeez, P. A. (2004). Vulture population decline, diclofenac and avian gout. *Current Science 87*:5 65-568.

Augliere B. (2015). Beyond GPS: *The Next High-Tech Frontier in Wild Animal Tracking.* Retrieved from https://blogs.scientificamerican.com/guest-blog/beyond-gps-the-next-high-tech-frontier-in-wild-animal-tracking/

Avery, M. (2004). *Managing Depredation and Nuisance Problems Caused by Vultures.* Washington: United States Department of Agriculture.

Avery, M. L. & Cummings, J. L. (2004). Livestock depredations by black vultures and golden eagles. *Sheep & Goat Research Journal 19*(1): 62-67.

Azevedo, F. C. (2008). Food habits and livestock depredation of sympatric jaguars and pumas in the Iguacu National Park area, South Brazil. *Biotropica, 40*(4): 494–500.

Bailey, T. N. (1993). *The African Leopard: A Study of the Ecology and Behavior of a Solitary Felid*. New York: Columbia University Press.

Barrios L. & Rodriguez, A. (2004). Behavioral and environmental correlates of soaring-bird mortality at on-shore wind turbines. *Journal of Applied Ecology 41*: 72–81.

Barron, D. G., Brawn, J. D. & Weatherhead, P. J. (2010). Meta-analysis of transmitter effects on avian behavior and ecology. *Methods in Ecology and Evolution 1*:180–187.

Bauer, H. & Van Der Merwe, S. (2004). Inventory of free-ranging lions *Panthera leo* in Africa. *Oryx 38(*1): 26–31. DOI: 10.1017/S00306 05304000055.

Bauer, H., Chapron, G., Nowell, K., Henschel, P., Funston, P., Hunter, L. T. B., Macdonald, D. W. & Packer, C. (2015). Lion (*Panthera leo*) populations are declining rapidly across Africa, except in intensively managed areas. *Proceedings of the National Academy of Sciences of the United States of America, 112* (48): 14894–14899. http://www.pnas.org/content/112/48/14894.

Beier, P. (1993). Determining minimum habitat areas and habitat corridors for cougars. *Conservation Biology 7*(1): 94-108.

Beier, P., Riley, S. P. D. & Sauvajot, R. M. (2010). Mountain lions (*Puma concolor*). In Gehrt, S. D., Riley, S. P. D. &and Cypher, B. L. (Eds.) *Urban Carnivores: Ecology, Conflict, and Conservation*, 140 - 155. Baltimore: John Hopkins University Press.

Belant, J. L. (1997). Gulls in the urban environment: landscape level management to reduce conflict. *Landscape and Urban Planning 38*: 245–258.

Benson E. P. (1998). The lord, the ruler: jaguar symbolism in the Americas. In Saunders, N. J. (Eds.) *Icons of Power: Feline Symbolism in The Americas*, 53 - 76. London: Routledge.

Benson, P. C. (1981). *Large Raptor Electrocution and Power Pole Utilization: A Study in Six Western States*. PhD. Thesis, Brigham Young University, Provo.

Berlanga, M. & Wood, P. (1992). Observations of the king vulture *Sarcoramphus papa* in Calakmul Biosphere Reserve, Campeche, Mexico. *Vulture News 26*:15-21.

Bird, D. M., Garland, D. E. & Negro, J. (1996). *Raptors in Human Landscapes Adaptations to Built and Cultivated Environments*. London: Academic Press.

Blackwell, B. F. & Wright, S. E. (2006). Collisions of red-tailed hawks (*Buteo jamaicensis*), turkey vultures (*Cathartes aura*), and black vultures (*Coragyps atratus*) with aircraft: implications for bird strike reduction. *Journal of Raptor Research 40*(1): 76-80.

Blokpoel, H. (1976). *Bird Hazard to Aircraft*. Buffalo, NY: Books Canada Inc.

Boron, V., Tzanopoulos, J., Gallo, J., Barragan, J., Jaimes-Rodriguez, L. & Schaller, G., et al., (2016). Jaguar densities across human-dominated landscapes in Colombia: The contribution of unprotected areas to long term conservation. *PLoS ONE 11*(5): e0153973. doi: 10.1371/journal. pone.0153973.

Braun, M. J., Finch, D. W., Robbins M. B. & Schmidt, B. K. (2007). *A Field Checklist of the Birds of Guyana, 2nd Ed*. Washington, D.C.: Smithsonian Institution.

Brodie, J. F. (2009). Is research effort allocated efficiently for conservation? Felidae as a global 343 case study. *Biodiversity Conservation 18*: 2927–2939.

Brown, D. E. & López-González, C. A. (2001). *Borderland Jaguars: Tigres de la Frontera*. Salt Lake City: University of Utah Press.

Brown, L. H., Urban, E. K. & Newman, K. (Eds.) (1982). *The Birds of Africa* Vol. 1. London: Academic Press.

Buckley, N. J. (1999). Black vulture (*Coragyps atratus*). Account 411. In Poole, A. & Gill, F. (Eds.) *The Birds of North America*. Philadelphia: National Academy of Sciences.

Burdett, C. L. Crooks, K. R., Theobald, D. M., Wilson, K. R., Boydston, E. E., Lyren, L. M., Fisher, R. N, Winston Vickers, I., Morrisson, A. & Boyce, W. M. (2010). Interfacing models of wildlife habitat and human development to predict the future distribution of puma habitat. *Ecosphere 1*:4.

Burggren, W. W. (1987). Invasive and noninvasive methodologies in ecological physiology: a plea for integration. In Feder, M.E., Bennett, A. F., Burggren, W. W., and Huey, R. B., (Eds.) *New Directions in Ecological Physiology*, 251–274. New York: Cambridge University Press.

Butler, P. J. (1989). Telemetric recording of physiological data from freeliving animals. In Grubb, P. J. & Whittaker, J. B. (Eds), *Towards a More Exact Ecology*, 63–84. Oxford: Blackwell Scientific Publications.

Campbell, M. & Lancaster, B. (2010). Public attitudes towards black bears (*Ursus americanus*) and cougars (*Puma concolor*) on Vancouver Island. *Journal of Society and Animals 10*(1): 47-51.

Campbell, M. & Torres-Alvarado, M. (2011). Public perceptions of jaguars (*Panthera onca*), pumas (*Puma concolor*) and coyotes (*Canis latrans*) in El Salvador. *Area 43*(3): 250-256.

Campbell, M. (1998). *Interactions between Biogeography and Rural Livelihoods in the Coastal Savanna of Ghana*. PhD Thesis. University of London, Imperial College.

Campbell, M. 2009. The impact of habitat characteristics on bird presence and the implications for wildlife management in the environs of Ottawa, Canada. *Urban Forestry and Urban Greening 8*: 87–95.

Campbell, M. (2012a). The relevance of age and gender for public attitudes to brown bears (*Ursus arctos*), black bears (*Ursus americanus*), and cougars (*Puma concolor*) in Kamloops, British Columbia. *Journal of Society and Animals 21*(4): 341-359.

Campbell, M. (2012b). Urban planning for cougar presence in North America: practices, challenges and benefits. In Amado, M. P. (Ed.) *Urban Planning: Practices, Challenges and Benefits*, 19 – 32. New York: Nova Science Publishers.

Campbell, M. (2015). *Vultures: Their Evolution, Ecology and Conservation*. New York: Taylor and Francis, CRC Press.

Campbell, M. (2017). *Biological Conservation in the 21st Century: A Conservation Biology of Large Wildlife*. New York: Nova Science Publishers.

Campbell, M. (2018). The ecology and conservation of the West African lion. In *Advances in Animal Science and Zoology*. New York: Nova Science Publishers.

Carette, M., Sanchez-Zapata, J.A., Benıtez, J. R., Lobon, M. and Donazar, J. A. 2009. Large scale risk-assessment of wind-farms on population viability of a globally endangered long-lived raptor. *Biological Conservation 142*: 2954–2962.

Carter, N., Jasny, M., Gurung, B. & Liu, J. (2015). Impacts of people and tigers on leopard spatiotemporal activity patterns in a global biodiversity hotspot. *Global Ecology and Conservation 3*, 149–162 DOI 10.1016/j.gecco.2014.11.013.

Caso, A., Lopez-Gonzalez, C., Payan, E., Eizirik, E., de Oliveira, T., Leite-Pitman, R., Kelly, M. & Valderrama, C. (2008). *Panthera onca. IUCN Red List of Threatened Species. Version 2011.1.* Gland: International Union for Conservation of Nature.

Cavalcanti, S. & Gese, E. M. (2010). Kill rates and predation patterns of jaguars (*Panthera onca*) in the southern Pantanal, Brazil. *Journal of Mammalogy 91*, 722-736.

Chardonnet, P., Soto, B., Fritz, H., Crosmary, W., Drouet-Hoguet, N., Mesochina, P., Pellerin, M., Mallon, D., Bakker, L., Boulet, H. & Lamarque, F. (2010). Managing the conflicts between people and lions. *Wildlife Management Working Paper 13*. Rome, FAO.

Chiari, C. G. & Seeland, K. (2004). Are urban green spaces optimally distributed to act as places for social integration? Results of a geographical information system GIS approach for urban forestry research. *Forest Policy and Economics 6*, 3–13.

Clark, P. E., Johnson, D. E., Kniep, M. A., Jermann, P., Huttash, B., Wood, A., Johnson, M., McGillivan, C. & Titus, K. (2006). An advanced, low-cost, GPS-Based animal tracking system. *Rangeland Ecology & Management 59*: 334–340.

Clinton-Eitniear, J. (1986). King vulture conservation and research program. *Endangered Species Technical Bulletin 3*:1-2. Ann Arbor: University of Michigan.

Cole, D. N. (1995). Disturbance of natural vegetation by comparing experimental applications of low level stress. *Environmental Management 19*(3): 405.

Conforti, V. L. & Azevedo, F. C. C. (2003). Local perceptions of jaguars Panthera once and pumas (*Puma concolor*) in the Iguacu National Park area south Brazil. *Biological Conservation 111*, 215-221.

Cooke, S. J. (2008). Biotelemetry and biologging in endangered species research and animal conservation: relevance to regional, national, and IUCN Red List threat assessments. *Endangered Species Research 4*:165–185.

Cooke, S. J., Hinch, S. G., Wikelski, M., Andrews, R.D., Kuchel, L. J., Wolcott, T. G. & Butler, P. J. (2004). Biotelemetry: a mechanistic approach to ecology. *Trends in Ecology and Evolution 19*: 334–343.

Cooke, S. J., Midwood, J. D., Thiem, J. D., Klimley, P., Lucas, M. C., Thorstad, E. B., Eiler, J., Holbrook, C. & Ebner, B. C. (2013). Tracking animals in freshwater with electronic tags: past, present and future. *Animal Biotelemetry 1*:5–5.

Cornelis, J. & Hermy, M. (2004). Biodiversity relationships in urban and suburban parks in Flanders. *Landscape and Urban Planning 69*: 385–401.

Costa, D. P. & Sinervo, B. (2004). Field physiology: physiological insights from animals in nature. *Annual Review of Physiology 66*: 209–238.

Craigie, I. D., Baillie, J. E. M., Balmford, A., Carbone, C., Collen, B., Green, R.E. & Hutton, J. M. (2010). Large mammal population declines in Africa's protected areas. *Biological Conservation 143*(9): 2221–2228.

Cronkite, G., Mulligan, T., Holmes, J. & Enzenhofer, H. (2007). Categorizing salmon migration behavior using characteristics of split-beam acoustic data. *Aquatic Living Resources 20*, 205–212. DOI: 10.1051/alr:2007033.

Daily Telegraph (2012). *Lion 'on a Rampage' Kills Two in Nigeria*. http://www.telegraph.co.uk/news/worldnews/africaandindianocean/nigeria/9095288/Lion-on-a-rampage-kills-two-in-Nigeria.html.

Davis, H., Wellwood, D. & Ciarniello, L. M. (2001). *"Bear Smart" Community Program: Background Report.* Victoria: British Columbia Ministry of Environment, Lands and Parks.

de Jong, Y. A., Cumming, D., d'Huart, J. & Butynski, T. (2016). *Phacochoerus africanus. (errata version published in 2017) The IUCN Red List of Threatened Species* 2016:e.T41768A1 09669842. http://dx.doi.org/10.2305/IUCN.UK.2016-2.RLTS.T41768A44140445.en. Downloaded on 01 September 2017.

De Lucas, M., Ferrer, M. & Janss G. F. E. (2012a). Using wind tunnels to predict bird mortality in wind farms: The case of griffon vultures. *PLoS ONE 7*(11): e48092. doi: 10.1371/journal.pone.0048092.

De Lucas, M., Ferrer, M., Bechard, M. J. & Muñoz, A. R. (2012b). Griffon vulture mortality at wind farms in southern Spain: Distribution of fatalities and active mitigation measures. *Biological Conservation 147*: 184–189.

De Lucas, M., Janss, G. F. E., Whitfield, D. P. & Ferrer, M. (2008). Collision fatality of raptors in wind farms does not depend on raptor abundance. *Journal of Applied Ecology 45*: 1695 – 1703.

Decker, D. J., Brown, T. L., Hustin, D. L., Clarke, S. H. & O'Pezio, J. (1981). Public attitudes toward black bears in the Catskills. *New York Fish and Game Journal 28*(1): 1-20.

DeMarco, E. (2015). *East, West Africa Could Lose 50% of Their Lions By 2035*. Retreived from http://www.sciencemag.org/news/2015/10/east-west-africa-could-lose-50-their-lions-2035.

Dickson, B. G. & Beier, P. (2002). Home-range and habitat selection by adult cougars in southern California. *Journal of Wildlife Management 66*: 1235–1245.

Dickson, B. G., Jenness, J. S. & Beier, P. (2005). Influence of vegetation, topography, and roads on cougar movement in southern California. *Journal of Wildlife Management 69*: 264–276.

Donadio, E. & Buskirk, S.W. (2006). Diet, morphology, and interspecific killing in Carnivora. *The American Naturalist 167*(4), 524–536 DOI 10.1086/501033.

Donazar, J. A., Ceballos, O. & Tella, J. L. (1996). Communal roosts of Egyptian vultures (*Neophron percnopterus*): dynamics and implications for the species conservation. In Muntaner, J., Mayol, J. (Eds.), *Biologıa y Conservacion de las Rapaces Mediterraneas, 1994* (Monograf ıas no 4), 189–201. Madrid: SEO.

Drewitt, A. L. & Langston, R. H. W. (2008). Collision effects of wind-power generators and other obstacles on birds. *Annals of the New York Academy of Science 1134*: 233–266.

Drewniak, Z., Finnegan, K., Miles, C., Miller, M. & Fox, G. (2012). National parks and protected areas in African countries: A free market environmentalism approach for social and environmental sustainability. *SURG 6*(1): 23 – 30.

Ellis, D. H., Glinski, R. L., Goodwin, J. G. & Whaley, W. H. (1983). New world vulture counts in Mexico, Central America, and South America. In Wilbur, S. R. & Jackson, J. A. (Eds.), *Vulture Biology and Management*, 124-132. Berkeley: University of California Press.

Emmons, L. H. (1987). Comparative feeding ecology of felids in a neotropical rainforest. *Behavioral Ecology and Sociobiology 20*: 271-283.

Emmons, L. H. (1991). Body size and feeding tactics. In Seidensticker, J. & Lumpkin, S. (Eds.) *Great Cats*, 62. Emmaus: Rodale Press.

Erickson, D. L. (2004). The relationship of historic city form and contemporary greenway implementation: a comparison of Milwaukie Wisconsin USA and Ottawa Ontario Canada. *Landscape and Urban Planning 68*: 199–221.

Farrell, L. E., Roman, J. & Sunquist, M. E. (2000). Dietary separation of sympatric carnivores identified by molecular analysis of scats. *Molecular Ecology 9*: 1583–1590.

Farve, R. (2018). *Demonstration of Satellite/GPS Telemetry for Monitoring Fine-Scale Movements of Lesser Prairie-Chickens*. Retrieved from https://www.fs.fed.us/t-d/programs/im/satellite_gps_telemetry/wildlifetrackingtelementry.htm.

Fennessy, J., Bidon, T., Reuss, F., Kumar, V., Elkan, P., Nilsson, M. A., Vamberger, M., Fritz, U. & Janke, A. (2016). Multi-locus analyses reveal four giraffe species instead of one. *Current Biology 26* (18): 2543. PMID 27618261. doi: 10.1016/j.cub.2016.07.036.

Ferguson-Lees, J. & Christie, D.A. (2001). *Raptors of the World*. Boston: Houghton Mifflin Harcourt.

Fernandez-Juricic, E. (2000). Avifaunal use of wooded streets in an urban landscape. *Conservation Biology 14*(2): 513–521.

Ferrer, M. (1993). *El Aguila Imperial*. Madrid: Quercus.

Ferrer, M. & Janss, F. E. (1999). *Birds and Power Lines*. Madrid: Quercus.

Fonturbel, F. E. & Simonetti, J. A. (2011). Translocations and human-carnivore conflicts: problem solving or problem creating. *Wildlife Biology 17*: 217-224.

Freedman, A. H., Portier, K. M. & Sunquist, M. E. 2003. Life history analysis for black bears (*Ursus americanus*) in a changing demographic landscape. *Ecological Modeling 167*(1): 47-64.

Garrett, K. & Dunn, J. (1981). *Birds of Southern California: Status and Distribution*. Los Angeles: Los Angeles Audubon Society.

Guggisberg, C. A. W. (1975). *Wild Cats of the World*. New York: Taplinger Publishing.

Gutiérrez-González, C. E. & López-González, C. A. (2017). Jaguar interactions with pumas and prey at the northern edge of jaguars' range. *Peer J 5*, e2886; DOI 10.7717/peerj.2886.

Guzmán, J. & Castaño, J. P. (1998). Electrocution of raptors on power lines in Sierra Morena Oriental and Campo de Montiel. *Ardeola 45*: 161-169.

Haas, D. & Nipkow, M. (2006). *Caution: Electrocution!* Bonn: NABU Bundesverband.

Haas, D. (2011). *Electrocution of Birds. Some Further Aspects of International High Significance*. Presentation at International Conference on Power Lines and Bird Mortality in Europe, Budapest.

Halleux, D. (1994). Annotated bird list of Macenta Prefecture, Guinea. *Malimbus 16*: 10–29.

Harmsen, B. J, Foster, R. J., Silver, S. C., Ostro, L. E. T. & Doncaster, P. (2009). Spatial and temporal interactions of sympatric jaguars (*Panthera onca*) and pumas (*Puma concolor*) in a Neotropical forest. *Journal of Mammalogy 90* (3), 612-620. DOI: https://doi.org/10.1644/08-MAMM-A-140R.1.

Harness, R.E. & Wilson, K.R. (2001). Utility structures associated with raptor electrocutions in rural areas. *Wildlife Society Bulletin 29*: 612–623.

Harris, R.J. (2013). The conservation of Accipitridae vultures of Nepal: a review. *Journal of Threatened Taxa 5*(2): 3603–3619.

Hastie, G.D. (2012). *Tracking Marine Mammals Around Marine Renewable Energy Devices Using Active Sonar*. SMRU Ltd report URN: 12D/328 to the Department of Energy and Climate Change. September 2012 (unpublished).

Hayward, M.W., Kamler, J.F., Montgomery, R.A., Newlove, A., Rostro-García, S., Sales, L.P. & Van Valkenburgh, B. (2016). Prey preferences of the jaguar *Panthera onca* reflect the post-Pleistocene demise of large prey. *Frontiers in Ecology and Evolution 3*: 1–19. DOI 10.3389/fevo.2015.00148.

Henschel, P., Bauer, H., Sogbohoussou, E. & Nowell, K. (2015). *Panthera leo (West Africa subpopulation). The IUCN Red List of Threatened Species 2015*: e.T68933833A54067639. http://dx.doi.org/10.2305/IUCN.UK.2015-2.RLTS.T68933833A54067639.en. Downloaded on 02 August 2017.

Henschel, P., Coad, L., Burton, C., Chataigner, B., Dunn, A., MacDonald, D., Saidu, Y. & Hunter, L.T.B. (2014). In Hayward, M. (Ed.) The Lion in West Africa is Critically Endangered. *PLoS ONE 9*: e83500. PMC 3885426 Freely accessible. PMID 24421889. doi: 10.1371/journal. pone.0083500.

Herrero, S. & Higgins, A. (1999). Human injuries inflicted by bears in British Columbia: 1960 - 1997. *Ursus 11*: 209-218.

Hill, J.R. & Neto, P.S. (1991). Black vultures nesting on skyscrapers in southern Brazil. *Journal of Field Ornithology 62*(2):173-176.

Höss, M., Kohn, M., Pääbo, S., Knauer, F. & Schröder, W. (1992). Excrement analysis by PCR. *Nature 359*: 199.

Houston, C.S., Terry, B., Blom, M. & Stoffel, M.J. (2007). Turkey nest success in abandoned houses in Saskatchewan. *Wilson Journal of Ornithology 119*(4):742-747.

Howard, B.C. (2014). *Lions Approach Extinction in West Africa.* Retrieved from https://news.nationalgeographic.com/news/2014/01/140108-west-african-lions-endangered-conservation-science/

Hunting, K. (2002). *A Roadmap for PIER Research on Avian Power Line Electrocution in California.* Sacramento: California Energy Commission.

Igl, L.D. & Peterson, S.L. (2010). Repeated use of an abandoned vehicle by nesting turkey vultures (*Cathartes aura*). *Journal of Raptor Research 44*(1): 73-75.

Iñigo Elias, E.E. (1987). Feeding habits and ingestion of synthetic products in a black vulture population from Chispas, Mexico. *Acta Zoologia: Mexico 22*: 1 - 15.

International Bank for Reconstruction and Development/The World Bank (2015). *Conservation and Development in Mozambique.* International Bank for Reconstruction and Development/The World Bank, Washington DC.

Iriarte, J.A., Franklin, W. L., Johnson, W. E. & Redford, K. H. (1990). Biogeographic variation of food habits and body size of the America puma. *Oecologia 85*: 185-190.

IUCN (1992). *Protected Areas of the World: A Review of National Systems. Vol. 3, Afrotropical*. IUCN, Gland, Cambridge.

IUCN SSC Antelope Specialist Group. (2008a). *Damaliscus lunatus* ssp. korrigum. *The IUCN Red List of Threatened Species 2008*:e. T6238 A12590995. http://dx.doi.org/10.2305/IUCN.UK.2008.RLTS.T6238A12590995.en. Downloaded on 04 September 2017.

IUCN SSC Antelope Specialist Group. (2008b). *Hippotragus equinus. The IUCN Red List of Threatened Species 2008*: e.T10167A3178972. http://www.iucnredlist.org/details/10167/0 Downloaded on 01 September 2017.

IUCN SSC Antelope Specialist Group. (2008c). *Syncerus caffer. The IUCN Red List of Threatened Species 2008*: e.T21251A9260904. http://dx.doi.org/10.2305/IUCN.UK.2008.RLTS.T21251A9260904.en. Downloaded on 01 September 2017.

IUCN SSC Antelope Specialist Group. (2008d). *Eudorcas rufifrons. The IUCN Red List of Threatened Species 2008*: e.T8973A12943749. http://dx.doi.org/10.2305/IUCN.UK.2008.RLTS.T8973A12943749.en. Downloaded on 04 September 2017.

IUCN SSC Antelope Specialist Group. (2008e). *Tragelaphus derbianus. The IUCN Red List of Threatened Species 2008*: e.T44172A10867579. http://dx.doi.org/10.2305/IUCN.UK.2008.RLTS.T44172A10867579.en. Downloaded on 05 September 2017.

IUCN SSC Antelope Specialist Group. (2016a). *Alcelaphus buselaphus. The IUCN Red List of Threatened Species 2016*: e.T811A50181009. http://dx.doi.org/10.2305/IUCN.UK.2016-2.RLTS.T811A50181009.en. Downloaded on 04 September 2017.

IUCN SSC Antelope Specialist Group. (2016b). *Nanger dama. The IUCN Red List of Threatened Species 2016*: e.T8968A50186128. http://dx.doi.org/10.2305/IUCN.UK.2016-2.RLTS.T8968A50186128.en. Downloaded on 04 September 2017.

IUCN SSC Antelope Specialist Group. (2016c). *Ourebia ourebi. The IUCN Red List of Threatened Species 2016*: e.T15730A50192202. http://dx.doi.org/10.2305/IUCN.UK.2016-1.RLTS.T15730A50192202.en. Downloaded on 04 September 2017.

IUCN SSC Antelope Specialist Group. (2016d). *Kobus kob. The IUCN Red List of Threatened Species 2016*: e.T11036A50189609. http://dx.doi.org/10.2305/IUCN.UK.2016-2.RLTS.T11036A50189609.en. Downloaded on 05 September 2017.

Jaksic, F. M., Greene, H. W. & Yáñez, J. L. (1981). The guild structure of a community of predatory vertebrates in central Chile. *Oecologia 49*: 21–28.

Janss, G. F. E. (2000). Avian mortality from power lines: a morphological approach of a species-specific mortality. *Biological Conservation 95*: 353–359.

Jenkins, A. (2010). Cape Vulture. *Africa - Birds & Birding 15*(2): 62-63.

Jokimaki, J. (1999). Occurrence of breeding bird species in urban parks: effects of park structure and broad-scale variables. *Urban Ecosystems 3*: 21–34.

Jokimaki, J. (2003). Spatial similarity of urban bird communities: a multiscale approach. *Journal of Biogeography 30*: 1183–1193.

Karanth, K.U. & Chellam, R. (2009). Carnivore conservation at the crossroads. *Oryx 43*: 1-2.

Kellert, S. R. (1994). Public attitudes towards bears and their conservation. *International Conference on Bear Research and Management 9*(1): 43-50.

Kellert, S. R., Black, M., Rush, C. R. & Bath, A. J. (1996). Human culture and large carnivore conservation in North America. *Conservation Biology 10*(4): 977-990.

Kertson, B. N., Spencer, R. D. & Grue, C. E. (2011b). Cougar prey use in a wildland-urban environment in western Washington. *Northwestern Naturalist 92*:175–185.

Kertson, B. N., Spencer, R. D. & Grue, C. E. (2013). Demographic influences on cougar residential use and interactions with people in western Washington. *Journal of Mammalogy 94*(2): 269–281.

Kertson, B. N., Spencer, R. D., Marzluff, J. M., Hepinstall-Cymerman, J. & Grue, C. E. (2011b). Cougar space use and movements in the wildland–urban landscape of western Washington. *Ecological Applications 21*: 2866–2881.

Kiff, L. F. 2000. The current status of North American vultures. In Chancellor, R. D. & Meyburg, B.-U. (Eds.) *Raptors at Risk*, 175–189. Berlin and Surrey, Canada: WWGBP/HancockHouse.

Kruger, R., Maritz, A. & van Rooyen, C. (2004). Vulture electrocutions on vertically configured medium voltage structures in the Northern Cape Province South Africa. In Chancellor, R. D. & Meyburg, B.-U. (Eds.) *Raptors Worldwide*, 437 - 441. Berlin: WWGBP/MME.

Kuhn, M. (2003). Greenbelt and greenheart: separating and integrating landscapes in European city regions. *Landscape and Urban Planning 64*: 19–27.

Land, M. F. & Nilsson, D.-E. (2002). *Animal Eyes*. Oxford: Oxford University Press.

Lasch, U., Zerbe, S. & Lenk, M. (2010). Electrocution of raptors at power lines in Central Kazakhstan. *Waldökologie, Landschaftsforschung und Naturschutz 9*: 95-100.

Lekuona, J. M. (2001). *Uso del espacio por la avifauna y control de la mortalidad de aves en los parques eolicos de Navarra*. Gobierno de Navarra, Pamplona. [*Use of space by avifauna and control of bird mortality in the wind parks of Navarra*. Government of Navarre, Pamplona]

Lekuona, J. M. & Ursula, C. 2007. Avian mortality in wind power plants of Navarra (North Spain). In De Lucas, M., Jonas, G. F. E. & Ferrer, M. (Eds.) *Birds and Wind Farms. Risk Assessment and Mitigation*, 177 – 192. Madrid: Quercus.

Lima, S. (1987). Vigilance while feeding and its relation to the risk of predation. *Journal of Theoretical Biology 124*: 303–316.

Lind, J. (2010). Vigilance and models of behavior. In Breed, M. D. & Moore, J., (Eds.), *Encyclopedia of Animal Behavior*, 506–510. Oxford: Academic Press.

Linnell, J. D. C., Swenson, J. E. & Anderson, R. (2001). Predators and people: conservation of large carnivores is possible at high human densities if management policy is favorable. *Animal Conservation 4*: 345-349.

Löe J. & Røskaft, E. (2004). Large carnivores and human safety: a review. *Ambio 33*(6): 283–288.

Logan, K. A. & Sweanor, L. L. (2001). *Desert Puma: Evolutionary Ecology and Conservation of an Enduring Carnivore*. Washington, D. C: Island Press.

Loomis, N. (2014). *West African Lions under Threat in Senegal*. Retrieved from https://www.voanews.com/a/west-african-lions-under-threat-in-senegal/1918938.html.

López-González, C. & González-Romero, A. (1998). A synthesis of current literature and knowledge about the ecology of the puma (*Puma concolor*). *Acta Zoologica Mexicana 75*, 171-190.

Lorenzo, J. A. (1995). Estudio preliminar sobre la mortalidad de aves por tendidos eléctricos en la isla de Fuerteventura (Islas Canarias). *Ecologıa 9*: 403–407. [Preliminary study on the mortality of birds by power lines on the island of Fuerteventura (Canary Islands).]

Loveridge, A. J., Valeix, M., Davidson, Z., Murindagomo, F., Fritz, H., & Macdonald, D. W. (2009). Changes in home range size of African lions in relation to pride size and prey biomass in a semi-arid savanna. *Ecography 32*(6): 953-962. doi: 10.1111/j.1600-0587.2009.05745.x.

Lowney, M. S. (1999). Damage by black and turkey vultures in Virginia, 1990 – 1996. *Wildland Society Bulletin 27*: 715–719.

Lynch, E., Northrup, J. M., McKenna, M. F., Anderson, C. R. Jr., Angeloni, L. & Wittemyer, G. (2014). Landscape and anthropogenic features influence the use of auditory vigilance by mule deer. *Behavioral Ecology 26*: 75–82.

Maehr, D. S., Shindle, D. B. & Hoctor, T. S. (2002). Florida panther dispersal and conservation. *Biological Conservation 106*:187–197.

Maffei, L., Polisar, J., Garcia, R., Moreira, J. & Noss, A. J. (2011). *Mesoamericana 15*: 49-49.

Martin, G. R. (2011a). Through birds' eyes: insights into avian sensory ecology. *Journal of Ornithology* doi: 10.1007/s10336-011- 0771-5.

Martin, G. R. (2011b). Understanding bird collisions with manmade objects: a sensory ecology approach. *Ibis 153*: 239 – 254.

Martin, G. R. (2012). Through birds' eyes: insights into avian sensory ecology. *Journal of Ornithology 153*: 23–48. http://dx.doi.org/10.1007/s10336-011-0771-5.

Martínez, J. E. (2003). *Impact of power lines on raptor populations in the Sierra Espuña Regional Park (Murcia).* Proceedings of the III International Conference on Prevention Strategies for Fires in Southern Europe, Barcelona.

Matthews, A., Ruykys, L., Ellis, B., Fitz Gibbon, S., Lunney, D., Crowther, M.S., Glen, A.S., Purcell, B., Moseby, K., Stott, J., et al. (2013). The success of GPS collar deployments on mammals in Australia. *Australian Mammalogy 35*(1) 65-83. https://doi.org/10.1071/AM12021.

McCain, E. B. & Childs, J. L. (2008). Evidence of resident jaguars (*Panthera onca*) in the southwestern United States and the implications for conservation. *Journal of Mammalogy 89*: 1-10.

McDougal, C. (1987). The man-eating tiger in geographical and historical perspective. In R. I. Tilson & U.S. Seal (Eds.) *Tigers of the World*, 435 - 448. Park Ridge: Noyes Publications.

Meikle, J. (2011). *Tracking equipment could be harming wild birds, experts warn.* Retrieved from https://www.theguardian.com/environment/2011/sep/20/tracking-equipment-birds-health.

Merkle, J. A., Robinson, H. S., Krausman, P. R. & Alaback, P. (2013). Food availability and foraging near human developments by black bears. *Journal of Mammalogy 94*: 378-385.

Merriam-Webster (2018). *Biotelemetry*. Retrieved from https://www.merriam-webster.com/dictionary/biotelemetry.

Mills, M. G. L., Broomhall, L. S. & du Toit, J. T. (2004). Cheetah *Acinonyx jubatus* feeding ecology in the Kruger National Park and a comparison across African savanna habitats: is the cheetah only a successful hunter on open grassland plains? *Wildlife Biology 10*: 177-186.

Mitani, Y., Andrews, R. D., Sato, K., Kato, A., Naito, Y. & Costa, D. P. (2010). Three-dimensional resting behaviour of northern elephant seals: drifting like a falling leaf. *Biology Letters 6*:163–166.

Mitchell, M. S. & Hebblewhite, M. (2012). Carnivore habitat ecology: integrating theory and application. In Boitani, L. & Powell, R. A. (Eds.) *Carnivore Ecology and Conservation*, 218–255. New York: Oxford University Press.

Moccia, J. & Arapogianni, A. (2011). *Pure Power: Wind Energy Targets for 2020 and 2030*. Brussels: European Wind Energy Association.

Morrison, J. C., Sechrest, W., Dinerstein, E., Wilcove, D. S. & Lamoreux, J. F. (2007). Persistence of large mammal faunas as indicators of global human impacts. *Journal of Mammalogy 88*; 1363-1380.

Morzillo, A. T., Mertig, A. G., Garner, N. & Liu, J. (2007). Resident attitudes toward black bears and population recovery in East Texas. *Human Dimensions of Wildlife 12*(6): 417-428.

Mosser, A. A., Avgar, T., Brown, G. S., Walker, C. S. & Fryxell, J. M. (2014). Towards an energetic landscape: broad-scale accelerometry in woodland caribou. *Journal of Animal Ecology 83*: 916–922.

Naidoo, V., Wolter, K., Espie, I. & Kotze, A. (2011). Vulture rescue and rehabilitation in South Africa: an urban perspective. *Journal of the South African Veterinary Association 82*(1): 24–31.

Nelson, M. W. & Nelson, P. (1976). Power lines and birds of prey. *Idaho Wildlife Review 28*: 3–7.

Nelson, M. W. (1979). *Power lines progress report on eagle protection research. Idaho Power Company, Boise, Idaho, USA*. Available from the U.S. Geological Survey, Richard R. Olendorff Memorial Library, 970 Lusk St., Boise, ID 83706.

Nelson, M. W. (1980). *Update on eagle protection practices. Idaho Power Company. Boise, Idaho, USA.* Available from the U.S. Geological Survey, Richard R. Olendorff Memorial Library, 970 Lusk St., Boise, ID 83706.

Niklaus, G. (1984). Large numbers of birds killed by electric power line. *Scopus 8*: 42.

Nilsson, C & Grelsson, G. (1995). The fragility of ecosystems: review. *Journal of Applied Ecology 32*(4): 677-692.

Norgaard, R. B, (1994). *Development Betrayed: The End of Progress and a Coevolutionary Revisioning of the Future*. London: Routledge.

Novack, A. J., Main, M. B., Sunquist, M. E. & Labisky, R. F. (2005). Foraging ecology of jaguar (*Panthera onca*) and puma (*Puma concolor*) in hunted and non-hunted sites within the Maya Biosphere Reserve, Guatemala. *Journal of Zoology 267*, 167–178. DOI 10.1017/ S0952836905007338.

Novaes, W. G. & Cintra, R. (2013). Factors influencing the selection of communal roost sites by the Black Vulture *Coragyps atratus* (Aves: Cathartidae) in an urban area in Central Amazon. *Zoologia* http://dx.doi.org/10.1590/S1984-46702013005000014.

Nowell, K. & Jackson P. (Eds.) (1996). *Wild Cats: Status Survey and Conservation Action Plan.* Gland: IUCN/SSC Cat Specialist Group.

Odden, M., Wegge, P. & Fredriksen, T. (2010). Do tigers displace leopards? If so, why? *Ecological Research 25*: 875–881 DOI 10.1007/s11284-010-0723-1.

Ogada, D. L. & Buij, R. (2011). Large declines of the Hooded Vulture Necrosyrtes monachus across its African range. *Ostrich: Journal of African Ornithology 82*(2): 101-113.

Oliveira, T. de (2002). Evaluación del estado de conservación del jaguar em el este de La Amazonia y noreste de Brasil. In Medellín, R. A., Equihua, C., Chetkiewitcz, C. L. B., Crawshaw Jr., P. G., Rabinowitz, A., Redford, K. H., Robinson, J. G., Sanderson, E. W. and Taber, A. B. (Eds.). *El Jaguar em El Nuevo Milenio*, 419-436. Mexico, New York: Fondo de Cultura Económica, México, Universidad Nacional Autónoma de México, Wildlife Conservation Society. [Evaluation of the state of conservation of the jaguar in the east of the Amazon and northeast of Brazil. In *The Jaguar in The New Millennium*]

Oliveira, T.G. de (1994). *Neotropical Cats: Ecology and Conservation.* São Luís: EDUFMA.

Packer, C., Ikanda, D., Kissui, B. & Kushnir, H. (2005). Lion attacks on humans in Tanzania. *Nature 436*: 927–928.

Packer, C., Ikanda, D., Kissui, B. & Kushnir, H. (2006). The ecology of man-eating lions in Tanzania. *Nature & Faune 21*(2). Accra: FAO.

Palacios, M. J. (2003). Power lines in Extremadura: conservation action and bird conservation. *Proceedings of the National Conference on Power lines and Bird Conservation in Protected Areas*. Murcia: Dirección General de Medio Ambiente.

Palmeira F. B. L, Crawshaw, P. G., Haddad, C. M., Ferrazd, K. M. P. M. B. & Verdadee, L. M. (2008). Cattle depredation by puma Puma concolor and jaguar Panthera onca in central western Brazil. *Biological Conservation 141*: 118 - 125.

Partenariat Régional pour la Conservation de la zone côtière et Marine en Afrique de l'Ouest (2017). *Guinea Bissau: protected areas coverage up from 15% to 26.3%.* Retrieved from http://www.prcmarine.org/en/guinea-bissau-protected-areas-coverage -15-263.

Paviolo, A., De Angelo, C. D., Di Blanco, Y. E. & Di Bitetti, M. S. (2008). Jaguar *Panthera onca* population decline in the Upper Paraná Atlantic Forest of Argentina and Brazil. *Oryx 42*: 554-561.

Paxinos, E., McIntosh, C., Ralls, K. & Fleischer, R. (1997). A non-invasive method for distinguishing among canid species: amplification and enzyme restriction of DNA from dung. *Molecular Ecology 6*: 483–486.

Pelton, M., Scott, C. & Burghardt, G. M. (1976). Attitudes and opinions of persons experiencing property damage and/or injury by black bears in the Great Smoky Mountains National Park. *International Conference of Bear Research and Management 3*: 157-167.

Perry, T. (1994). Living in cougar country. *Los Angeles Times*, December 19, A3, A38. A39.

Pirnat, J. 2000. Conservation and management of forest patches and corridors in suburban landscapes. *Landscape and Urban Planning 52*,135–143.

Polisar, J., Maxit, I., Scognamillo, D., Farrell, L., Sunquist, M. & Eisenberg, J.F. (2003). Jaguars, pumas, their prey base, and cattle ranching: ecological interpretations of a management problem. *Biological Conservation 109*: 297–310 DOI 10.1016/S0006-207(02)00157-X.

Ponganis, P. J. (2007). Bio-logging of physiological parameters in higher marine vertebrates. *Deep Sea Research Part II: Topical Studies in Oceanography 54*: 183–192.

Prance, G. T. (1995). Systematics, Conservation and Sustainable Development. *Biodiversity and Conservation 4* (5): 49-50.

Prinsen, H. A. M., Boere, G. C., Píres N. & Smallie, J. J. (2011). Review of the conflict between migratory birds and electricity power grids in the African-Eurasian region. *CMS Technical Series No. XX, AEWA Technical Series No. XX Bonn.*

Quiroga, V. A., Boaglio, G. I., Noss, A. J. & Di Bitetti, M. S. (2013). Critical population status of the jaguar *Panthera onca* in the Argentine Chaco: camera-trap surveys suggest recent collapse and imminent regional extinction. *Oryx 48*: 1–8. doi: 10.1017/S0030605312 000944.

Rabinowitz, A. & Nottingham, B. (1986). Ecology and behavior of the jaguar in Belize, Central America. *Journal of Zoology 210*: 149–159.

Rabinowitz, A. & Zeller, K. A. (2010). A range-wide model of landscape connectivity and conservation for the jaguar, *Panthera onca. Biological Conservation 143*: 939-945.

Ramesh, T., Kalle, R., Sankar, K. & Qureshi, Q. (2012). Spatio-temporal partitioning among large carnivores in relation to major prey species in Western Ghats. *Journal of Zoology 287*(4): 269–275 DOI 10.1111/j.1469-7998.2012.00908.x.

Reid, S. B. 1989. Flying behavior and habitat preferences of the king vulture *Sarcorarnphus papa* in the western Orinoco Basin of Venezuela. *Ibis 131*: 301-303.

Richards, D. K. 1982. The birds of Conakry and Kakulima, Democratic Republic of Guinee. *Malimbus 4*: 93–104.

Romero-Muñoz, A., Maffei, L., Cuéllar, E. & Noss, A. J. (2010). Temporal separation between jaguar and puma in the dry forests of southern Bolivia. *Journal of Tropical Ecology 26*(3): 303–311 DOI 10.1017/S0266467410000052.

Rondeau, G., Condeé, M.M., Ahon, B., Diallo, O. & Pouakouyou, D. 2008. Survey of the occurrence and relative abundance of raptors in Guinea subject to international trade. *JNCC Report No. 412*. Peterborogh: Joint Nature Conservation Committee.

Rosas-Rosas, O. C. & Valdez R. (2010). The role of landowners in jaguar conservation in Sonora, Mexico. *Conservation Biology 24*(2): 366–371 DOI 10.1111/j.1523-1739. 2009.01441.x.

Rosas-Rosas, O. C. & Bender, L.C. 2012. Population status of jaguars (*Panthera onca*) and pumas (*Puma concolor*) in northeastern Sonora, Mexico. *Acta Zoologica Mexicana 28*(1): 86 - 101.

Rosas-Rosas, O. C., Bender, L. C. & Valdez, R. (2008). Jaguar and puma predation on cattle calves in northeastern Sonora, Mexico. *Rangeland Ecology and Management 61*: 554-560.

Russell, K. R. & Gauthreaux, Jr., S. A. (1998). Use of weather radar to characterize movements of roosting purple martins. *Wildlife Society Bulletin 26*: 5-16.

Ruth, T. K. & Murphy, K. (2010). Competition with other carnivores for prey. In M. Hornocker & S. Negri (Eds.), *Cougar Ecology & Conservation*, 163–172. Chicago: University of Chicago Press.

Ruth, T. K. (1991). *Mountain Lion Use of an Area of High Recreational Development in Big Bend National Park Texas*. MS. Thesis, Texas A&M University, College Station.

Rutz, C. & Hays, G. C. (2009). New frontiers in biologging science. *Biology Letters 5*:289–292.

Saenz, J. C. & Carrillo, E. (2002). Jaguares depredadores de ganado en Costa Rica: Un problema sin solucion? *El Jaguar en el Nuevo Milenio*, 127–138. Mexico: A Taber Universidad Nacional Autonoma de Mexico and Wildlife Conservation Society. [Jaguars predators of cattle in Costa Rica: A problem without solution? *The Jaguar in the New Millennium*, 127-138. Mexico: A Taber National Autonomous University of Mexico and Wildlife Conservation Society.]

Salom-Pérez, R., Camillo, E., Sáenz, J. C. & Mora, J. M. (2007). Critical condition of the jaguar Panthera onca population in Corcovado National Park, Costa Rica. *Oryx 41*: 51-56.

Sanderson, E. W., Redford, R. H., Chetkiewicz, C. B., Medellin, R., Rabinowitz, A., Robinson, J. & Taber, A. (2002). Planning to save a species: the jaguar as a model. *Conservation Biology 16*: 58-72.

Santos, F. R., Jácomo, A. T. A. & Silveira, L. (2008). Humans and jaguars in five Brazilian biomes: same country, different perceptions. *Cat News 4*: 21-25.

Sarmiento, R. (2004). *Population estimation methods for jaguar (Panthera onca), Corcovado National Park, Costa Rica.* Master of Science Thesis. Universidad Nacional, Heredia, Costa Rica.

Satheesan, S. M. & Satheesan, M. (2000). Serious vulture-hits to aircraft over the world. *International Bird Strike Committee IBSC25/WP-SA3*. Amsterdam: IBSC.

Satheesan, S. M. (1992). *Ecology and Behaviour of Pariah Kite Milvus Migrans Govinda Sykes as a Problem Bird at Some Indian Aerodromes*. PhD. Thesis. University of Bombay. Bombay.

Satheesan, S. M. 1994. The more serious vulture hits to Military Aircraft in India between 1980 and 1994. In *Proceedings of the 22 BSCE Meeting Vienna, Austria*, 163 – 168. 29 August–2 September, WP 23.

Satheesan, S. M. (1996). Raptors associated with airports and aircraft. In Bird, D., Varland, D. and Negro, J. (Eds.) *Raptors in Human landscapes*, 315 – 323. Academic Press, U.K. and Raptor Research Foundation.

Satheesan, S. M. (1998a). Need for imparting training at national level to bird controllers at Civil and Military Airports. In *Proceedings of the IBSC 24*, 3 – 8. Stara Lesna, Slovakia, 14-18 September. WP1.

Satheesan, S. M. (1998b). Sound and light can control bird activity at Indian Airports. In *Proceedings of the IBSC 24*, 223 - 226. Stara Lesna, Slovakia, 14-18 September. WP19. Ibid, WP 19.

Saunders, N. J. (1998). *Icons of Power: Feline Symbolism in the Americas*. London: Routledge.

Savard, J. P. L., Clergeau, P. & Mennechez, G. (2000). Biodiversity concepts and urban ecosystems. *Landscape and Urban Planning 48*,131–142.

Scantlebury, D. M., Mills, M. G. L., Wilson, R. P., Wilson, J. W., Mills, M. E. J., Durant, S. M., Bennett, N.C., Bradford, P., Marks, N. J. and Speakman, J. R. (2014). Flexible energetics of cheetah hunting strategies provide resistance against kleptoparasitism. *Science 346*: 79–81.

Schwartz, M. (2017). *The Lion and the Cow: Conservation, Pastoralism, and Conflict. Cat Watch.* https://voices.nationalgeographic.org/%202017/04/28/the-lion-and-the-cow-conservation-pastoralism-and-conflict/

Scognamillo, D., Maxit, I.E., Sunquist, M. & Polisar, J. (2003). Coexistence of jaguar (*Panthera onca*) and puma (*Puma concolor*) in a mosaic landscape in the Venezuelan Llanos. *Journal of Zoology (London) 259*: 269–279.

Shepard, E. L., Wilson, R. P., Rees, W. G., Grundy, E., Labertucci, S. A. & Vosper, S. B. (2013). Energy landscapes shape animal movement ecology. *American Naturalist 182*: 298–312.

Shuey, M. L. (2008). *The Influence of Landscape on Cougar-Human Encounter Hazard in Boulder and Colorado Springs, Colorado*. ETD Collection for Texas State University. Retrieved from https://digital.library.txstate.edu/handle/10877/3004.

Sigismondi, A. & Politano, E. (1996). Unusually high concentrations of the Egyptian vulture *Neophron percnopterus* in a border area of the Dancalia region of Ethiopia. It may be one of the most important wintering areas known, 88–89. Abstracts *2nd International Conference on Raptors, Urbino, Italy 2–5 October 1996*. Raptor Research Foundation—University of Urbino, Urbino.

Silver, S. C., Ostro, L. E. T., Marsh, L. K., Maffei, L., Noss, A. J., Kelly, M. J., Wallace, R. B., Gómez, H. & Ayala, G. (2004). The use of camera traps for estimating jaguar *Panthera onca* abundance and density using capture-recapture analysis. *Oryx 38*(2), 148 - 154.

Slater, L. E. (1965). (Introductory editorial). *Bio Science 15*(2): 81-82.

Smallie, J. & Virani, M. (2010). A preliminary assessment of the potential risks from electrical infrastructure to large birds in Kenya. *Scopus 30*: 32-39.

Soisalo, M. K. & Cavalcanti, S. M. C. (2006). Estimating the density of a jaguar population in the Brazilian Pantanal using camera-traps and capture–recapture sampling in combination with GPS radio-telemetry. *Biological Conservation 129*: 487 – 496.

Sollmann, R., Gardner, B., Belant, J. L. (2012). How Does Spatial Study Design Influence Density Estimates from Spatial Capture-Recapture Models? *PLoS ONE 7*: e34575.

Sollmann, R., Tôrres, N. M. & Silveira, L. (2008). Jaguar conservation in Brazil: the role of protected areas. *Cat News 4*, 15-20.

Spencer, R. D., Pierce, D. J., Schirato, G. A., Dixon, K. R. & Richards C. B. (2001). *Mountain Lion Home Range, Dispersal, Mortality and Survival in the Western Cascade Mountains of Washington*. Olympia: Washington Department of Fish and Wildlife.

Ssemmanda, R. & Plumptre, A. J. (2011). *Declining Populations: What is the Status of Vultures in Queen Elizabeth Park-Uganda?* New York; Wildlife Conservation Society.

Ssemmanda, R. & Pomeroy, D. (2010). Scavenging birds in Kampala since 1973. *Scopus 30*: 26–31.

Ssemmanda, R. (2005). An apparent increase in Hooded Vulture *Necrosyrtes monachus* numbers in Kampala, Uganda. *Vulture News 53*: 10 - 14.

Stephens, P. A. (2015). Land sparing, land sharing, and the fate of Africa's lions. *PNAS 112*(48): 14753–14754.

Stiles, F. G. (1985). Conservation of forest birds in Costa Rica: problems and perspectives. In Diamond, A. W. & Lovejoy, T. E. (Eds.), *Conservation of Tropical Forest Birds. Vol. 4*, 141-168. Cambridge: ICBP Technical Publications.

Sunquist, M. & Sunquist, F. (2002). *Wild Cats of the World*. Chicago: The University of Chicago Press.

Telleria, J. L. (2009a). Overlap between wind power plants and griffon vultures *Gyps fulvus* in Spain. *Bird Study 56*: 368 – 271.

Telleria, J. L. (2009b). Wind power plants and the conservation of birds and bats in Spain: a geographical assessment. *Biodiversity Conservation 18*:1781–1791.

Thiollay, J.-M. (2006a). The decline of raptors in West Africa: long-term assessment and the role of protected areas. *Ibis 148*: 240-254.

Thornton, C. T. & Quinn, M.S. (2009). Coexisting with cougars: Public perceptions, attitudes and awareness of cougars on the urban-rural fringe of Calgary, Alberta, Canada. *Human–Wildlife Conflicts 3*(2); 282–295.

Tobler, M. W., Carrillo-Percastegui, S. E., Zúñiga Hartley, A. & Powell, G. V. N. (2013). High jaguar densities and large population sizes in the core habitat of the southwestern Amazon. *Biological Conservation 159*: 375–381. doi: 10.1016/j.biocon. 2012.12.012.

Tomkiewicz, S. M., Fuller, M. R., Kie, J. G. & Bates, K. K. (2010). *Philosophical Transactions of the Royal Society of London B 365*: 2163–2176 doi: 10.1098/ rstb.2010.0090.

Tôrres, N. M., De Marco Jr, P., Diniz-Filho, J. A. F. & Silveira, L. (2008). Jaguar distribution in Brazil: Past, present and future. *Cat News Special Issue 4*: 4-8.

Tôrres, S. G., Mansfield, T. M., Foley, J. E., Lupo, T., and Brinkhaus, A. (1996). Mountain lion and human activity in California: testing speculations. *Wildlife Society Bulletin 24*: 451– 460.

Tortato, F. R., Layme, V. M. G., Crawshaw, Jr P. G. & Izzo, T. J. (2015). The impact of herd composition and foraging area on livestock predation by big cats in the Pantanal of Brazil. *Animal Conservation 18*(6): 539–547 DOI 10.1111/acv.12207.

United States Fish and Wildlife Service (2017). *Endangered and Threatened Wildlife and Plants; Listing Two Lion Subspecies.* Federal Register Retrieved from https://www.federalregister.gov/documents/2015/12/23/2015-31958/endangered-and-threatened-wildlife-and-plants-listing-two-lion-subspecies.

Unitt, P. (1984). The birds of San Diego County. *San Diego Society of Natural History Memoir 13.*

Van Dyke, F. G., Brooke, R. H., Shaw, H. G., Ackerman, B. B., Hemker T. P. & Lindzey, F. G. (1986). Reactions of mountain lion to logging and human activity. *Journal of Wildlife Management 50*: 95–102.

Vanak, A. T., Fortin, D., Thaker, M., Ogden, M., Owen, C., Greatwood, S. & Slotow, R. (2013). Moving to stay in place: behavioral mechanisms for coexistence of African large carnivores. *Ecology* 94(11): 2619–2631. DOI 10.1890/13-0217.1.

Virani, M.Z. (2006). In Steep Decline. *SWARA (Magazine of the East African Wildlife Society) April – June 2006.*

Viviant, M., Monestiez, P. & Guinet, C. (2014). Can we predict foraging success in a marine predator from dive patterns only? Validation with prey capture attempt data. *PLoS ONE 9*(3): e88503.

Wacher, T., Newby, J., Houdou, I., Harouna, A. & Rabeil, T. (2013). Vulture observations in the Sahelian zones of Chad and Niger. *Bull ABC 20*(2): 186 - 199.

Wallace, R. B., Gomez, H., Ayala, G. & Espinoza, F. (2003). Camera trapping capture frequencies for jaguar (*Panthera onca*) in the Tuichi Valley, Bolivia. *Mastozoologia Neotropical 10*(1): 133-139.

Walsh, J. F. (1987). Records of birds seen in north-eastern Guinea in 1984–1985. *Malimbus 9*: 105–122.

Wasser, S. K., Houston, C. S., Koehler, G. M., Cadd, G. G. & Fain, S.R. (1997). Techniques for application of faecal DNA methods to field studies of Ursids. *Molecular Ecology 6*: 1091–1097.

Watanabe, Y. Y. & Takahashi, A. (2013). Linking animal borne video to accelerometers reveals prey capture variability. *Proceedings of the National Academy of Sciences USA 110*: 2199–2204.

Weaver, J. L., Paquet, P. C. & Ruggiero, L. F. (1996). Resilience and conservation of large carnivores in the Rocky Mountains. *Conservation Biology 10*: 964–976.

Weise, F. J., Vijay, V., Jacobson, A. P., Schoonover, R. F., Groom, R.J., Horgan, J., Keeping, D., Klein, R., Marnewick, K., Maude, G., Melzheimer, J., Mills, G., van der Merwe, V., van der Meer, E., van Vuuren, R.J., Wachter, B. & Pimm, S.L. (2017). The distribution and numbers of cheetah (*Acinonyx jubatus*) in southern Africa. *PeerJ 5*:e4096; DOI 10.7717/peerj.4096.

Whitacre, D. F., Baker, A. J., Jones, J. E., Villegas Patraca, R., Sutter, J. & Swartz, C. M. (1991). Results of census efforts in three units of the Maya Biosphere Reserve/ Calakmul Biosphere Reserve Complex. In Whitacre, D. F., Burnham, W. A. & Penny, J. P. (Eds.) *Maya Progress Report*, 43-58. The Peregrine Fund, Boise.

White, C. R., Cassey, P., NSchimpf, G., Halsey, L. G., Green, J. A. & Portugal, S. J. (2013). Implantation reduces the negative effects of bio-logging devices on birds. *Journal of Experimental Biology 216*: 537–542.

White, K. R., Koehler, G. M., Maletze, B. T. & Wielgus, R. B. (2011). Differential prey use by male and female cougars in Washington. *Journal of Wildlife Management* 75: 1115–1120.

Wildlife, A.C.T. (2018). *Monitoring and Tracking Technology*. Retrieved from https://wildlifeact.com/about-wildlife-act/monitoring-tracking-technology/

Williams, T. M., Wolfe, L., Davis, T., Kendall, T., Richter, B., Wang, Y., Bryce, C., Elkaim, G. H. & Wilmers, C. C. (2014). Instantaneous energetics of puma kills reveal advantage of felid sneak attacks. *Science 81*: 81–85.

Wilmers, C. C., Nickel, B., Bryce, C. M., Smith, J. A., Wheat, R. E., Yovovich, V. (2015). The golden age of bio-logging: how animal-borne sensors area advancing the frontiers of ecology. *Ecology 96*: 1741–1753.

Wilson, A. M., Lowe, J. C., Roskilly, K., Hudson, P. E., Golabek, K. A. & McNutt, J. W. (2013). Locomotion dynamics of hunting in wild cheetahs. *Nature 498*:185–189.

Wilson, R. P., Grundy, E., Massy, R., Soltis, J., Tysse, B., Holton, M., Cai, Y., Parrott, A., Downey, L. A., Qasem, L., Butt, T. (2014). Wild state secrets: ultra-sensitive measurement of micro-movement can reveal internal processes in animals. *Frontiers in Ecology and the Environment 12*: 582–587.

Wolch, J., Gullo, A. & Lassiter, U. (1997). Changing Attitudes toward California's Cougars. *Society and Animals 5*(2): 95 - 116.

Wood, H. B. (1945). The history of bird banding. *Auk 62*: 256 – 265.

Yáñez, J. L., Cárdenas, J. C., Gezelle, P. & Jaksic, F. M. (1986). Food habits of the southernmost mountain lions (*Felis concolor*) in South America: natural versus livestocked ranges. *Journal of Mammalogy 67*: 604–606.

Zeller, K. (2007). *Jaguars in the New Millennium Data Base Update: The State of the Jaguar in 2006*. New York: Wildlife Conservation Society-Jaguar Conservation Program.

Zeller, K. A., Nijhawan, S., Salom-Perez, R. Potosme, S. H. & Hinnes, J. E. (2011). Integrating occupancy modeling and interview data for corridor identification: A case study for jaguars in Nicaragua. *Biological Conservation 144*: 892–901.

Zilio, F., Bolzan, A., de Mendonça-Lima, A., da Silva, A. O., Verrastro, L. & Borges-Martins, M. (2013). Raptor assemblages in grasslands of Southern Brazil: species richness and abundance and the influence of the survey method. *Zoological Studies 52*: 27.

Zimmermann, A., Walpole, M. J. & Leader-Williams, N. (2005). Cattle ranchers' attitudes to conflicts with jaguar (*Panthera onca*) in the Pantanal of Brazil. *Oryx 39*: 406–412.

In: Geomatics and Conservation Biology
Editor: Michael O'Neal Campbell
ISBN: 978-1-53614-468-0

Chapter 7

GEOMATICS FOR MICRO-SPACES OF AVIAN FORAGING, SUBJECTIVE BEHAVIOR AND HUMAN PRESENCE

Michael O'Neal Campbell
Camosun College, Victoria, Canada

ABSTRACT

The study of avian foraging in micro-spaces, that is small areas such as gardens, shrubs, and parks, is an important aspect of conservation biology. However, many studies use human observation or 'eye balling' to record bird behavior in such micro-spaces. Recent developments in geomatics technology have created possibilities for the investigation of such micro-spaces. These would include satellite images and aerial photographs at very sharp resolutions for the recording of habitats, and global positioning systems (GPS) and motion activated cameras for the recording of bird movements. These will enable more accurate measurements of ecological relations of birds, especially in relation to their reactions to human encroachment. Particularly problematic are the measurements of the more dynamic actions of birds, that may in some cases escape the generalizations of species level studies. In this respect the methodologies of the sciences that investigate individual or group-based bird behavior may guide the application of geomatics techniques. This chapter examines the recent developments in geomatics techniques that enable measurement of the details of bird behavior in dynamic contexts and links these applications to the methodology of animal geography, focusing on those strands that study the subjective individualism of animal behavior. It also examines the use of alert and flight distances in measuring the reactions of birds to human approaches. The chapter concludes that current advancements in geomatics techniques are adequate for the measurement of bird behavior at micro-levels, but more experimentation and comparison with older methodologies will hone these applications further.

INTRODUCTION

Geomatics techniques are applied to biological, ecological and geographical topics to ensure the derivation of more authoritative and reliable information on the parameters of environmental dynamics (Elliott & Campbell, 2002). Sharper resolutions and more flexible techniques increasingly enable the measurement of details overlooked by older, less advanced techniques (Campbell, 2017). These points are particularly important when smaller species, such as common urban avian species are studied (Marzluff et al., 2001). The application of such detailed geomatics techniques may have a stronger impact if they are applied within the theoretical framework of methodological developments related to the study of small-scale ecological relations and possible individual and group behaviors, which may be difficult to detect with more generalized studies (Campbell, 2006a,b, 2009a,b,c, 2010a). These theoretical and methodological approaches may be used to test the geomatics techniques to derive findings that may contribute to the deeper understanding of the small-scale contexts of animal behavior (Campbell, 2007, 2008).

Three particularly important developments have occurred within the fields of biogeography and ecology that take new looks at ecological relationships (Campbell, 1998; Wolch & Emel 1998; Fernández-Juricic et al. 2001). One of these is the development of disequilibrium ecology, which challenges the generalized, positivist approach of classical ecology, and acknowledges the possibility that non-equilibrium dynamics may be fundamental to ecosystems, in that an ecosystem may exhibit constantly changing dynamism in ecological relations, making generalization difficult (Campbell, 1998). Another development concerns certain strands of animal geography, a discipline or subdiscipline related to zoogeography, which explores evidence of individualism in animal species. Like disequilibrium ecology, this approach challenges generalization in favor of unpredictability, non-generalization and animal individualism (Wolch & Emel, 1998, Lorimer, 2007). Another development concerns the measurement of avian responses to human approaches, using the bird's first awareness to human approach (alert distance) or when it takes flight (flight distance) (Fernández-Juricic et al., 2001). This is illustrated in Figure 1.

Increased detail in the observation of ecological relations was the hallmark of these developments (Campbell, 2017). Parallel developments concerned the appearance of increasingly sophisticated techniques for ecological measurements, including more powerful and longer-range cameras, and the remote measurement of animal movements, bodily reactions and even physiology (Bowlin et al., 2005; Cagnacci et al., 2010; Bouten et al., 2013). Difficulties in the application of these techniques are acknowledged and critiqued, but the current direction is towards the successful measurement of avian behavior, of sufficient magnitude to contribute to greater detail in ecological studies. Lameris & Kleeheeg (2017: 24) argue

> The development of lightweight electronic tracking devices, which can transmit or log global positions and are small enough to deploy on free-ranging wild birds, has unchained a revolution in ornithology. Such devices enable tracking of birds during their entire annual cycle and assessing fine-scale movement behavior and individual behavioral strategies, which is difficult or impossible using conventional marking methods (metal rings or color bands), especially in long-distance migrants. The ability to remotely track bird movements with high resolution and accuracy… has greatly improved our understanding of their behavior and ecology, with important implications for their conservation.

The remainder of this chapter examines the theoretical frameworks of disequilibrium ecology, strands of animal geography and alert distances, and links these findings to the development of relevant geomatics techniques. In this it draws heavily on review and research pieces I have published elsewhere (book chapters Campbell, 2013, 2014a, 2015; and articles on the alert distances of birds in the green spaces of Stirling, Scotland (Campbell, 2006a) and Peterborough, Canada (Campbell, 2010a), and two papers on the utility of a strand of animal geography (emphasizing avian individualism) in feeding spaces of Glasgow, Scotland (2007) and Peterborough, Canada (Campbell, 2008). The key point concerns the possibility of extended knowledge on the behavior of birds at micro-levels, and the current enabling and constraining paradigms and techniques that frame this development. Individualism one of the themes of animal geography. Alert distances (Figure 1) are one way of measuring this phenomenon, but any observation is also suitable. Individualism rests partly on the ideas of ecological disequilibrium, as it precludes over-generalization and assumes that ecological equilibrium may be challenged by micro-studies that can differentiate individual behavioral traits. Such behavior can be measured by human appraisal, but the geomatics techniques can impart the quantification required for more definitive assessment.

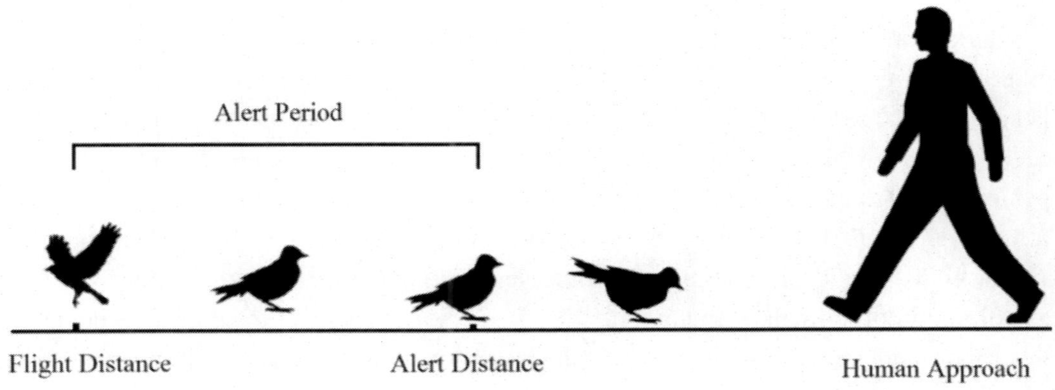

Figure 1. The bird was in a relaxed position until awareness of human approach. The alertness ensued for a period before flight.

GPS Applications

Small bird size is an important problem of GPS supported studies of such dynamics. Most such equipment is only suitable for larger animals. For example, Williams et al. (2014) developed a GPS–accelerometer collar on caged cougars, testing the apparatus on the cats' movement on a treadmill and walking on flat surfaces, also measuring the energetic costs of locomotion, linked to gaits. In another example, Augliere (2015) examined the GPS tracking of wolves, arguing new technology that not only documents the animals' movements, but also the metabolic costs of the different activities of the wolves, based on accelerometers which recorded the burning of calories during athletics or running. This technology was tested on larger animals, including wolves, polar bears, narwhals, and mountain lions.

In another example, Wildlife ACT (2018) developed life- saving tracking collars, used to track larger animals such as cheetahs or rhinoceros. This uses tracking collars, examples being radio, GPS and satellite collars. These monitors track animals daily, to monitor injuries, illegal trapping, movement patterns and inter-species interactions (Wildlife ACT, 2018). This company also developed remote camera trapping, were remote-activated camera traps take snapshots through infrared or motion sensors after detecting movements. However, this usually records locations of elusive species, rather than the dynamic, second by second movements of small species (Wildlife ACT (2018). This issue was noted by Clark et al. (2006, 334):

> An improved global positioning system (GPS)–based animal tracking system is needed to meet quickly evolving demands of ecological research, range livestock production, and natural resource management. Commercially available tracking systems lack the data storage capacity needed to frequently collect animal location data (e.g., 15-minute intervals or less) over long-term deployment periods (e.g., 1 year or more).

Clark et al. (2006) therefore mention a new system, the Clark ATS, which enables detailed examination of target animal behavior, including reactions to short-term disturbances (Clark et al., 2006: 334). Other recent developments include tri-axial accelerometers, which can be calibrated to movements as miniscule as individual fish opening and closing their mouths (Viviant et al. 2014) and recording of fungi on insects (Wilson et al., 2014). There are also observations of eating animals (Watanabe & Takahashi 2013) and tri-axial digital compasses, or magnetometers, which can record the gaze direction of animals (Mitani et al., 2010). Some sensors can measure the metabolic rates, body and jaw acceleration, and hunger and satiation indicators (Ponganis, 2007; Rutz & Hays, 2009; Barron et al., 2010; White et al., 2013). Wilmers et al. (2015, 1741) note that "this technology is revolutionizing our ability to study animal ecology by allowing researchers to …answer questions about the physiology, behavior, and ecology of wild animals that would have previously been limited….”

Studies using GPS on smaller birds have recently been developed. Hill (2017) gives the example of upland sandpipers (*Bartramia longicauda*, Bechstein, 1812) which were tracked using very small back-mounted satellite tags, to document their migratory patterns. Jones et al. (2014) used the targeting of male Bachman’s sparrow (*Peucaea aestivalis,* Lichtenstein, 1823) using plastic color bands which were tracked using a handheld Trimble Geo XM global positioning system (GPS), which was accurate to within 5 m. The data was later analyzed using GIS ESRI software. Morris (2016) describes a study of urban herring gulls (*Larus argentatus*, Pontoppidan, 1763) using solar-powered GPS tags, which recorded their distributional patterns and feeding habits. Harris (2012) used back-mounted GPS loggers for tracking Atlantic puffins (Linnaeus, 1758) breeding at a colony near the proposed site for a wind farm. They argue that:

> Until recently, Global Positioning System (GPS) loggers that provide detailed data on location have been too large to deploy on many of the seabird species breeding in the UK, but devices now available are potentially small enough to be used on birds weighing as little as 300–400 g…despite it having long been recognized that bird-borne devices can affect the flight and foraging behavior of seabirds, particularly auks and other diving species (Harris et al., 2012, 43).

Studies of avian behavioral dynamics would require high resolution images to record the habitats surrounding the bird action, as the base upon which GPS-based tracking and behavioral data would be based. For smaller species, the resolution would necessarily be in millimeters or centimeters. For this reason, aerial photographs or various types of field cameras would be better than satellite images (Campbell, 2017). Tables 1 and 2 show the resolutions, dates and origins of some of the satellite type imagery which has been published over the last 20 years. Most resolutions range between 30 cm and 5 meters. Most are multispectral images, which include several bands in each image produced, where bands refer to representations of visible and invisible parts of electromagnetic spectrum, i.e., the ultraviolet, visible and infrared bands. Each image usually has three to ten bands; for example, the multispectral IKONOS image has four bands: blue, green, red and near infrared. The Landsat TM multispectral image has seven bands: blue, green, red, near-IR bands, two SWIR bands, and a thermal IR band.

Boyle et al. (2014) note that with time, the resolutions may improve, as technology advances. This is however a difficult undertaking, as most changes in technology are expensive, and as large numbers of images may be required for many applications, the cost of some projects may be prohibitive. For example, Boyle et al. (2014) examined the possibility that imagery of high spatial resolution (≤5 m, 4 m multispectral IKONOS) was better than lower-resolution imagery (≥30 m, 30 m multispectral Landsat), using a study of fragmented Interior Atlantic Forest in Paraguay. The higher-resolution imagery was more accurate for delineating cover classes and identifying patch shapes and locations. They found that despite these advantages, leading conservation journals (Biological Conservation, Biotropica, Conservation Biology) used high-resolution imagery less (10.4% of the 26.8% of the land cover studies that used analysis of satellite imagery used imagery of ≤5 m resolution). The conclusion was that high-resolution imagery was neglected in conservation research, and that there should be greater access to low-cost, multiband, high-resolution satellite imagery.

Some studies have used satellite imagery with coarse resolutions to determine avian spatial ecology. For example, Cannon et al. (1982) used Landsat MultiSpectral Scanner (MSS), with 80 m spatial resolution dated 1978 for the classification of shinnery oak *Quercus havardii* (Rydberg) dominated landcover in Western Oklahoma, US. The results found a positive correlation between the presence of male lesser prairie chickens (Tympanuchus pallidicinctus, Ridgway, 1873) and the percentage of grassland habitat. Palmeirim (1988) also used a map derived from 30m spatial resolution Landsat Thematic Mapper (TM) imagery to estimate the number of avian species in northeastern Kansas. Another older study used Normalized Difference Vegetation Index (NDVI) data, which at a coarse resolution (1 km) (this can record vegetation attributes including biomass, productivity and photosynthetic activity) related vegetation landcover variation to breeding habitat of the Red-billed quelea (*Quelea quelea*, Linnaeus, 1758) (Wallin et al., 1992). Another such study used 1-km NDVI imagery (Jørgensen & Nøhr, 1996), and demonstrated correlations between avian species presence and vegetative biomass in Senegal. LIDAR, or Light Detection and Ranging (National Ocean Service, 2018) is a more advanced technique that uses light pulses to generate detailed three-dimensional information about features on the Earth's surface. LIDAR has a sharper resolution and hence would be more useful for the study of the micro-habitats of small avian species.

Table 1. Recent satellite launched by the United States

Name	Nation	Active Dates	Spatial Resolution (m)
EarlyBird-1	USA	Dec. 24, 1997 - 28 Dec.1997	3 m PAN, 15 m MS
Ikonos	USA	Sept. 24, 1999 -31 Mar. 2015	82 cm PAN 3.2 m MS
QuickBird	USA	Oct. 18, 2001 – Dec.17, 2014	60 cm PAN 2.4 m MS
OrbView-3	USA	Jun. 2003 - March 13, 2011	1 m PAN 4 m MS
WorldView-1	USA	Sept. 18, 2007 - 2018	0.5 m PAN
GeoEye-1	USA	Sep. 6, 2008 - 2018	0.41-46 m PAN, 1.65 m MS
WorldView-2	USA	Oct. 8, 2009 - 2018	46 cm PAN 1.84 m MS
WorldView-3	USA	Aug. 13, 2014 - 2018	31cm PAN, 1.24m MS, 3.7m SWIR, 30m CAVIS

Sources: Livingston (2015, 15-16), Indian Space Research Organization (2018); NASA (2018d). NB: PAN refers to panchromatic, MS refers to multispectral and SAR to synthetic aperture radar. CAVIS refers to Clouds, Aerosols, Vapors, Ice, and Snow. SW refers to short wave, IR to infrared, NIR to near infrared and SWIR to shortwave infrared imagery.

Table 2. Recent satellite launched by nations other than the United States

Name	Nation	Dates (Launch to Retirement)	Spatial Resolution (m)
RADARSAT-1	Canada	Nov. 4, 1995 - Mar. 28, 2013	10 – 100 SAR
IRS-1C	India	Dec. 28, 1995 - Sept. 21, 2007	6 m PAN
EROS-A	Israel	Dec. 5, 2000 - May 2016	1.8 meter
Spot 5	France	May 4, 2002 -Mar. 31, 2015.	5 m PAN 10 m MS
Carosat-1	India	May 5, 2005 - 2018	2.5 m PAN
EROS-B	Israel	Apr. 25, 2006 - 2018	70 cm
Cartosat-2	India	Jan. 10, 2007 - 2018	< 1 m
EROS-B	Israel	Apr. 25, 2006 - 2018	70 cm
Cartosat-2	India	Jan. 10, 2007 -2018	< 1 m
Cosmo SkyMed-1	Italy	Jun. 8, 2007- 2018	SAR
TerraSar-X	Germany	Jun. 15, 2007	Variable, to 1 m SAR
SAR-Lupe 1 -5	Germany	Dec. 19, 2006 - Jul. 22, 2008 +	< 1 m, SAR
Cosmo SkyMed-2	Italy	Dec. 9, 2007 - 2018	SAR
RADARSAT-2	Canada	Dec. 14, 2007 - 2018	3-100, <1 m SAR
Cartosat-2A	India	April 28, 2008	< 1 m PAN
Cosmo SkyMed-3	Italy	Oct. 25, 2008 - 2018	SAR
TanDEM-X	Germany	Jun. 21, 2010 - 2018	2 – 4m
Cartosat-2B	India	Jul. 12, 2010 - 2018	< 1 m PAN
Cosmo SkyMed-4	Italy	Nov. 6, 2010 - 2018	SAR
NigComSat-1R	Nigeria	Dec. 19, 2011 - 2018	2.5 m PAN 5 m MS
Spot 6	France	Sept. 9, 2012 - 2018	PAN - 1.5 m, MS– 6 m, (BGR, NIR)
Spot 7	France	Jun. 30, 2014 – 2018	PAN 1.5m MS- 6m (BGR, NIR)
Pleiades-1A	France	Dec. 16, 2011 – 2018	PAN 0.5 m, 2 m MS
Pleiades-1B	France	Dec. 2, 2012 – 2018	PAN 0.5 m 2-m MS

Sources: Livingston (2015, 15-16), Indian Space Research Organization (2018); NASA (2018d). NB: PAN refers to panchromatic, MS refers to multispectral and SAR to synthetic aperture radar. SW refers to short wave, IR to infrared and SWIR to shortwave infrared imagery. BGR refers to blue, green and red. Note for the Spot satellites, Spot 1 – 5 were launched 1986 – 2002, and deactivated 1990 – 2015.

In the application of GPS and image-based technologies to small-scale ecological studies, several parameters are particularly important. For the strands of animal geography examined (the interspecies competition for food at the individual levels), the main parameters for the application of geomatics techniques include: species differences in competition; manifestations of interspecies fear and dominance; learning behavior of birds in human feeding contacts; possibilities of determining aggression and fear physiologically; detection of anticipatory behavior; and avian abilities to distinguish human feeders from non-feeding humans. For the alert distances, the main parameters for the application of geomatics techniques include: possibilities for application to small species; detection of exact distances between birds and approaching humans; scanning of the vegetation configuration near the studied bird; recording of the details of birds after alert, flight and the period in between (the alert period); and bird physiological changes during the entire period of study. This chapter concludes that for effective fuller understanding of the avian behavior, there must be measurement of bird (intra- and inter-species) adaptation and habituation to human presence and each other. This is where the application of new geomatics-based techniques is vital. This would represent an important contribution to zoogeography, ecology and geomatics, and consequently to conservation biology.

THE DEVELOPMENT OF ANIMAL GEOGRAPHY

This section examines the development of the relevant strands of animal geography, and the contribution of this emerging subdiscipline to the understanding of individualism in animals. The theory and development of zoogeography is also considered, to illustrate the differences between this discipline and that of animal geography (Campbell, 2014a). Zoogeographical theory largely developed with a global focus, with relative neglect of the micro-spaces of interaction between people and animals. As such, aerial photographs and satellite images covering large areas would be the best applications of geomatics (Campbell, 1998, 2002). The classical version of zoogeography gave a more generalized assessment of smaller scale studies. This position may be contrasted with the considered strand of animal geography, which offers a challenging perspective, focused on the unpredictable, subjective, individualist behavior of animals which may complement the more classical, large scale approach of zoogeography. For this branch of animal geography, different geomatics applications would be needed, focused on the more dynamic parameters considered.

The Development of Zoogeography

Biogeography studies the distribution of living species in geographic space and through geological time, and as such links ecology, evolutionary biology and palaeontology (McDonald, 2003: 3). Branches of biogeography include ecological, historical, analytical conservation biogeography. Zoogeography studies the geographic distribution (present and past) of animal species; and phytogeography studies the geographic distribution of plant species and their influence on the earth's surface. Zoogeography includes methodologies from both zoology and geography (McDonald, 2003; Campbell, 2009a). Zoogeography began as a regional or global science; "the science of mapping out the surface of the earth with reference to its faunæ"; and the "study of the distribution of animal life," this concerned with "local peculiarities in animals and plants" in the "various regions of the earth's surface" (Ingersoll, 1920). Some have argued that zoogeography is less studied than phytogeography, as animals are a "paradox in biogeography"; they are popular, but there is a "relative scarcity of texts in animal biogeography" (Ganderton & Coker, 2005: 123). Micro-scale interfaces between zoogeography and human geography are also neglected (Campbell, 1998, 2009, 2013a,b,c; McDonald, 2003). Two developments that encourage the integration of the global and regional analysis of traditional zoogeography and the micro-scale interactions between animals and human influenced landcover are disequilibrium ecology and animal geography.

Disequilibrium Ecology or 'New Ecology'

Disequilibrium ecology or the 'New' ecology differs from the classical ecology at the root of biogeography in that the latter was based on Newtonian physics, which analyzed ecology as comprised of "deterministic dynamics," caused by a "few simple laws" (Solbrig, 1994:17), generalization, scientific universalism, vegetation climaxes, ecological succession and equilibrium, and human/nature separation (Pickett et al., 1992; Zimmerer, 1994; Begon et al., 2005). In disequilibrium ecology, the emphases shifted to continual flux, unpredictable outcomes and trajectories, non-linear, non-equilibrium and multi-directional ecological dynamics, subject to "multiple interpretations" (Nicolis, 1994; Scoones, 1997, 162; Hobbs et al., 2009). Such trajectories are regarded as internally, rather than externally derived and are part of constantly changing "non-equilibrium or multi-state systems" (Young & Solbrig, 1993: 324).

Scoones (1999) argues that the emergence of the new ecology has had some implications for the linking of social and natural science methods in the investigation and documentation of environmental change. There were however limitations on the impact on socioenvironmental analysis, possibly "a consequence of the lag times of cross-disciplinary communication: different languages, frames of reference and methodological approaches are evident across the disciplinary divides" (Scoones, 1999: 489). Nevertheless, there were three relevant impacts that the new ecology had on integrated socio-environmental studies. Firstly, there was increased attention to environmental history, with offshoots towards paleoecology and evolutionary ecology (Williams, 1994; O'Connor, 1997). This allowed appraisals of nature-society interactions to be studied in more detail. Secondly, the promotion of variability in both space and time, enabled the description of environmental dynamics to move "beyond the simple assumptions of equilibrial regulation to a wider appreciation of complex dynamics, uncertainty, and surprise" (Scoones, 1999: 483). Thirdly, there was a greater attention to scale, structure and agency, allowing the documentation of "nonlinear interactions" (Scoones, 1999: 483) and the contribution of these to small and large-scale patterns (Allen & Starr, 1982; Turner, 2003).

Animal Geography

During the 1960s, certain strands within 'animal geography', studied animals as subjects with active abilities (termed actancy, like human action) rather than as mechanistic objects (Wolch & Emel, 1998; Lorimer, 2007; Barua, 2013). Actancy is defined as a behavior, in which free choices are made in response to different events (Murdoch, 1998). Like the disequilibrium ecology paradigm, these strands argued for a more flexible perspective to uncover subjectivities undetected by the more positivist science of zoogeography, with more effective links to the human geographies of wildlife (Anderson, 1997; Philo & Wilbert, 2000; Yarwood & Evans, 2000; Emel et al., 2002). This approach included a greater theoretical focus on animal behavioral subjectivity and comprised perspectives of 'cultural human geography', the hypothesis of shared actancy between people and animals and the integration of these with quantitative zoogeographical field methods (Wolch & Emel, 1998; Johnston et al., 2000). The contribution of these to zoogeographical paradigms has been validated by some recent studies (Campbell 2008, 2009a,b). Therefore, for avian studies, animal geography and avian ecology/zoogeography may represent two ways of looking at avian relations with the environment and human behavior; animal geography sees birds as actants; active agents creating relations with people as individuals and groups, while avian ecology/zoogeography focusses on statistical generalization, at inter- and intra- species levels (Campbell, 2013).

Urban areas are important for the testing of the animal geography paradigm, as animal food access is dependent on human behavior and artificial landcover and animals may adapt at inter- and intra-species levels (Campbell, 2013). Birds are an important and behaviorally variable urban animal group, as they relate to humans who may be both deliberate and non - deliberate food providers (Belant, 1997; Burger, 1981; Marzluff et al. 2001; Mitchell et al., 2004; Vuorisalo et al., 2003). The key question concerns the extent to which animal actancy may emerge in such intensive contexts and the methods available to determine and measure such behaviors. To develop this line, closer attention must be placed on the behavior of birds. As argued by Johnston et al. (2000: 25) there are instances where "animals emerge as active agents fashioning the environment they inhabit or as experimental objects in the social relationships which bind them in different ways to people."

In Campbell (2013), the methodology of such an actancy-based approach was detailed, applied to birds being fed by people, and the inter- and intra-species competition and other behaviors exhibited by these birds. Individualism was determined from observing bird waiting behaviors prior to feeding, especially for food provided by humans, the links between such behavior and individualist behavior in human feeders, and interspecies conflicts, especially where there was evidence of strategic behavior for food access. Such individualism required individual bird identification, by recording features such as size, feather patterns, broken, dirty or missing feathers, and soiled feet. Other parameters were species presence correlated with land-cover variation and human presence/behavior and the impacts of human presence. Bird behavior was also measured by recording actions such as flapping, pecking, chasing, dodging, hovering and moving around human feeders. These actions at the inter-species and intra-species levels usually resulted in winners and losers, where the loser retreated from the feeding space and and/or lost its food by dropping it or by losing it to the bill of another bird. There were also inter- and intra-species pursuits (on land, water or through the air) and attacks by pecking and/or wing flapping. Such behaviors were recorded as occurring in three periods: the pre-feeding phase, before human presence, when birds were perching, or foraging in individual or group spaces; the human presence phase, as birds clustered around people (non-feeding or feeding) and avian conflicts began, including the feeding phase, when conflicts linked to dropped food, and food dragging; and the post-feeding phase, after most food was consumed, and conflicts continued as birds remained and disputed for spaces around potential human feeders.

In Campbell (2013), dominant avian scavengers were studied and individualist actancy among different species was compared contrasted. The urban areas were Glasgow and Brighton in the UK, and Peterborough in Canada. Parameters clustered under the headings of the relations between the dominant scavengers and other scavengers (including actant behavior in interspecies conflict) and individualism in bird-human relations. In the case of Glasgow, the dominant avian scavenger was the lesser black-backed gull (*Larus fuscus*, Linnaeus, 1758), the others being the rock dove (*Columba livia*, Gmelin, 1789) and the mallard duck (*Anas platyrhynchos*, Linnaeus, 1758) (Campbell, 2007). Gulls dominated ducks more than doves. Ninety five percent of the observed conflicts between gulls and ducks ended with the ducks' withdrawal. Dominance behavior consisted of flapping, pecking, chasing and hovering. Gulls drove ducks away and even removed food from their bills. Gulls dominated doves, winning ninety two percent of the observed conflicts. Dominance consisted of wing-flapping, but doves were too fast to be pecked. The doves' superior agility resulted in gulls being outmaneuvered in 46 percent of the conflicts (Campbell, 2007, 2013).

Actant behavior was revealed by occasional dove/duck grouping of individual aggression on gulls, and doves using their superior agility to snatch food from under gulls. Identifiable doves in small groups prevailed over gulls through aggressive actions, including two or three flapping around the gull's heads will others snatched food from the gull's feet. This was successful in about 30 percent of attempts, but no other doves except those identified in these actions employed this technique. More frequently, groups of doves would simply crowd onto solitary feeding gulls, which either lost its food, or flapped aggressively and usually drive the doves back, where individual skills were apparent. Ducks used three main actions against gulls. One was group attacks, where the gulls either dropped the food, increased flapping and calling, or flew with the food. Another was picking contested food and moving within a group away from the gulls, with gulls either flying over the ducks and diving for the food, attacking ducks closest to the food, or holding onto the food and flapping. Ducks could also react by sitting on the food like an egg, but gulls attacked these ducks. Using individual identification techniques, revealed these to be individual, sometimes the same individuals taking such action. On water, ducks prevailed against gulls more often, as they could put their heads under water, with the food in their bills, or use their wings or bodies to cover the food on the surface. Gulls attacked these birds, prevailing in about 65% of the fights.

Concerning bird-human interactions in feeding situations, gull, duck and dove numbers increased with human presence even when there was no feeding, with human presence positively correlated with increased inter- and intra- species conflict. Although some people deliberately fed ducks and avoided other species, gulls still dominated feeding areas. Gulls grouped on ducks or doves being fed and 'stole' their food, intercepted food in mid-air or attacked people; sometimes gull attacks drove people from the feeding area. Individual gull attacks on people included perching and defecating on people's heads, flapping and diving at people, attacking ducks and doves both near and far from people, which made the latter species avoid the feeding area. Ducks reacted by approaching feeding humans by flight from the water rather than swimming, sitting between peoples' shoes or near people, or leaving the feeding area and returning if or when the gull flock left. Dove reactions to such gull aggression including hiding under park benches, if these were near the feeding area, or agile flight near people, which could escalate in aerial conflicts with gulls (Campbell, 2007).

In the study by Campbell (2010b) in Brighton, UK, the dominant scavenger was the herring gull (*Larus argentatus*, Pontoppidan, 1763, Denmark). Other common scavengers were black- headed gulls (*Chroicocephalus ridibundus*, Linnaeus, 1766), carrion crows (*Corvus corone*, Linnaeus, 1758), rooks (*Corvus frugilegus*, Linnaeus, 1758), jackdaws (*Coloeus monedula*, Linnaeus, 1758), mallard ducks and rock doves. Gull dominance over other species included diving when the other bird was on the ground, hopping and flapping when both were on the ground, aerial chasing, and pecking during contestations for food. Actant behavior consisted of reactions (greater agility, closer flights to people, faster reactions to feeders) of the smaller, lighter black-headed gulls against the herring gull attacks, which were successful for the smaller birds in about 15% of the conflicts. Herring gull reactions included temporary avoidance, flight tumbling, diving to the water surface, dropping food with immediate pickup or bullying of the smaller gull. Gulls also attacked the crows, rooks and jackdaws, which resulted in the gulls chasing the corvids, beyond the feeding area. Corvids only reacted by individual avoidance. There were no records of corvids chasing gulls or other species, only minor aggression to doves, with avoidance and leaving behavior on the part of doves (Campbell, 2010b).

Bird-human relations were like those of the Glasgow study. Bird numbers were positively correlated with human numbers, whether the people were feeding birds or not. Human presence was positively correlated with bird conflicts only when a gull species was present. Herring gulls were dominant and obtained the most food, even when people preferentially ducks and doves, and when people sought to protect ducks and doves by placing food nearer their bills, stopped feeding when gulls began to dominate, tossed food to preferred target birds, moved from one feeding area to another, or waved and shouted at the gulls. Gulls reacted to human actions by observational perching, hovering and diving, and attacking other species. As in Glasgow (Campbell, 2007), doves reacted to gulls more successfully than ducks did, due to their greater agility and speed. Ducks actions included faster ingestion of food, anticipation of food throwing by feeders and approaching feeders with open bills. Gulls also reacted by swooping at such ducks (Campbell, 2010b).

Interactions between gulls and corvids in human feeding presence, were similar in that people never fed corvids preferentially. Gulls used observational perching or hovering, mid-air interception of food and attacks on corvids, driving of corvids from feeding areas and attacks on people. Gulls were significantly more successful in these actions than corvids. Corvids reactions included dodging, perching in trees with small branches, which gulls avoided and flying around people. Herring gulls were very successful against the mostly solitary carrion crows and jackdaws, but less successful against the more social rook. The smaller, more mobile black headed gull was more successful than the herring gull against the similarly mobile jackdaw, but slightly less so against the larger crow and rook. The results were not statistically significant, due to the strongly individual nature of the actions (Campbell, 2010b).

In the study by Campbell (2008) in Peterborough, Canada, the dominant scavengers were the ring-billed gull (*Larus delawarensis*, Ord, 1815), Canada goose (*Branta canadensis*, Linnaeus, 1758), American crow (*Corvus brachyrhynchos*, Brehm, 1822) and mallard duck. In this case, there was no single dominant species, and this complicated the interspecies relations. The ring-billed gull is a small gull, like the black headed gull, and the American crow is only slightly larger than the British jackdaw. The large and aggressive Canada goose is slow moving and does not contest for food on the ground or in the air, only in some cases on water surfaces. Mallards and rock doves had similar behavior to those in Glasgow and Brighton. Ring-billed gulls were fast and aggressive, while crows and doves never landed on water. Gulls and geese were the main competitors, the gulls winning the most victories (44.6%) compared to geese (34.4%), with several indecisive results (21%). Gulls dominated ducks more often than geese did, due to superior mobility, and aerial fighting of which the geese were incapable. Gulls chased ducks on land, water and in air for long distances, while geese did so on land and water for one or two meters. The large size of the Canada goose was sometimes insufficient against group attacks by gulls and the greater mobility of crows, gulls and doves. Gull and geese numbers were negatively correlated with duck numbers near feeders. Duck numbers had no effect on other species numbers (Campbell, 2008).

Actant behaviors could be discerned in duck actions against gulls and geese, and competitive mobility between crows, gulls and doves. Although ducks won only about 30 percent of food pickings, even when the food was nearer them than the geese or gulls, some ducks used spurts of mobility to obtain food when attacked by other species. Some ducks, in some cases identifiable as consistently aggressive and innovative, and used sudden attacks on large food pieces being broken up and entered spaces dominated by flocks of gulls and geese. Other ducks relied on the gulls and geese to break up food, which they then picked. Crows and doves had fewer conflicts with ducks but could 'steal' food already taken by ducks by diving from observational posts, snatching and flying behavior, and rapid flight in which their only challengers were gulls and ducks only rarely challenged in the air and geese never did. In non-riverine areas, gulls, crows and doves were the main competitors, as geese and ducks were fewer and were much slower on land than in water. The main issues were gull aggression, crow deviousness and dove mobility. Individually aggressive gulls employed diving, flapping, low flights, calling and pecking to dominate other birds, while gulls merely watched or took were there was little competition. Some crows robbed slow gulls. Fast moving doves sometimes snatched food from gulls, ducking under the gulls' bodies, or exploiting crowded situations. All the bird species were positively correlated to human presence, even non-feeders. People deliberately fed ducks, and sometimes placed food closer to the ducks than other species. This would be initially successful, but eventually the ducks could be overpowered by geese on the ground, gulls and crows on the ground and in the air with low swoops. Gulls could even remove food from people's hands, defecate on human clothes, flap about people's heads and contribute to people leaving feeding areas (Campbell, 2008).

Single Species Applications of Animal Geography: The Case of Turkey Vultures

This section draws on my published book chapter on the behavioral ecology of turkey vultures (*Cathartes aura*, Linnaeus 1758) in urban contexts (Campbell (2015). It seeks to abstract those behavioral traits that may be observed and detected by GPS and other geomatics techniques, for more authoritative documentation and analysis of the foraging ecology of this species. It also points to the parameters to be measured to illustrate the possibilities of individual actancy, for an animal geography perspective. The turkey vulture is an extremely adaptive species, ranging across the Americas; from the northern forests of Canada and the Northern United States, through the deciduous forests of Central and South America, to the Pampas plains of southern South America. Kiff (2000; 175) described the turkey vulture as "the most widespread and differentiated cathartid species, occurring from southern Canada to Tierra del Fuego and the Falkland Islands, ranging from sea level to 4.300 m." It has adapted to deforestation, urbanization and declining, natural food sources. Their sense of smell, unique among vultures, allows their detection of hidden food sources. Possible intra-species, individualist adaptations are currently neglected in ecological studies, but this could be ameliorated by detail, GPS led surveys. This would assist in the assessment of the adaptation and habituation of turkey vultures to human-spaces, a common phenomenon across its range. Campbell (2015) examined three research points: (1) the evidence of individualist vulture behavior from foraging, communication, competition and adaptation activities, through the discernment of outliers in statistical generalization; (2) the possible or detectable intra-species and/or intra-subspecies behavioral differences, partly through records of different behaviors, in different studies of the same species or subspecies; and (3) new or previously unreported behaviors.

Four issues may be examined to determine any individual actancy in the behavior of the turkey vulture, and the related parameters that be measured using advanced geomatics. One concerns its recent range expansion and its adaptation to new, possibly challenging environments (Campbell, 2015). Kiff (2000, 175 – 176) notes that the turkey vulture has "undergone dramatic northward range extensions since the 1920s." The turkey vulture also has the "most catholic habitat preferences of any member of the Cathartidae, perhaps owing in part to their ability to use olfactory clues in locating carcasses" (Kiff 2000: 175 – 176), with habitats including cultural landcover (including gardens, pastures, towns, roadsides, agricultural lands, disturbed forests), grasslands, savanna, forests, scrub or brush and habitats altered by humans, such as and forest edge (Carrette et al. 2009). They may also roost in groups or singly, on natural (trees or cliffs) or artificial features (roofs, ledges, old concrete structures) (Kiff 2000; Houston et al. 2007). Communal roosts promote energy savings through thermoregulation, social interactions and reduce predation risks (Devault et al. 2004).

A second issue concerns their foraging ability. They feed on a wide variety of foods, including garbage, coconuts, oil-palm fruit, salt, dead and live fish, young wading birds, marine animals, amphibians, reptiles, snails, shrimps and even insects, plant matter and young domesticated sheep (Woods & Woods 2006; Cringan, 2007; Campbell, 2015). Cringan 2007). Stolen (1996: 177) argued that "turkey tultures locate carcasses primarily using olfaction and must remain close to the ground where odors are concentrated." The main actancy issue is related to the issue of sense of smell or olfactory ability; the individual abilities in food detection and the comparison with other competing species with inferior food detection abilities. Measurable parameters would include individual differences including: (1) the changing flight patterns such as low flight over the canopy and maneuvering within the canopy and understory when food is scented; (2) variable abilities in the detection of carrion covered by leaves or uncovered, or carcasses of different variable age. Stolen (1996) found that first discovery of a food source by one turkey vulture attracted other Turkey vultures from more than 200 meters distant. Houston (1986) found that "there was no clear correlation between the percentage of open sky above a carcass, as measured from photographs of the canopy, and the time taken in daylight hours for vultures to find the food"; hinting at individual abilities in food discovery (Houston 1986: 319). Individualism in food detection rates, would be indicated by outliers that would prevent a significant correlation between open sky and food detection rates.

Smith et al. (2002) described a turkey vulture that dug up and partially consumed as a deliberately buried woodchuck (*Marmota monax*, Linnaeus 1758) carcass. They argued that this incident "suggests that this bird had knowledge of the location of the carcass and that it had previous experience in this behavior" (Smith et al. 2002: 144). The observed bird may have learning observational ability: either from its own experience digging out buried carcasses, observing other vultures doing so, or it had a particularly acute sense of smell, or it merely exhibited normal foraging behavior. It may also have observed the experimenters burying the carcasses. As the carcasses were buried late, after the roosting time for turkey vultures the latter scenario was discounted.

A third issue concerns the turkey vultures' competitive foraging with other species (Campbell, 2015). Turkey vultures lead other species to food sources, largely due to the unique olfactory ability, which helps were carrion may be hidden by vegetation or earth structures (Houston 1986). Interspecies competition involves the larger king vulture (*Sarcoramphus papa*, Linnaeus 1758) dominating turkey vultures and sometimes tearing open carcasses beyond the ability of the turkey vulture (Snyder & Snyder, 2006). The social, similarly sized black vulture (*Coragyps atratus*, Bechstein 1793) may use flock behavior to dominate turkey vultures (Campbell, 2015). The main actancy issue is related to individual differences in turkey vulture reactions to other species; and the extent to which such individual reactions would vary from generalized inter-species studies. As turkey vultures have a sense of smell and lead Black vultures to food sources, they lead the latter to food in forested areas, while both species may arrive together in open areas (Houston, 1986).

Measurable parameters would include individual abilities during site competition with competitors. Turkey vultures sometimes prevail over black vultures in one to one confrontations, but this may be reversed in some situations. Large numbers of social black vultures usually dominate the solitary turkey vulture (Buckley, 1996). Wallace and Temple (1987: 293) observed that in "one-on-one situations, turkey vultures won only slightly over half of their aggressive encounters with black vultures, demonstrating very little dominance over black Vultures" This was despite the slightly larger size of the black vulture (adults 1.5 kg, with a range of 1.2 to 1.8 kg; juveniles 1.3 kg, with a range of 1.1. to 1.5 kg) than turkey vultures (adults 1.4 kg, with a range of 0.9 to 1.8; juvenile 1.3 kg with a range of 1.0 – 1.5) (Wallace & Temple 1987: 293). Black vultures nevertheless dominated turkey vultures in situations where black vultures were more numerous, some flocks numbering over 100 (Campbell, 2015). Turkey vultures were about evenly matched with greater yellow-headed vultures (Hilty, 1977; Schulenberg, 2010). Turkey vultures were dominated by Andean and California condors, golden eagles, king vultures and caracaras (Polyborinae, Lafresnaye 1839) (Wallace & Temple,

1987). Adult caracaras prevailed in 91 percent of the interactions with turkey vultures, turkey vultures were slightly dominant over ravens (Koford, 1953; Wallace & Temple 1987; Snyder & Snyder, 2000). In extreme cases, turkey vultures, on an individual basis may regurgitate their food, possibly to distract the attacker (Campbell, 2015).

Other studies recorded varying results. Red-tailed hawks (*Buteo jamaicensis*, Gmelin 1788) dominated turkey vultures, which were dominant over the American crow (*Corvus brachyrhynchos*, Brehm 1822). Individual behavior was illustrated by the individual vultures dragging food from feeding red-tailed hawks. Vultures retreated from approaching red-tailed hawks, but some bolder vultures returned and perched close to the feeding hawks. In most cases, turkey vultures were dominated by domestic cats (*Felis catus*, Linnaeus 1758), but in one case a turkey vulture aggressively confronted a cat (Prior & Weatherhead, 1991). Some studies recorded the striated caracara (*Phalcoboenus australis*, Gmelin 1788) dominating turkey vultures (Strange 1996; Catry et al. 2008), but another (Dwyer & Cockwell (2011) recorded the turkey vulture as the dominant of the two species. Dwyer and Cockwell (2011) studied interactions between turkey vultures, variable hawks (*Geranoaetus polyosoma*, Quoy and Gaimard 1824), southern caracaras (*Caracara plancus*, Miller 1777) and striated caracaras. Variable hawks ranked first in competitive dominance, being the aggressors with striated caracaras (the weakest species, 98%), southern caracaras (80%) and turkey vultures (82%). Southern caracaras were the aggressors with striated caracaras (100%) and Turkey vultures (90%). Turkey vultures were more often the aggressors with Striated caracaras (71%). Turkey vultures were sometimes the aggressors with variable hawks (18%) and southern caracaras (10%) (Dwyer & Cockwell 2011).

Another issue concerns competition between the different turkey vulture subspecies, which may occur were their ranges overlap, or one subspecies migrates seasonally to the range of the other subspecies. Larger subspecies are usually dominant of the smaller subspecies. (Campbell, 2015). For example, migrating northern turkey vultures (*Cathartes aura meridionalis*) originating from the northern North American winter, into the parallel dry season (October to April) of northern South America, dominated the smaller resident turkey vultures (*Cathartes aura ruficollis*) in central Venezuela, displacing the resident subspecies from the open savanna into the denser forest and consequently affecting their health which improved when the migrants moved back north (Kirk & Gosler 1994).

Concerning actancy, measurable parameters would concern how the different Turkey vulture subspecies variably adapted to ecological opportunities and constraining factors, how they related to each other at the individual level and the evidence of individual vulture behavioral outliers in this relationship. Outliers in the dominance of larger species (how individuals of the weaker subspecies varied in response to this dominance) may examined. For example, in the example quoted above, the weaker resident subspecies

(*Cathartes aura ruficollis*) ate food faster than the larger subspecies (C. *a. meridionalis*). Kirk and Gosler (1994) note that similar actions may occur between the stronger and weaker members within a subspecies. Faster eating could be due to both fear of stronger birds displacing the weaker birds from food, or the greater hunger of the weaker birds from earlier displacements (Kirk & Gosler, 1994). Adult Turkey vultures also dominated juveniles: "during aggressive encounters between juvenile and adult turkey Vultures, 231 (68%) of 339 interactions were won by adults; the remaining encounters were won by juveniles" (Wallace & Temple 1987: 292). This indicates some outliers and exceptions (32%) were the juvenile dominated, which could be measured with advanced research techniques.

Another issue concerns roosting among turkey vultures. This is not so much about their competition for roosting spaces, but their possible ability to share foraging information from the previous days foraging, to enable the next day's food search (Ward & Zahavi, 1973; Rabenold, 1983). This is a hypothesis, which would require more research using advanced techniques for tracking, individual identification and roost locations of target birds (in close pairs or groups). These arguments center on why competitors would lead others to food and the possibility of needing help to open carcasses (Alexander, 1974; Jackson, 1975; McHargue, 1977; Rabenold 1983). The possibility of information sharing envisions the individualized one to one communication. Such ideas have been touted for other species, such as bank swallows (*Riparia riparia*, Linnaeus 1758), great blue herons (*Ardea herodias*, Linnaeus 1758), and reed cormorants (*Microcarbo africanus*, Gmelin 1789) (Emlen, 1971; Ward & Zahavi, 1973; Krebs, 1974; Erwin, 1978; Jones, 1978).

Evidence of such shared information from the previous days foraging would rely on evidence that vulture departures from communal roosts at the beginning of the next day's foraging "deviate from random both in direction taken and time" (Rabenold 1983, 303). Vultures with knowledge of food source from the previous day tend to fly in a straight route to the food source the next day, using directional, flapping flight. In comparison, without such knowledge the vultures begin foraging later in the morning soar in circular flight reliant on the thermals over the heating land. Directional flight begins only with the sighting of a food source (Rabenold 1983). Additional evidence of information sharing might be a greater amount of group behavior in departing than arriving vultures (Rabenold 1983). To record this, it is necessary to identify the roosting birds that sighted food on the previous day (termed the primary birds) and determine if other birds (termed the secondary birds) follow the primary birds to food sources the on the second day, before the secondary birds found new food sources.

Crucially, are the primary birds followed by the secondary birds *before* they sight food on the second day? This is important because after the primary bird sights food, the secondary birds would follow them, regardless of any shared information (Rabenold

1983). However, some authors discount these possibilities, but the possibilities of inter-species differences make a conclusion difficult. For example, Prior and Weatherhead (2004: 385) argued that "feeding trials failed to provide evidence of recruitment due to information transfer" in a study of roosting Turkey vultures in southern Ontario. They acknowledged however that, in possible contrast to other species, Turkey vultures perhaps would not transfer information due to "intraspecific competition, preferential selection of small-sized carrion, and a low degree of kin association" (Prior & Weatherhead, 2004: 385).

Another issue concerns vulture adaptation to human presence and human modified landcover. Within this field, the actancy issues would include individual vulture behavioral differences in adaptation to human presence, the evidence of learned behavior between individual birds and evidence of behavioral differences between in natural habitats and those in more artificial habitats. There is no doubt that at the group level Turkey vultures can adapt to the new contexts they encounter. Devault et al. (2004: 710) argue that "Turkey vultures are able to adjust their home ranges, movement patterns, and flight behavior to local conditions, especially those related to habitat structure and foraging opportunities," defined as "plasticity in ranging behavior," an important factor for their current range expansion. However, Devault et al. (2004: 710) acknowledge that more research is required "concerning the interplay of ranging behavior, foraging ecology, and flight behavior would identify key elements that control plasticity of vulture movements and provide a more complete assessment."

Evidence of individualism has been documented in a few studies. Turkey vultures have been found to nest in abandoned buildings and even in an abandoned car (Houston et al. 2007; Igl & Peterson 2010). Adaptive, individualist behaviors would include intrusion into areas such as food and market stalls, doors and windows, roofs and window ledges, butcheries and garbage dumps, individual tolerance of human presence and any variations in these activities at the individual level (Campbell 2014b). In this study, turkey vultures, were recorded in downtown areas (41%), town suburbs (22%), fallow farm and grassland (15%), savanna (11%), open forest (8%) and closed forest (4%) (Campbell 2014b). Turkey vultures often used observational and possibly anticipatory perching, even when no food was available, hinting learned behavior. This behavior did not occur in more natural environments. In a similar study of the similarly sized Hooded vulture (Necrosyrtes monachus, Temminck, 1823), Campbell (2009) vulture presence was strongly correlated with human presence in urban contexts, manifested by individually variable intrusion into human dominated spaces, such as butchers' tables, windows and doors, and aggressive interactions such as flapping over people's heads, approaching people engaged in culinary activities (e.g., cooking or cutting meat, even discarding non-meat refuse), observational perching on the ground within meters of people, and forays to snatch food, sometimes at between human legs (Campbell, 2009).

The Use of Alert and Flight Distances

As mentioned above, alert distances (Figure 1) refer to the distance between a bird and an approaching human when the bird shows awareness of the person. Flight distances refer to the distance between the bird and the person when the bird takes flight. Alert periods are the period between first alertness and flight (Burger & Gochfield, 1998; Campbell 2006a, 2010). All these variables may vary according individual bird behavior, species size, vegetation cover structure, elevation and proximity, and human and road proximity. In line with the approach of the strand of animal geography discussed above, there must be evaluation of the relative importance of the generalizable factors, such as species, size and environmental structures) and the individual animal behavior that illustrates actancy in reaction to variable human behavior (e.g., noise, hand movements, approaches with and without focused intent) (Kalof & Fitzgerald 2007; Steinberg, 2010; Campbell, 2014a). Such encounters principally occur in urban green spaces, including parks, treed avenues, gardens, greenbelts, suburban forest patches, greenhearts in city centers (Fernandez-Juricic et al., 2001; Campbell 2006a, 2014a).

Campbell (2014a) collated the results of Campbell (2006a) and Campbell (2010), to answer the hypotheses concerning the factors variation in alert and flight distances. The first hypothesis was that larger birds are less tolerant than smaller species of human presence (Cooke, 1980; Humphrey et al. 1987; Skagen et al. 1991; Holmes et al. 1993; Gutzwiller et al. 1998). Fernández-Juricic et al. (2001: 264) selected four species: house sparrow (*Passer domesticus*, Linnaeus, 1758), blackbird (*Turdus merula*, L. 1758), wood pigeon (*Columba palumbus*, L., 1758) and magpie (*Pica pica*, L., 1758) as "representative model species" for a study of alert and flight distances. The larger species (magpie and wood pigeon) were less tolerant (hence, had longer alert distances) of human approaches than the smaller species. The blackbird was less tolerant than the smaller house sparrow. Fernández-Juricic et al. (2001: 267) speculated about the reasons for these results: possibly, "small species have higher energy expenditure than large ones due to greater area/body mass ratios, so small species are expected to be more tolerant of people to diminish energy costs associated with fleeing." Possibly also, the larger species have a greater perceptual ability and perhaps sharper vision over a larger radius (see also Kiltie 2000).

Campbell (2006a) used 13 species (including the large carrion crow (*Corvus corone*, Linnaeus 1758) and rook (*Corvus frugilegus*, Linnaeus 1758), the medium-sized magpie (*Pica, pica,* Linnaeus, 1758), jackdaw (*Corvus monedula*, L.1758), and the smaller starling (*Sturnus vulgaris*, L. 1758) and house sparrow (*Passer domesticus*, (L. 1758). Campbell (2010) used 10 species (including the large American crow (*Corvus brachyrhynchos*, C.L. Brehm 1822), medium-sized American robin (*Turdus migratorius*, L. 1766) and the smaller house sparrow (Passer domesticus, L.1758). Campbell (2016) used seven species including those used for Campbell (2010). These allowed a greater size range than Fernandez-Juricic et al. (2001). Campbell (2006a), found very strong correlation between alert distance and bird size. Campbell (2010) and Campbell (2016) had similar findings. These results contradicted other studies which focused on human feeding of birds, which found that larger species, moved closer than smaller than smaller species to both feeding and non-feeding people (Campbell 2007, 2008). Possible reasons anticipatory behavior from birds expecting food from even non-feeding people. Also, human approaches when feeding sometimes led to bird leaving behavior, but only temporarily; commonly anticipatory perching resulted, with turns when the person retreated.

Several measurable parameters were related to alert and flight distances. Fernández-Juricic et al. (2001) found cover proximity and density affected bird reactions to human presence. Possibly grass cover was positively correlated with alert distances for some species as bird's felt more vulnerable in open areas. Denser cover increased some species tolerance of people, as did shrub height, perhaps of the perception of lowered predation risk. Tree height also increased alert distances for some, perhaps as with the increased risk of predation, birds took longer to reach tree canopies (Fernández-Juricic et al. 2001). In Campbell (2006a,b), grass and shrub percentage cover reduced alert distances In some species alert distances were reduced by tree proximity more than by shrub proximity. Shrub density reduced the alert distances of some species.

In Campbell (2010), vegetation height was a significant predictor of alert and flight distances for some species, as shrub height, and tree height (combined with grass and undergrowth), correlated with shorter alert distances, while tree height combined with shrubs was a predictor for longer alert distances. Vegetation cover also affected both alert and flight distances, for several species. For some species increased grass, shrub and tree cover resulted in reduced alert distances, while for some species increased grass and undergrowth cover were linked with shorter alert distances. For others, tree cover was linked to shorter alert distances, while grass cover was a significant predictor of longer alert distances, and shrub and undergrowth were insignificant in effect. For the animal geography studies, alert and flight distances were reduced in human proximity in open grassed areas (Campbell, 2007, 2008). This indicated that the association of food with people overrode the association of open areas with predation risk. Species size was not significantly correlated with human presence, which was the only significant predictor for bird presence. In Fernández-Juricic et al. (2001) distance from pathways was negatively related to some alert distances. Similarly, in Campbell (2006a), road and path proximity reduced the alert distances of some but increased the alert distances for the others. In Campbell (2010) road and path proximity were not predictors of alert and flight distances for any species. In Campbell (2008) some species were attracted to roads, paths and pavements.

An important, measurable parameter of alert distances concerns the behavior of birds in response to human movements. These human movements include not only direct, frontal approaches to birds, but also hand movements and sideways approaches. In the more structured study of Fernández-Juricic et al. (2001: 265), the person approaching the bird at a steady speed of one step per second; wearing "light blue jeans and grey shirts to avoid differential bird reaction." Campbell (2006a) used a more complex approach, based on the premise that animals can detect frontal human attention. Hence, the approaching person used: (1) a straight approach; (2) a sideways approach (walking sideways, with the face at right angles to the bird; and (3) a straight approach moving arms by a regular swing up to 45° for each step. This assumed that the sideways approach would create the least alarm for the bird, hand movements would create the maximum alarm. Campbell (2010) went further; the approach included the first three approaches and a new, fourth approach where the arms were swung to a vertical position above the head (90°).

The findings of this methodology were that the arm swing and directness of approach (facing forward) were the most important factors for longer alert and flight distances for all the studied bird species. Campbell (2006a), found that hand movements affected larger species such as crows, more than the smaller species. The sideways approach, by comparison resulted in the shortest alert distances. For larger bird species, aggressive hand waving approaches doubled or nearly doubled the alert distances compared with the sideways approaches. For smaller species such as starlings, American robins and house sparrows, the increase was 20 to 30 percent. Alert and flight distances for straight approaches were about 10 – 35 percent, and 20 to 48 percent shorter than those of approaches using arm waving.

Campbell (2010), had similar findings, as arm movements through 90° created the longest alert and flight distances, followed by arm movements through 45°, straight approaches with no arm movement, and sideways approaches. The straight and sideways approaches with no hand movements had the largest differences, while the two forms of hand swinging (45° and 90°) had the smallest differences. Campbell (2016) found that the human hand movements were the most important factors for alert and flight distances, with bird size and tree cover being the only other predictors for alert and flight distances, with larger species having long flight distances. An important finding was that larger species were both attracted and disturbed by human attract approach and presence, therefore some alert periods might be anticipatory perching for possible food. Most species exhibited strongly individual behavior, resulting in much of the non-significant status of many predictors.

These findings did not indicate the exclusion of bird species due to human presence. For example, Campbell (2006b and 2009c) high correlations between human presence/modified landscapes and some bird species. These included American crow, starling, American robin and house sparrow in Canada and the carrion crow, rook, jackdaw, magpie, common starling and house sparrow in the UK. Most of the birds recorded in these cases showed individual behavior such as individual variations in property invasion, accessing of human eating spaces, observational perching, food competition and the blending of alert periods from observational perching.

The measurement of bird reactions to human approaches is the main parameter for the application of geomatics techniques. Bird behavior was highly variable and in some cases individualized, hence more quantified measurements beyond mere 'eyeballing' would lend greater reliability. Another important justification for the greater precision and quantification of the geomatics techniques, especially recent advanced techniques in global positioning systems is the examination of the possibility of bird actancy or individual adaptation to both human presence and the survey methodology. Fernández-Juricic et al. (2001: 264) point out that to avoid bird adaptation to the approach methods, the study methods were conducted a "20-ha park not included in this study, to avoid conditioning birds in our five study sites." In Campbell's (2006a, 2010, 2016) studies, this was also used; bird anticipatory perching might also have influenced the results. To avoid confusing alert periods, observational perching, and elastic retreat and advance behavior, more observation would be required (Campbell 2007, 2008), but the lack of quantification prevents a definitive result on the matter.

Conclusion

This chapter has examined the need for quantification in the study of small-scale, ecological relations of avian species. The points raised would also apply to mammals and other animal species. The main point examined concerns the possibilities of individualism in such animals, which is difficult to record with generalized data or monitoring tools with coarse resolution. The strand of animal geography discussed, which looks at the unique behaviors of individual animals, and examines animals as adaptive subjects rather than passive objects, is established in the literature. However, it requires more support from geomatics technology that quantities the miniscule variations in behavior that characterize individual animals. Intra- and inter-species interactions and animal-human relations were cited as examples where important ecological information could be derived from rudimentary observation methods. More advanced GPS, image analysis and other associated technologies would assist derivation of more accurate results, that may make an important contribution to conservation biology.

References

Alexander, R. D. (1974). The evolution of social behavior. *Annual Review of Ecological Systems 5*: 325–383.

Allen, T. F. H. & Starr, T. B. (1982). *Hierarchy: Perspectives for Ecological Complexity.* Chicago: University of Chicago Press.

Anderson. K. (1997). A walk on the wild side: a critical geography of domestication. *Progress in Human Geography 21*: 463 - 485.

Barron, D. G., Brawn, J. D. & Weatherhead, P. J. (2010). Meta-analysis of transmitter effects on avian behavior and ecology. *Methods in Ecology and Evolution 1*:180–187.

Barua, M. (2013). Volatile ecologies: towards a material politics of human-animal relations. *Environment and Planning A*: doi: 10.1068/a46138.

Begon, M., Townsend, C. R. & Harper, J. L. (2005). *Ecology: From Individuals to Ecosystems*. Hoboken: Wiley-Blackwell.

Belant, J. L. (1997). Gulls in urban environments: landscape-level management to reduce conflict. *Landscape and Urban Planning 38*: 245-258.

Bouten, W., Baaij, E. W., Shamoun-Baranes, J. & Camphuysen, K. C. J. (2013). A flexible GPS tracking system for studying bird behavior at multiple scales. *Journal of Ornithology 154*: 571. https://doi.org/10.1007/s10336-012-0908-1.

Bowlin, M. S., Cochran, W. W. & Wikelski, M. C. (2005). Biotelemetry of New World thrushes during migration: Physiology, energetics and orientation in the wild. *Integrative and Comparative Biology 45*(2), 295–304, https://doi.org/10.1093/icb/45.2.295.

Buckley, N. J. (1996). Food finding and the influence of information, local enhancement, and communal roosting on foraging success of North American vultures. *Auk 113*: 473–488.

Burger, J. (1981). Feeding competition between Laughing Gulls and Herring Gulls at a sanitary landfill. *Condor 83*: 328–335.

Burger, J. & Gochfield, M. (1998). Effects of eco-tourists on bird behaviour at Loxahatchee National Wildlife Refuge, Florida. *Environmental Conservation 25*: 13–21.

Cagnacci, F., Boitani, L., Powell, R. A., & Boyce, M. S. (2010). Animal ecology meets GPS-based radiotelemetry: a perfect storm of opportunities and challenges. *Philosophical Transactions of the Royal Society B: Biological Sciences 365*(1550), 2157–2162. http://doi.org/10.1098/rstb.2010.0107.

Campbell, M. (1998). *Interactions Between Biogeography and Rural Livelihoods in The Coastal Savanna of Ghana*. Ph.D. Thesis, University of London.

Campbell, M. (2006a). Urban parks as shared spaces? The utility of alert distances as indicators of avian tolerance of humans in Stirling, Scotland. *Area 38*(3): 301- 311.

Campbell, M. (2006b). *Impact of Landscape Features on Corvids in Stirling and Lewes.* Unpublished Certificate Thesis, University of Lancaster, Lancaster.

Campbell, M. (2007). An animal geography of avian ecology in Glasgow. *Applied Geography 27*(2): 78-88.

Campbell, M. (2008). An animal geography of avian feeding habits in Peterborough, Ontario *Area 40*(4): 472–480.

Campbell, M. (2009a). A new zoogeography of domestication and agricultural planning in Southern Ghana. *Area 41*(2): 139–148.

Campbell, M. (2009b). Repositioning zoogeography within the nature–culture borderlands: An animal geography of reptiles in southern Ghana. *Applied Geography 29*: 260–268.

Campbell, M. (2009c). The impact of habitat characteristics on bird presence and the implications for wildlife management in the environs of Ottawa, Canada. *Urban Forestry & Urban Greening* 8(2): 87–95.

Campbell, M. (2009d). Factors for the presence of avian scavengers in Accra and Kumasi, Ghana. *Area 41*(3): 341–349.

Campbell, M. (2010a). Passerine reactions to human behaviour and vegetation structure in Peterborough, Canada. *Urban Forestry & Urban Greening 10*: 47–51.

Campbell, M. (2010b). An animal geography of avian foraging competition on the Sussex coast of England. *Journal of Coastal Research 26*(1): 44 – 52.

Campbell, M. (2013). An animal geography of the dominant urban avian scavenger in contrasting case studies. In Ruiz, L. & Iglesias, F. (Eds.) *Birds: Evolution and Behavior, Breeding Strategies, Migration and Spread of Disease*, 67 - 82. Nova Science Publishers, New York.

Campbell, M. (2014a). Advances in zoogeography: Micro-spaces of avian foraging and human presence. In Jenkins, O. P. (Ed.) *Advances in Zoology Research 6*: 51 – 64. New York: Nova Science Publishers.

Campbell, M. (2014). The impact of urbanization and agricultural development on vultures in El Salvador. *Vulture News 66*: 16–28.

Campbell, M. (2015). The adaptation of the turkey vulture (Cathartes aura, Linnaeus 1758) to changing circumstances and ecologies. In Jenkins OP, (Ed.), *Advances in Animal Science and Zoology 7*, 99–117. New York: Nova Science Publishers.

Campbell, M. (2016). Avian reactions towards human approaches in different urban greenery structures in Nanaimo. *Urban Forestry & Urban Greening 19*: 47–55.

Cannon, R. W., Knopf, F. L. & Pettinger, L. R. (1982). Landsat data to evaluate lesser prairie chicken habitat in west Oklahama. *Journal of Wildlife Management 46*: 916-922.

Carrete, M., Sánchez-Zapata, J. A., Benítez, J. R., Lobón, M. & Donázar. J. A. (2009). Large-scale risk-assessment of wind-farms on population viability of a globally-endangered long-lived raptor. *Biological Conservation 142*: 2954–2961.

Catry, P., Lecoq, M. & Strange, I. J. (2008). Population growth and density, diet and breeding success of striated caracaras *Phalcoboenus australis* on New Island, Falkland Islands. *Polar Biology 31*: 1167–1174.

Clark, P. E., Johnson, D. E., Kniep, M. A., Jermann, P., Huttash, B., Wood, A., Johnson, M., McGillivan, C. & Titus, K. (2006). An advanced, low-cost, GPS-Based animal tracking system. *Rangeland Ecology & Management 59*: 334–340.

Cringan, A. T. (2007). An observation of apparent 'fishing' by Turkey Vultures. *Colorado Birds 41*: 258–259.

Devault, T. L., Reinhart, B. D., Brisbin, I. L. & Rhodes, O. E. (2004). Home ranges of sympatric Black and Turkey vultures in South Carolina. *The Condor 106*: 705 – 711.

Dwyer, J. F. & Cockwell, S. G. (2011). Social hierarchy of scavenging raptors on the Falkland Islands, Malvinas. *Journal of Raptor Research 45*(3): 229–235.

Elliott, J. A. & Campbell, M. (2002). The environmental imprints and complexes of social dynamics in rural Africa: cases from Zimbabwe and Ghana. *Geoforum 33*(2): 221-237.

Emel, J., Wilbert, C. & Wolch, J. (2002). Society and animals. *Journal of Human Animal Studies 10*(4): 1.

Emlen, S. T. (1971). Adaptive aspects of coloniality in the Bank Swallow. *American Zoology 11*: 628.

Erwin, R. M. (1978). Coloniality in terns: The role of social feed. *Condor 80*: 211–215.

Fernández-Juricic, E., Dolores, M. & Lucas, E. (2001). Alert distance as an alternative measure of bird tolerance to human disturbance implications for park design. *Environmental Conservation 28*: 263–269.

Ganderton, P. & Coker, P. (2005). *Environmental Biogeography*. Harlow: Pearson/ Prentice Hall.

Harris, M. P. Bogdanova, M. I., Daunt, F. & Wanless, S. (2012). Using GPS technology to assess feeding areas of Atlantic Puffins Fratercula arctica. *Ringing & Migration 27*:1: 43-49, DOI: 10.1080/03078698.2012.691247.

Hill, J. (2017). *GPS Tags Reveal a Hidden Side to a Grassland Bird*. Retrieved from https://vtecostudies.org/blog/gps-tags-reveal-a-hidden-side-to-a-grassland-bird/.

Hilty, S. L. (1977). *A Guide to the Birds of Colombia*. Princeton: Princeton University Press.

Hobbs, R. J., Higgs, E. & Harris, J. A. (2009). Novel ecosystems: implications for conservation and restoration. *Trends in Ecology and Evolution 24*: 599–605.

Houston, D. C. (1986). Scavenging efficiency of Turkey Vultures in Tropical Forest. *Condor 88*: 318 - 323.

Houston, C. S., Terry, B., Blom, M. & Stoffel, M. J. (2007). Turkey Vulture nest success in abandoned houses in Saskatchewan. *Wilson Journal of Ornithology 199*: 742–747.

Igl, L. D. & Peterson, S. L. (2010). Repeated Use of an Abandoned Vehicle by Nesting Turkey Vultures (Cathartes aura). *Journal of Raptor Research 44*(1): 73–75.

Ingersoll, E. (1920). Zoogeography. In: *Encyclopedia Americana*. New York: Grolier.

Jackson, J. A. (1975). Reguritative feeding of young black vultures in December. *Auk 92*: 802–803.

Johnston, R. J., Gregory, D., Pratt, G. & Watts, M. (2000). *Human Geography (4th ed.)*. Malden: Blackwell Publishing.

Jones, P. J. (1978). A possible function of the 'wing-drying' posture in the Reed Cormorant *Phalacrocorax africanus*. *Ibis 120*(4): 540–542.

Jones, C. D., Cox, J. A. & Cooper, R. J. (2014). Bachman's Sparrow (*Peucaea aestivalis*) response to variation in the extent of burns conducted during the nesting season. *Avian Conservation and Ecology 9*(1): 3. http://dx.doi.org/10.5751/ACE-00632-090103.

Jørgensen, A. F. & Nøhr, H. 1996. The use of satellite images for mapping of landscape and biological diversity in the Sahel. International. *Journal of Remote Sensing 17*: 91–109.

Kiff, L. F. (2000). The current status of North American vultures. In Chancellor, R. D. & and Meyburg, B. U. (Eds.), 175 - 189. *Raptors at Risk*, 175-189. Berlin/Blaine, WA: World Working Group on Birds of Prey/Hancock House.

Kiltie, R. A. (2000). Scaling of visual acuity with body size in mammals and birds. *Functional Ecology 14*: 226–234.

Kirk, D. A. & Gosler, A. G. (1994). Body condition varies with migration and competition in migrant and resident South American vultures. *Auk 111*: 933–944.

Koford, C. B. (1953). *The California Condor*. Toronto: General Publishing Company.

Krebs, J. R. (1974). Colonial nesting and social feeding as strategies for exploiting food resources in the Great Blue Heron (Ardea Herodias). *Behavior 51*: 99–134.

Lameris, T. K. & Kleyheeg, E. (2017). Reduction in adverse effects of tracking devices on waterfowl requires better measuring and reporting. *Animal Biotelemetry* 5: 24. DOI 10.1186/s40317-017-0139-6.

Lorimer, J. (2007). Nonhuman charisma. *Environment and Planning D: Society and Space 25*: 911-932.

McDonald, G. (2003). *Biogeography: Introduction to Space, Time and Life*. New York: John Wiley and Sons.

Marzluff, J. M., Bowman, R. & Donnelly, R. (Eds.) (2001). *Avian Ecology and Conservation in an Urbanizing World*. Norwell, MA: Kluwer Academic.

McHargue, L. (1977). Nesting of turkey and black vultures in Panama. *Wilson Bulletin 89*(2): 328–329.

Mitani, Y., Andrews, R. D., Sato, K., Kato, A., Naito, Y. & Costa, D. P. (2010). Three-dimensional resting behaviour of northern elephant seals: drifting like a falling leaf. *Biology Letters 6*: 163–166.

Mitchell, I. P., Newton, S. F., Radcliffe, N. & Dunn, T. E. (2004). *Seabird Populations in Britain and Ireland*. London: T. and A. D. Poyser.

Morris, S. (2016). *GPS Tags Reveal the Secret Life of Urban Seagulls*. Retrieved from https://www.theguardian.com/environment/2016/jul/14/gps-tags-reveal-the-secret-life-of-urban-seagulls.

Murdoch, J. (1998). The spaces of actor network theory. *Geoforum 29*: 357–374.

National Ocean Service (2018). *What is LIDAR?* Retrieved from https://oceanservice.noaa.gov/facts/lidar.html.

Nicolis, G. (1994). Dynamical systems, biological complexity and global change. In Solbrig, O. T., Van Emden, H. M. & Oordt, P. G. W. J. (Eds.), *Biodiversity and Global Change*, 21-31, Wallingford: CAB International.

O'Connor, J. (1997). What is environmental history? Why environmental history? *Capitalism, Nature, Socialism 8*: 3-29.

Palmeirim, J. (1988). Automatic mapping of avian species habitat using satellite imagery. *Oikos 52*(1): 59-68. doi: 10.2307/3565982.

Philo, C. & Wilbert, C. (Eds.) (2000). *Animal Spaces Beastly Spaces: New Geographies of Human–Animal Relations.* London: Routledge.

Pickett, S. T. A., Parker, V. T. & Fiedler, P. L. (1992). The New Paradigm in Ecology: Implications for Conservation Biology above the Species Level. In Fiedler, P.L. & Jain, K. (Eds.) *Conservation Biology: The Theory and Practice of Nature Conservation, Preservation and Management*, 66-87. New York: Chapman and Hall.

Ponganis, P. J. (2007). Bio-logging of physiological parameters in higher marine vertebrates. *Deep Sea Research Part II: Topical Studies in Oceanography 54*: 183–192.

Rabenold, P. P. (1983). The communal roost in black and turkey vultures—an information center? pp. 303–321. In Wilbur, S. R. & Jackson, J. R. (Eds.), *Vulture Biology and Management.* Berkley: University of California Press.

Rutz, C. & Hays, G. C. (2009). New frontiers in biologging science. *Biology Letters 5*: 289–292.

Schulenberg, T. S., Stotz, D. F. Lane, D. F., O'Neill, J. P. & Parker III, T. A. (2010). *Birds of Peru*. Princeton: Princeton University Press.

Scoones, I. (1997). The dynamics of soil fertility change: historical perspectives on environmental transformation from Zimbabwe. *The Geographical Journal 63* (2): 161–169.

Scoones, I. (1999) New ecology and social sciences: What prospects for a fruitful engagement? *Annual Review of Anthrolopology 28*: 479-507.

Smith, H. R., DeGraaf, R. M. & Miller, R. S. (2002). Exhumation of food by Turkey vulture. *Journal of Raptor Research 36*(2):144-145.

Snyder, N. F. R. & Snyder, H. (2006). *Raptors of North America: Natural History and Conservation*. London: Voyageur Press.

Stolen, E. D. (1996). Black and turkey vulture interactions with bald eagles in Florida. *Florida Field Naturalist 24*(2): 43–45.

Strange, I. J. (1996). *The Striated Caracara Phalcoboenus australis in the Falkland Islands*. Author's Edition, Falkland Islands.

Turner, W. R. (2003). City biological monitoring as a tool for ecology and conservation in urban landscapes: The case of the Tucson bird count. *Landscape and Urban Planning 65*: 149–166.

Viviant, M., Monestiez, P. & Guinet, C. (2014). Can we predict foraging success in a marine predator from dive patterns only? Validation with prey capture attempt data. *PLoS ONE 9*(3): e88503.

Vuorisalo, T., Andersson, H., Hugg, T., Lahtinen, R., Laaksonen, H. & Lehikoinen, E. (2003). Urban development from an avian perspective: causes of hooded crow Corvus corone cornix urbanisation in two Finnish cities. *Landscape and Urban Planning 62*: 69–87.

Wallace, M. P. & Temple, S. A. (1987). Competitive interactions within and between species in a guild of avian scavengers. *The Auk 104*: 290-295.

Wallin, D. O., Elliott, C. C. H., Shugart, H. H., Tucker, C. J., Wilhelmi, F., 1992. Satellite remote sensing of breeding habitat for an African weaver-bird. *Landscape Ecology 7*: 87–99.

Ward, P. & Zahavi, A. (1973). The importance of certain assemblages of birds as "information centres" for food–finding. *Ibis 115*: 515–545.

Watanabe, Y. Y. & A. Takahashi. (2013). Linking animal borne video to accelerometers reveals prey capture variability. *Proceedings of the National Academy of Sciences USA 110*: 2199–2204.

Wilmers, C. C., Nickel, B., Bryce, C. M., Smith, J. A., Wheat, R. E. & Yovovich, V. (2015). The golden age of bio-logging: how animal-borne sensors area advancing the frontiers of ecology. *Ecology 96*: 1741–1753.

White, C. R., Cassey, P., NSchimpf, G., Halsey, L. G., Green, J. A. & Portugal, S. J. (2013). Implantation reduces the negative effects of bio-logging devices on birds. *Journal of Experimental Biology 216*: 537–542.

Wildlife ACT (2018). *Monitoring and Tracking Technology.* Retrieved from https://wildlifeact.com/about-wildlife-act/monitoring-tracking-technology/.

Williams, M. (1994). The relations of environmental history and historical geography. *Journal of Historical Geography 20*: 3-21.

Williams, T. M., Wolfe, L., Davis, T., Kendall, T., Richter, B., Wang, Y., Bryce, C., Elkaim, G. H. & Wilmers, C. C. (2014). Instantaneous energetics of puma kills reveal advantage of felid sneak attacks. *Science 81*: 81–85.

Wilson, A. M., Lowe, J. C., Roskilly, K., Hudson, P. E. Golabek, K. A. & McNutt, J. W. (2013). Locomotion dynamics of hunting in wild cheetahs. *Nature 498*:185–189.

Wolch, J. & Emel, J. 1998. *Animal Geographies: Place Politics and Identity in the Nature-Culture Borderlands*. London: Verso.

Woods, R. W. & Woods, A. (2006). *Birds and Mammals of the Falkland Islands*. Hampshire: WILDGuides.

Yarwood, R. & Evans, N. (2000). Taking stock of farm animals and rurality. In Philo, C. & Wilbert, C. (Eds.) *Animal Spaces, Beastly Spaces: New Geographies of Human–Animal Relations*, 98–114. London: Routledge.

Young, M. D. & Solbrig, O. T. (1993). Providing an environmentally sustainable, economically profitable and socially equitable future for the world's savannas. In Young, M. D. & Solbrig O. T. (Eds.), *The World's Savannas: Economic Driving Forces, Ecological Constraints and Policy Options for Sustainable Land Use*, 321 – 344. Paris: Pantheon Publishing Group.

Zimmerer, K. S. (1994). Human geography and the 'New Ecology': the prospect of integration. *Annals of the Association of American Geographers 84*(1): 108-125.

In: Geomatics and Conservation Biology
Editor: Michael O'Neal Campbell
ISBN: 978-1-53614-468-0

Chapter 8

APPLICATION OF GEOGRAPHICAL TECHNOLOGY IN THE MODELLING OF MIGRATORY BIRD HABITATS IN MANGROVE FORESTS: A REVIEW

Azimah Abd Rahman[1,*] and Tarmiji Masron[2]
[1]School of Humanities, Universiti Sains Malaysia, Malaysia
[2]Faculty of Social Sciences, Universiti Malaysia Sarawak, Malaysia

ABSTRACT

Migratory bird habitat modelling is the best supportive method for the implementation of preservation and conservation programs for such species. Such habitat models help to accurately identify the birds' main area of concentration, population distribution pattern, habitat requirements, and potential locations for the future habitats. Field information can be analysed to identify the habitat requirements of migratory birds and determine and predict their population distributions. Geographical technologies such as geographic information systems (GIS) and remote sensing can be used for habitat modelling, using the spatial parameters of habitat variation. The GIS applications are helpful in mapping out the data gathered in the field, while remote sensing can be used to model the parameters of the study that have been identified as affecting the migratory birds' choice of habitat. This chapter refers to the example of avian migratory presence in the Larut Matang Mangrove Forest Reserve in Malaysia and examines relevant habitat model-constructions using geographical technology, for analysis that may determine the study parameters for migratory birds' choice of habitat, population distribution in the current year, and future habitat choices of habitat. This visual mapping may simplify consumers' interpretation of such ecological dynamics and support more successful preservation and conservation programmes of migratory birds' habitat concentrations in sea mangrove areas and possibly wider habitat types. This study makes an important contribution to habitat analysis, geomatics, conservation biology and wildlife policy and planning.

[*] Corresponding Author Email: azimahrahman@usm.my.

Keywords: geographic information systems (GIS), remote sensing, migratory bird, habitat model, conservation

INTRODUCTION

Migratory bird populations have been declining in recent years, this occurrence attracting the attention and reaction of various interested stakeholders throughout the world (Jonzén et al., 2002). Various reasons for this decline have been identified in most Asian countries, including Thailand (Barbier & Cox, 2002), Vietnam (Yen et al., 2005), Hong Kong (Corlett, 1993), and Malaysia (Lum, 2005), as well as in western countries such as the United States (Burke & Erica, 1998), Australia (Greenberg & Marra, 2005), and New Zealand (Pulido et al., 2001). Disturbances to the migratory bird habitats are the main factors for the decline in migratory bird population sizes (Berthold, 2001; Howes et al., 2003). Natural disturbances such as tidal waves, tsunamis and storms (Kathiresan & Rajendran, 2005), as well as human actions such as agricultural activities, fish farming, mangroves logging and so on (Pfister et al., 1992) have indirectly affected migratory birds' habitats and food sources. These factors have reduced the arrival of the migratory birds to preferred habitats, as the birds are sensitive to such environmental changes. Environmental development for human settlement, agricultural activities, fish farming and the like cause water pollution, while deterioration of water quality, changes in land use, erosion of the banks and logging disturb the habitat of migratory birds (Eng et al., 1989).

Mangrove areas provide food sources, sanctuary, and rest areas for migratory birds before they continue their flying journey (Holguin et al., 2001). The importance of mangrove areas can be seen in the distribution pattern of the migratory birds (Martin & Finch, 1995). However, conventional methods for monitoring and managing mangrove areas, as performed by the responsible agencies, are usually time-consuming and relatively ineffective (Bishop, 1999). A more systematic and effective approach is therefore required to overcome the decline in the populations of these migratory bird species. Geographic technology applications such as geographic information systems (GIS) and remote sensing offer such an enhanced method (Peterson, 2001). Combining these geographical technologies will enhance the analytical accuracy of data by taking into consideration the spatial aspects of habitat parameters and analyses. A pixel-based, remote sensing technology approach will also allow the spatial mapping and analysis of all the environmental parameters that affect bird behavior in a more comprehensive manner (Tankersley & Orvis, 2003). This chapter examines the issues of migratory bird ecology, including the merits and problems of GIS and remote sensing applications, relevant ecological parameters and the syntheses of these approaches to the analysis of migratory bird habitat selection and conservation. Although several geomatics methods are considered, a pixel-based methodology is cited as relevant to the case discussed. The case of avian migration through the Larut Matang Mangrove Forest Reserve in Malaysia is referred to as a study area for the testing of these technologies. This area is an important area for bird migration (Ahmad, 2009).

THE MIGRATORY BIRDS' HABITAT MANGROVE FOREST

Migratory bird species are often associated with mangrove forest, which is their choice of habitat for foraging and nesting during migration (Wilson & Twedt, 2003). Mangrove forest (*Rhizophora*, *Avicennia*, *Sonneratia* and *Bruguiera* spp.) is part of the tropical forest vegetation found in the tropics and subtropical areas between 30° N and 30° S latitudes. Mangrove forest is found in India and Bangladesh and down to Southeast Asia. Across the world, it is estimated that 123 countries in the tropical and subtropical areas have mangrove swamp forest, with a total area of 150,000 square kilometres. These ecosystems provide food chains for marine and other life forms such as small fish species and invertebrate animals, which form part of the complex ecosystems of the mangrove swamp areas (Holguin et al., 2001; Valiela et al., 2001). Kathiresan and Rajendran (2005) explain that mangrove plants adapt to the challenging environmental conditions of surrounding ecosystems, through natural adjustments found in their roots, leaves and seeds (Table 1).

Table 1. Natural adaptation for mangrove plant

Species	Descriptions
Rhizophora • *Rhizophora mucronata* spp. • *Rhizophora apiculata* spp.	Roots  Leaf  Seeds 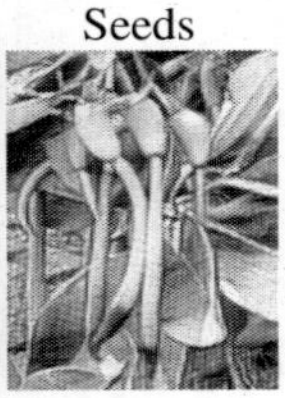 Flower & buds 

Table 1. (Continued)

Species	**Descriptions**
Avicennia • *Avicennia alba* spp. • *Avicennia marina* spp. • *Avicennia officinalis* spp.	Roots 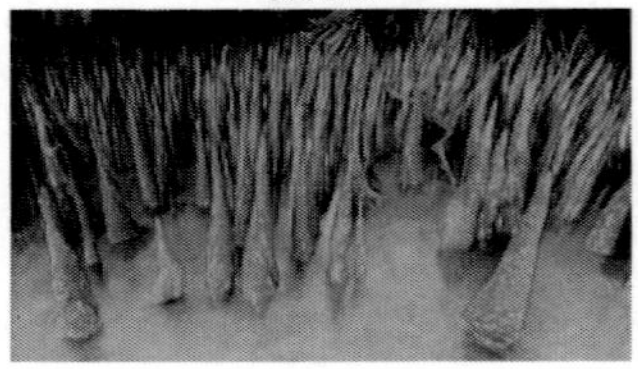 Leaf 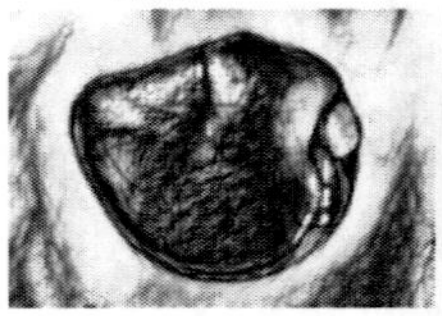 seeds 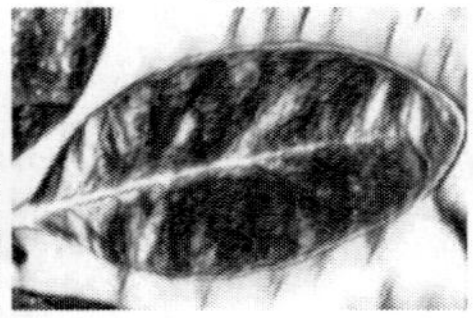 Flower & buds 
Sonneratia • *Sonneratia Alba* spp. • *Sonneratia caseolaris* spp. • *Sonneratia ovate* spp.	Roots  Leaf  Seeds  Flower & buds 
Bruguiera • Bruguiera gymnorhiza spp. • Bruguiera cylindrical spp. • Bruguiera parviflora spp.	Roots 

Species	Descriptions
	Leaf seeds 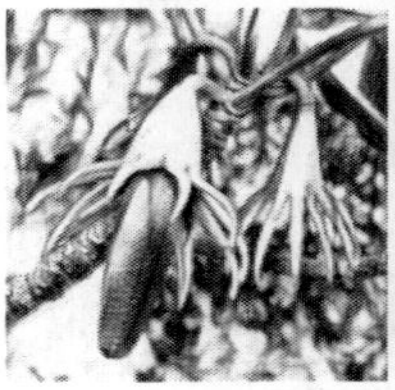  Flower & buds 

A study by Robertson (1992) found that plant species such as *Avicennia sonneratia* spp., *Rhizophora apiculata* spp., *Rhizophora mucronata* spp., *Sonneratia alba* spp., *Bruguiera Gymnorhiza* spp., *Bruguiera parviflora* spp. and *Bruguiera parviflora* spp. can survive change in a mangrove environment through their natural genetic adaptation. Mangrove plant adaptation is important for survival, by acquiring and providing nutrients, and enabling reproduction and circulation in the plant cells (Valiela et al., 2001). According to Kathiresan and Rajendran (2005), mangrove swamps can produce the mechanism for restoring the nutrients required for the growth of mangrove plants. The rate of nutrient dispersion in a swamp area varies according to the characteristics of deposition and the movement of nutrients in the whole area (Medina et al., 1990). This condition can impact on the rate of distribution and location of the habitats of marine life, including those of migratory birds, which differ according to the rates of high and low tides, or based on the highest and lowest levels of sea water (Pearson & Dawson, 2003).

Importance of Mangrove Swamps to Migratory Birds

According to Alerstam and Christien (1993), the total cumulative time spent by migratory birds in habitat areas is more than the time spent in flying during the migrating period. Therefore, habitat is crucial for migratory birds during their migrating period. The inshore mangrove forests, with thick mud areas are rich with various mangrove plants and marine life forms and are the location of choice for the migratory bird species (Lum, 2005).

The importance of a habitat for migratory birds is now the main focus of studies on migratory birds. This research largely concerns the suitability of the habitat type for the requirements of life supports and protection from any disturbances (Dunn & Bul, 2002). Migratory bird species have differing reasons for their choice of habitat location. Reasons may include the changing of feathers or the stock of lipid supply in their bodies and resting, while waiting for nightfall (Colwell, 2010). Milsom et al. (2000) argue that the choice of habitat for migratory birds is correlated with wider geographical areas and the associated flying time limit. Canterbury et al. (2000) note that the location of temporary rest areas is also related to the suitability of the area as a habitat. According to Moore et al. (2005), migratory birds will usually stop at an area for a few hours or a few days, depending upon the suitability of the area and the availability of resources that can fulfil their necessities, which include food supplies, protection and resting space for continuing their flight journey.

CHRONOLOGY IN CONSTRUCTING HABITAT MODEL

Developing a habitat model using geographic information systems and remote sensing approaches has captured the interest of many researchers, especially with increased interest in the applications of geographical technology. These researchers have articulated the value of such techniques in the successful development of conservation programs for migratory bird habitat conservation. In such cases, the analyses of the field data creates study parameters and visual mappings which facilitate the interpretation of data (Table 2a, b, c).

Table 2a. Chronology of Habitat Modelling (1988 - 2000)

Year	Progress	References
1988	Digital Elevation Model (DEM) analysing elevation.	Jenson & Domingue (1988)
1989	Determining the scale in building habitat model.	Wiens, 1989; Eng et al. (1989)
1991	Habitat model analyses niches for migratory birds.	Horne & Wiens (1991)
1993	Remote sensing migratort flight routes.	Homer et al. (1993)
1995-1996	Geographic Information system (GIS) for conservation of mangrove swamp forests.	Kairo et al. (1995); Câmara et al. (1996)
1996	Pixel data in place of point data for precision	Gatrell et al. (1996)
1998	Simulated habitat models, for presence prediction.	Pearson & Carroll (1998)
2000	Habitat modelling of avian migratory distribution.	Guisan & Zimmermann (2000)
2000	Habitat models need to consider parameters such as food, climate, landscape, land use, human disturbances and natural causes of environment.	Schaub & Jenni (2000); Lennon et al. (2000)
2000	Numerical representation of data in the form of statements for analysis using GIS applications.	Allen & O'Connor (2000)
2000	The change of data points to topographic structure to guarantee precision of analysis using GIS applications.	Buckley (2000)
2000-2004	Technological improvement in elevation analysis for habitat model using GIS applications and remote sensing using Light *Detection and Ranging* (LIDAR) and *Shuttle Radar Topography Mission* (SRTM) which provides clear data in three-dimensional representation.	Marks & Bates (2000); Smith & Sandwell (2003); Dowding et al. (2004)

Table 2a. Chronology of Habitat Modelling

(2001 - 2003)

Year	Progress	References
2001	Habitat modelling for migratory bird species using combination of GIS and remote sensing technologies.	Valiela et al. (2001)
2001	Habitat model is important for looking at distribution of living species, identifying condition of biodiversity of an area and development of strategy for spatial conservation.	Peterson (2001)
2001	GIS Applications can help identify the relationship between migratory bird species and their habitat. GIS is the most perfect method for evaluating the environment.	Atauri & de Lucio (2001)
2001	GIS applications are also used in building habitat model that focuses on the effects of the relationship between the habitat and biodiversity in an area of study.	Putz et al. (2001)
2001	GIS is a useful method for quantitative analysis and visual presentation of the analysis results	Keane et al. (2001)
2001	Remote sensing applications presents pixel-based methods important for identifying habitat requirement.	Osborne et al. (2001)
2001-2003	With GIS applications the conditions of the landscape structure and land use in a given area can be identified.	Keane et al. (2001); Appleton et al. (2002); Smith (2003)

(2003 - 2009)

Year	Progress	References
2003	Remote sensing technology using pixel analyses better than point analysis.	Tankersley & Orvis (2003); Store & Jokimäki, (2003); Anselin (2003)
2004	Remote sensing analysis using pixel-based approach shows the role of a landscape	DeFries et al. (2004)
2005	Satellite data used to record migratory paths and heir resting locations	Yen et al. (2005)
2005	Habitat model construction requires knowledge on the relationship between the landscape structure and species continuity.	Lum (2005)
2005	Construction of pixel-based model is appropriate for analysing the distribution of moving live species using image-based data	Heikkil & Ainen (2005)
2005	Remote sensing technology requires large investment and satellite data is limited.	Yen et al. (2005)
2006	GIS applications can analyze and process data in various formats and in different scales.	Yasue (2006)
2006	Remote sensing analyses and integrates species data and environmental parameters for predicting new areas for future studies.	Oyana et al. (2006)
2006	Pixel size in analysis using remote sensing is appropriate with the study model.	Phillips et al. (2006)
2006	Pixel-based analysis, e.g., Kernel Density in GIS software can analyse parameter impacts on the distribution of migratory birds.	Buyong (2006)
2006	In remote sensing analyses, SRTM data has high resolution values for large scale studies.	Gorokhovich & Voustianiouk (2006)

Table 2a. (Continued)

Year	Progress	References
2007	Global elevation data is uneven as elevation mapping used varied scales and resolutions in different countries.	McKee et al. (2007)
2007	Satellite images (e.g., *Landsat TM)* processed using three dimensional and coloured presentations, analyzes size proportion and shape of the forest, topological diversity, altitude rate, and forest group relationships.	Wade & Hickey (2007)
2008	Remote sensing analyses landscape elements by determining the distribution pattern and changes in the landscape.	Shao & Wu (2008)
2009	GIS applications use recorded data to identify choice of location to lay eggs.	Bonter et al. (2009)

HABITAT MODELLING CHOOSING MODEL PARAMETERS

Habitat model construction requires an understanding of the ecosystem, to determine the relationship between the environmental landscape structure and the sustainability of life for a living species (Lum, 2005). Rotenberry (1985) opined that the landscape diversity within a habitat influences the distribution of living species, including migrant birds in an area. In fact, over half (55%) of the bird species composition may be determined by the local flora, whereas only 35% of such avian interspecies variation is the result of the other environmental and physical conditions of the migratory birds' habitat. Furthermore, according to Dupré and Ehrlén (2002), the relationship between ecology and animal habitat enables a species to live, whether temporarily or permanently, in a certain habitat. Rosenfeld et al. (2000) notes that habitat is the basis that explains the animal's requirements for sustainable living and reproduction. Therefore, analysis of the density of an animal species can show the habitat of their choice. The natural characteristics of migratory birds, which are sensitive to environmental changes, show that this bird species group has its own criteria for habitat choice, even for the areas of temporary occupance (Schaub & Jenni, 2000; Schaub et al., 2001; Desholm, 2009; Calvert et al., 2009). Therefore, Schaub and Jenni (2000) argued that a habitat model must include certain factors such as food, climate, landscape, land use, human disturbances and influences of the natural environment (Figure 1).

All these components fulfil the habitat requirements of migratory birds in their preparation for continued migration (Piersma & Lindström, 2004; Deppe & Rotenberry, 2008). Lennon et al. (2000) also found that in most previous studies, the common factors cited in migratory bird habitat model construction are variables such as food or climatic changes or a combination of the two. Both these factors are closely interrelated in influencing the pattern of density of species population in a studied area (Figure 2).

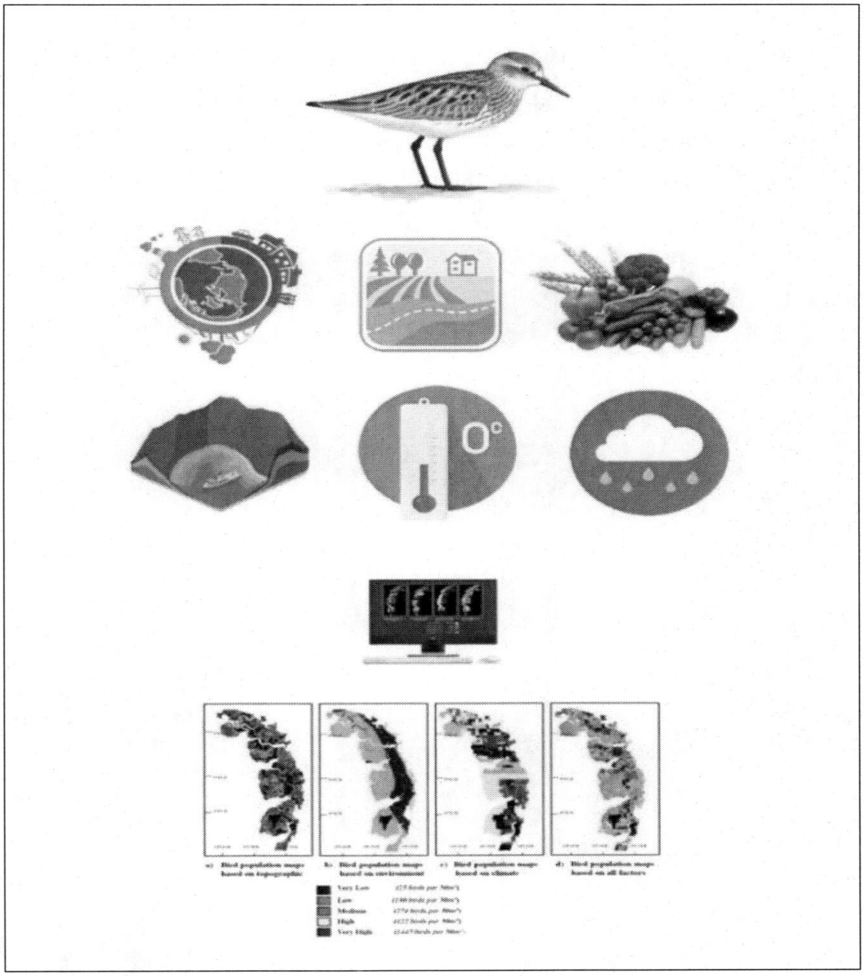

Figure 1. The overall workflow for habitat modelling studies with parameters identified from field sampling and literature review. The research parameters included are [left to right]: (i) land cover, (ii) landuse, (iii) food source, (iv) elevation, (v) temperature, (vi) rainfall. The symbol of laptop describe about habitat modelling that show mapping of parameter as a result of model analysis.

In addition, environmental factors, such as land type, land use, altitude of the area, local economic activities and landscape conditions are relevant parameters. These factors are also interrelated with one another and affect the species distribution of migratory birds in the surroundings of the study area (Phillips et al., 2005). Many studies have taken them into consideration in determining the suitability of an area as a habitat for migratory birds, due to the realization of the importance of such factors in influencing the distribution of migratory birds, (Bonter et al., 2009; Philpott et al., 2008). Altitude is an extra influence that has often been cited, especially in previous studies that indicate that the density of migratory birds is inversely proportional with changes in altitude of an area (Figure 3).

Gathering data for a such studies is a global issue that is often discussed (Bonter et al., 2009). Data on altitudes at the global level show mixed results because many altitude maps were created using nationally variable scales and resolutions (McKee et al., 2007). The scarcity of quality digital altitude data affects the accuracy of the analyses for some areas. However, current technological developments such as Digital Elevation Models (DEMs) (Jenson & Domingue, 1988), Synthetic Aperture Radar (SAR), Light Detection and Ranging (LIDAR) (Marks & Bates, 2000) and Shuttle Radar Topography Mission (SRTM) (Smith & Sandwell, 2003) are capable of providing better digital data on the altitudes suitable for conducting such studies.

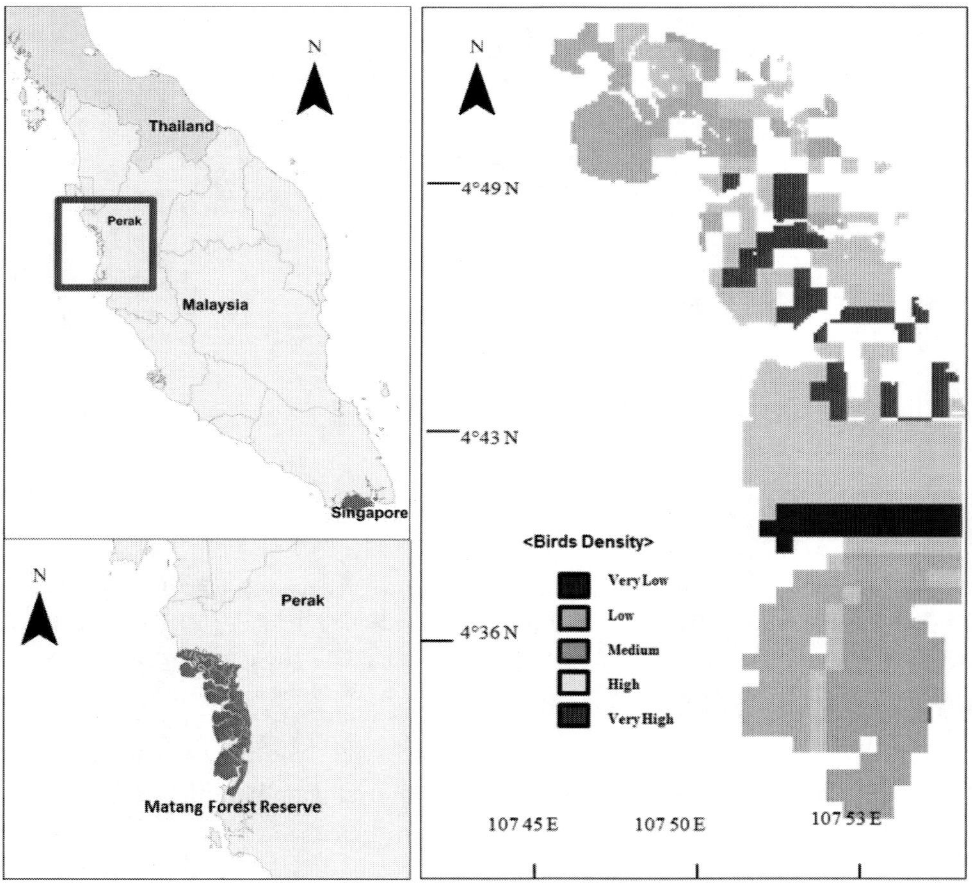

Figure 2. Map of climate changes impact on bird population and distribution in the Larut Matang Forest Reserve.

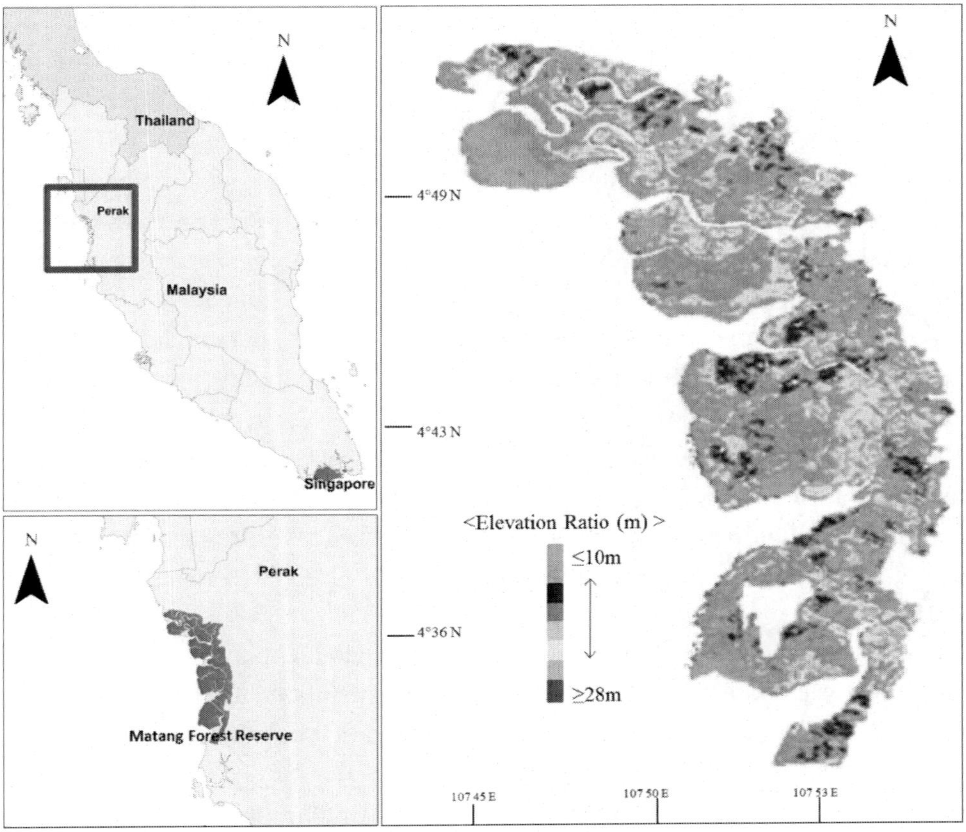

Figure 3. Map of elevation that influence birds distribution in the Larut Matang Mangrove Forest Reserve.

Geographic Information System (GIS) in Habitat Modelling

Geographic technologies such as geographic information systems (GIS) applications are not limited to geographical mappings but may also be applied to various fields including biological studies that focus on conserving the biodiversity of a habitat (Bakker et al., 2002; Câmara et al., 1996; Pennington et al., 2008). Many studies on migratory birds have used GIS applications as for background analyses and quanitification, especially for the purpose of preservation and conservation of mangrove forests, which have become the main habitat for most migratory bird species in some areas (Kairo et al., 1995; Dahdouh-Guebas, 2002; Zharikov et al., 2005). In determining the choice of habitat for migratory birds, the landscape ecology principle associated with the dynamic change in population needs to be taken into consideration.

GIS applications assist in the identification and quantification of relationships between the migratory bird species and their habitat (Atauri & de Lucio, 2001; Tian et al., 2008). GIS analyses support the spatial division of study areas and can predict presence based on known habitat preferences, for the identification of habitat choices for a migratory bird species (Atauri & de Lucio, 2001). Areas lacking the known parameters for bird presence may be classified as unsuitable habitat substrates (Ellner et al., 2001). Mapping of the potential habitas for such species is a vital requirement for habitat conservation, as it allows the demarcation of such areas especially in regions with intensive landcover change and competing human landuse (Ellner et al., 2001; Larson et al., 2004). Habitat modelling using GIS applications helps to identify areas that are suitable for migratory birds to migrate, lay eggs, rest and get food (Brotons et al., 2004; (Bonter et al., 2009). Using GIS applications, classifications for mapping habitat, detection of changes and updates for habitat database can be obtained.

GIS models also examine and analyze the impacts of human landuse, such as agricultural activities, commercial logging activities and urban development for an area contribute to the shrinking of the original avian habitat areas, contributing to spcies decline or even extirpation (Jensen et al., 2008). Landscape change processes influence the continuity of life of migratory bird species (Kouki et al., 2001). Information related to landscape ecology helps to show the population patterns of a bird species dynamically and highlights the distribution and sustainability of life of certain organisms (Andries et al., 1994).

GIS applications are also used for constructing habitat models that focus on the relationship between habitats and general biodiversity in an area under study (Putz et al., 2001). Peterson et al. (2001) stated that the construction of a habitat model is important to determine the distribution of a living species, identify the biodiversity condition of an area, and develop a conservation strategy for the preservation of an area. A habitat model also provides the tools for a practical framework for the identification of complex evaluation factors, while producing visual images from the analysis at the same time (Horne & Wiens, 1991). In addition, habitat models analyse ecological niches for migratory birds and identify rare and common species population densities, while showing the location of the zone that becomes the area of preference for the conservation of the rare species (Horne & Wiens, 1991). According to Broton et al. (2004), the construction of habitat model is important because it enables the identification of potential area for breeding of a living species in an area that has never been explored before.

GIS models also analyze forest distribution changes and the relevance of these to avian ecology. Currently, changes in forest fragmentation are the main problems for the habitat of migratory birds (Alongi, 2002). This problem occurs as a result of human economic activities (housing, transportation, agriculture, and industrial development), which indirectly disturb the conditions of the natural biodiversity of forested landcover. These activities have contributed to the reduction of landcover favoured by migratory avian species (Sweitzer et al., 1996). GIS applications can identify the structural conditions and land use of such areas for landcover analyses (Appleton et al., 2002). Spatial statements and numerical representations measure, represent and facilitate such studies (Allen & O'Connor, 2000).

GIS models also depict and analyze scale as an important environmental parameter. According to Lennon et al. (2000), the diversity pattern of species presence is related to the differences in scales of a habitat. Determination of the scale in building a habitat model is important for documenting the distribution diversities of migratory bird species (Store & Jokimäki, 2003; Wiens, 1989). Most of the ecological components, such as animal species and population density of animals and geographical conditions in an area, differ according to the scale of the area (Cuervo & Møller, 2013). A GIS application is the most accurate method to evaluate ecological effects on the distribution of bird species measured using different scales (Atauri & de Lucio, 2001; Greenberg & Marra, 2005). This is because the GIS application is capable of analysing and processing data in various formats and in different scales (Yasue, 2006). GIS techniques are low cost, with high effectiveness in mapping and updating data, and very useful for performing quantitative analysis and displaying results visually for a vast area using the migratory bird habitat parameters (Keane et al., 2001; Li et al., 2005).

Remote Sensing in Migratory Bird's Habitat

Remote sensing is a recent and powerful technology that has many applications for determining the movements of migratory bird species, especially for the identification of bird presence and flight routes (Homer et al., 1993; Desholm, 2006). It is also the best application for monitoring migratory birds' habitats, especially those involving mangrove swamps (Berthold, 2001; Keane et al., 2001). The remote sensing method considers the study area holistically in its analysis, including everything that is thought to affect the distribution and the movement of migratory birds in an area (Anselin, 2003; Muzein, 2006).

In remote sensing, satellite images are the main data source for monitoring the movements and foraging cluster zones for long distance migratory birds. For example, Landsat TM satellite images may be processed using a three-dimensional colour presentation to enable the analysis of forest fragmentations, shapes of forests, topological variations, altitude rate, and relationships among forest groups (Wade & Hickey, 2007). These images allow the documentation and analysis of the conditions of mangrove swamps, including features such as the mariqueter, mud zone and stream zones, and other factors such as altitude, types of land coverage, and height of sea water during high and low tides, and biomass of micro-organisms (Howes et al., 2003).

The construction of habitat models using the remote sensing approach acknowledges the migratory avian habitat choices and correlations between avian species presence and population density and mangrove landscape patterns (Tian et al., 2008). Habitat model simulations illustrate the influence of the environment on the distribution of migratory birds (Guisan & Zimmermann, 2000). Landscape elements and other environmental variables can be easily analysed using remote sensing (Shao & Wu, 2008). Such landscape analysis using remote sensing technology is equivalent to and supplements the field observations, through presentations in the form of geometry, combinations of pixel points and results from resampling (Gottschalk et al., 2005; Homer et al., 1993). Remote sensing also enables habitat monitoring and evaluation of landscape at different scales (Keane et al., 2001). Landscape evaluation is required to ensure that the various landscape functions can be sustained (McGuckin & Brown, 1995). Remote sensing analyses using pixel-based approaches also contribute to the implementation of in-depth analyses of the role of landscape units in supporting communities of plants or ecological units that are structured according to related functions, geomorphology, and disturbances (DeFries et al., 2004).

Habitat model simulations are often applied to studies that aim to identify problems contributing to negative environmental change (Larson et al., 2004; Pearson & Carroll, 1998; Weng, 2008). These models use the real-world methods to predict animal presence and population distributions (Pearson & Carroll, 1998). Image-based data is supported by field numerical data for presence/absence maps (Banker & Morey, 1986; Stockwell & Peterson, 2002; Yen et al., 2005; Wisz et al., 2008). Osborne et al. (2001) argue that remote sensing applications have now moved beyond the simple mapping of habitats to include the identification of parameters that influence the distribution of wildlife, such as food sources, protection, water and space. Remote sensing applications have been successful in analysing and integrating data for predicting the environmental factors that may occupy future studies (Oyana et al., 2006; Gottschalk et al., 2005). Although such methods are based on the real world, in that they are based on data obtained directly through satellite and aerial images, large investments are required, and the data gathered by satellites are limited (Yen et al., 2005).

Pixel-Based Approach in Habitat Modelling

Pixel or grid-based analysis facilitates the identification of the parameters that influence the distribution of animal species in an area of study. The presentation of data on migratory birds using the pixel-based method is important for identifying the habitat requirements for the migratory birds, based on the characteristics of each pixel which represents a habitat unit with bird presence (Osborne et al., 2001; Tian et al., 2008). However, the pixel size used must be appropriate to the study model constructed (Phillips et al., 2006; Tuttle et al., 2006). Additionally, the construction of the pixel-based model must be suited for analysing the distribution of the mobile birds and include the entire range of the studied animals included in the image-based data (Heikkil & Ainen, 2005). Examples of such pixel-based methods include the Kernel Density Analysis found in ArcMap the GIS software package, which can analyse environmental parameters related to the distribution of migratory birds (Buyong, 2006).

The pixel-based approach is appropriate for showing the influence of environmental parameters on the migratory birds that cannot be shown accurately using point data analysis, which focuses only on coverage of a small area (Anselin, 2003). Furthermore, to ensure the accuracy of the analysis results, the point data must be transformed into topological structures (Buckley, 2000). This hinders the accuracy of the results produced, especially considering the distribution of migratory birds, which are sensitive to environmental changes such as temperature, rainfall distribution and other environmental elements (Piersma & Lindström, 2004). Point data is also usually unsuited for more complex analyses, like those that include several types of parameters. Smith (2003) argues that the use of point data in an analysis complicates the estimation of multifarious data distributions, because there is a possibility that the distribution patterns produced are almost the same. This complicates the data analysis, especially when comparing different data from the same area to determine the types of distribution pattern. This is especially true with grouped data distributions (Smith, 2003).

Pixel data also captures parameters that influence avian distributions and are useful for study and analysis of the distribution of migratory birds (Lum, 2005). When using point data, determination of the most important parameters (weights) is difficult (Anselin, 1992; Gatrell et al., 1996). Therefore, an alternative method is to use pixel data that classifies data in the form of cells, the size of these cells being linked to the resolution or detail required and the needs of the studied area (Benz et al., 2004; Gatrell et al., 1996; Im et al., 2008). By using pixel data, the whole area of study, which contains different parameters, will be taken into consideration in the analysis performed (Im et al., 2008; Store & Jokimäki, 2003).

LIMITATIONS OF GEOGRAPHIC TECHNOLOGY IN HABITAT MODELLING TECHNOLOGY

Advanced geographical technologies, such as GIS and remote sensing for habitat modelling in the mangrove area of the Larut Matang Mangrove Forest Reserve, can be applied to other areas of similar size or acreage to obtain more accurate results. However, such methodologies must be modified for larger areas, as the pixel size and scale must adapted to the areal coverage, parameters and avian population densiies to be considered for the habitat modelling. The study of the Larut Matang Mangrove Forest Reserve used the pixel-based approach, which is only appropriate for studies with pixel data that can be modified easily. Point data are not suitable for this model, because the use of point data requires the influence of each parameter to be considered within the scope of the area studied. For this reason, the influence of each parameter under study cannot be fully shown.

Some field data derived from either field or secondary sources are unsuited for modelling, as elevations of river banks indistinct. Supporting data such as Shuttle Radar Topography Mission (SRTM) are useful in such cases. SRTM data show the altitude of the study area because they provide three-dimensional data, with x, y and z values, and can be used with remote sensing and GIS applications in habitat modelling (Dowding et al., 2004). SRTM data have large resolution values (like many image based data sets) with errors up to 90 m, but they are still suitable for studies involving large scale areas (Gorokhovich & Voustianiouk, 2006; Siart et al., 2009). Modelling may also require data with certain formats and structures, such as numerical data. This may limit the utility of studies involving many data attributes such as land use data, land coverage, food sources, and so forth. In such cases, the form of data attributes must be changed into detailed numerical forms for suitable analyses.

CONCLUSION

In this review, the importance of the avian migratory habitat has been discussed, particularly in reference to mangrove swamps, where the main issues are avian foraging and resting in preparations for migration resumption. Therefore, the conservation and preservation of migratory avian habitats is justified to support the presence of these migratory birds. The construction of habitat models is a step towards ensuring the success of such research and programmes. The use of geographical technology-based approaches such as geographic information systems (GIS) and remote sensing can produce suitable habitat models, which are very useful for identifying the areas of high avian migratory presence during the migration season, and avian habitat requirements. Moreover, modelling the habitat for migratory birds also helps to predict avian presence, particularly in cases where the original habitat is destroyed by natural or human factors.

Relevant agencies and organizations may utilize the results of habitat modelling to monitor, control and conserve the migratory bird species and their habitats. Such conservation policies have increased in importance, partly due to their value as tourist visitation sites. Information retrieved from the analyses performed through these models can be the best reference sources for the preparation of development plans for the areas outside the mangrove forests, with important economic implications for the local human populations. In addition, it is also important for plant and animal species, which depend on mangrove forests for sustenance. These results and positions make important contributions to the fields of geomatics and conservation biology.

ACKNOWLEDGMENTS

The authors wish to express their heartiest appreciation to Universiti Sains Malaysia (USM), Penang, Malaysia, for providing the short-term research grant (304/

PHUMANITI/6313187). We would also wish to acknowledge H. P. S Abdul Khalil, who helped us through his invaluable advice and guidance in the writing of this article.

REFERENCES

Ahmad, S. (2009). Recreational values of mangrove forest in Larut Matang, Perak. *Journal of Tropical Forest Science*, *21*(2), 81–87.

Alerstam, T. & Hedenström, A. (1998). The development of bird migration Theory. *Journal of Avian Biology Optimal Migration*, *29*(4), 343-369.

Alerstam, T. & Christie, D. A. (1993). *Bird Migration*. London: Cambridge University Press.

Allen, A. P. & O'Connor, R. J. (2000). Interactive effects of land use and other factors on regional bird distributions. *Journal of Biogeography*, *27*, 889–900.

Alongi, D. M. (2002). Present state and future of the world's mangrove forests. *Environmental Conservation*, *29*(3), 331–349.

Alongi, D. M., Sasekumarb, A., Tirendia, F. & Dixona, P. (1998). The influence of stand age on benthic decomposition and recycling of organic matter in managed mangrove forests of Malaysia. *Journal of Experimental Marine Biology and Ecology*, *225* (2), 197–218.

Andries, A. M., Gulinck, H. & Herremans, M. (1994). Spatial modelling of the barn owl *Tyto alba* habitat using landscape characteristics derived from Spot data. *Ecography*, *17* (3), 278–287.

Anselin, L. (2003). *An Introduction to Point Pattern Analysis using CrimeStat. Spatial Analysis Laboratory. Department of Agricultural and Consumer Economics*. Urbana-Champaign; University of Illinois.

Anselin, L. & Getis, A. (1992). Spatial statistical analysis and geographic information systems. *Annals of Regional Science*, *26*, 19-33.

Appleton, K., Lovett, A., Sünnenberg, G. & Dockerty, T. (2002). Rural landscape visualization from GIS databases: A comparison of approaches, options and problems. *Computers, Environment and Urban Systems.*, *26*, 141–162.

Atauri, J. A. & de Lucio, J. V. (2001). The role of landscape structure in species richness distribution of birds, amphibians, reptiles and Lepidopteron in Mediterranean Landscapes. *Landscape Ecology*, *16*, 147–159.

Bakker, K. K., Naugle, D. E. & Higgins, K. F. (2002). Incorporating landscape attributes into models for migratory grassland bird conservation. *Conservation Biology*, *16*, 1638-1646.

Banker, R. D. & Morey, R. (1986). The use of categorical variables in data envelopment analysis. *Management Sciences*, *32* (12), 1613-1627.

Barbier, E. & Cox, M. (2002). Economic and Demographic Factors Affecting Mangrove Loss in the Coastal Provinces of Thailand, 1979–1996. *AMBIO: A Journal of the Human Environment*, *31* (4), 351-357.

Berthold, P. (1991). Recent advances in studies of bird migration. *Annual Review of Ecology and Systematics*, *22*, 357-378.

Berthold, P. (1996). *Control of Bird Migration*. London: Chapman & Hall.

Berthold, P. (2001). *Bird Migration: A General Survey*. United Kingdom: Oxford University Press.

Berthold, P. & Helbig, A. (1992). The genetics of bird migration: stimulus, timing, and direction. *Ibis*, *34*, 35-40.

Bishop, J. T. (1999). *Valuing Forests: A Review of Methods and Applications in Developing Countries*. London: International Institute for Environment and Development: London.

Bonter, D. N., Gauthreaux, Jr. S. A. & Donovan, T. M. (2009). Characteristics of important stopover locations for migrating birds: Remote sensing with radar in the Great Lakes Basin. *Conservation Biology*, *23* (2), 440–448.

Brotons, L., Thuiller, W., Arau´jo, M. B. & Hirzel, A. H. (2004). Presence-absence versus presence-only modelling methods for predicting bird habitat suitability. *Ecography*, *27*, 437- 448.

Buckley, D. J. (2000). *Vector and Raster - Advantages and Disadvantages. Biographical Sketch*. Portland: Pacific Meridian Resources, Inc.

Burke, D. M. & Erica. (1998). Influence of food abundance, nest-site habitat, and forest fragmentation on breeding ovenbirds no l. *The Auk*, *115*(1), 96-104.

Buyong, T. (2006). *Spatial Statistics for Geographic Information Science*. Skudai: UTM Publisher.

Câmara, G., Souza, R. C. M., Freitas, U. M., Garrido, J. & Mitsuo Ii, F. (1996). *Spring: Integrating Remote Sensing and GIS by Object Oriented Data Modelling*. Retrieved from http://www.dpi.inpe.br/geopro/papers/spring.pdf.

Canterbury, G. E., Martin, T. E., Petit, D. R., Petit, L. J. & Bradford, D. F. (2000). Bird communication and habitat as ecological indicators of forest condition in regional monitoring. *Conservation biology*, *14*(2), 544-558.

Colwell, M. A. (2010). *Shorebird Ecology, Conservation, and Management*. Berkeley: University of California Press.

Corlett, R. T. (1993). Reproductive phenology of Hong Kong shrubland. *Journal of Tropical Ecology*, *9*(4), 501-510.

Cuervo, J. J. & Møller, A. P. (2013). Temporal variation in population size of european bird species: effects of latitude and marginality of distribution. *PLoS ONE*, *8*(10), e77654. https://doi.org/10.1371/journal.pone.0077654.

Dahdouh-Guebas, F. (2002). Remote Sensing and GIS in the Sustainable Management of Tropical Coastal Ecosystems. *Environment, Development and Sustainability*, *4* (2), 93-112.

DeFries, R. S., Asner, G. P. & Houghton, R. A. (2004). *Geophysical Monograph*, *153*. Washington, DC: American Geophysical Union.

Desholm, M. (2006). *Wind Farm Related Mortality Among Avian Migrants – A Remote Sensing Study and Model Analysis*. PhD Thesis. Dept. of Wildlife Ecology and Biodiversity, NERI, and Dept. of Population Biology, University of Copenhagen. National Environmental Research Institute, Denmark.

Desholm, M. (2009). Avian sensitivity to mortality: Prioritising migratory bird species for assessment at proposed wind farms. *Journal of Environmental Management*, *90*, 2672-9. 10.1016/j.jenvman.2009.02.005.

Dowding, S., Kuuskivi, T. & LI, X. (2004). Void fill of SRTM elevation data – principles, processes and performance. *Images to Decisions: Remote Sensing Foundations for GIS Applications*, ASPRS, Fall Conference, September 12-16, Kansas City, MO, USA.

Dunn, E. H. & Bul, W. (2002). A cross-Canada comparison of mass change in birds during migration stopover. *Wilson Bulletin*, *114*(3), 368–379.

Dupré, C. & Ehrlén, J. (2002). Habitat configuration, species traits and plant distributions. *Journal of Ecology*, *90*(5), 796-805.

Ellner, S. P., McCauley, E., Kendall, B. E., Briggs, C. J., Hosseini, P. R., Wood, S. N. & Murdoch, W. W. (2001). Habitat structure and population persistence in an experimental community. *Nature*, *412*(6846), 538-543.

Eng, C. T., Paw, J. N. & Guarin, F. Y. (1989). The environmental impact of aquaculture and the effects of pollution on coastal aquaculture development in Southeast Asia. *Marine Pollution Bulletin*, *20*(7), 335–343.

Gatrell, A. C., Bailey, T. C., Diggle, P. J. & Rowlingson, B. S. (1996). Spatial Point Pattern Analysis and Its Application in Geographical Epidemiology. *Transactions of the Institute of British Geographers*, *21*, 256–274.

Gorokhovich, Y. & Voustianiouk, A. (2006). Accuracy assessment of the processed SRTM based elevation data by CGIAR using field data from USA and Thailand and its relation to the terrain characteristics. *Remote Sensing of Environment*, *104*, 409-415.

Gottschalk, T. K., Huettmann, F. & Ehlers, M. (2005). Thirty years of analysing and modelling avian habitat relationships using satellite imagery data: a review. *International Journal of Remote Sensing*, *26*, 2631–2656.

Greenberg, R. & Marra, P. P. (2005). *Birds of Two Worlds: The Ecology and Evolution of Migration*. Baltimore: Johns Hopkins University Press.

Guisan, A. & Zimmermann, N. E. (2000). Predictive habitat distribution models in ecology. *Ecological Modelling*, *135*, 147–186.

Heikkilä, M. & Ainen, M. P. (2005). *A Texture-Based Method for Modelling the Background and Detecting Moving Objects.* Retrieved http://www.ee.oulu.fi/mvg/files/pdf/pdf_662.pdf.

Holguin, G., Vazquez, P. & Bashan, Y. (2001). The role of sediment microorganisms in the productivity, conservation, and rehabilitation of mangrove ecosystems: an overview. *Biology and Fertility of Soils*, *33*, 265–278.

Homer, C. G., Edwards, T. C., Ramsey, R. D. & Price, K. P. (1993). Use of remote sensing methods in modelling sage grouse winter habitat. *Journal of Wildlife Management*, *57*(1), 78-84.

Horne, B. V. & Wiens, J. A. (1991). Forest Bird Habitat Suitability Models and the Development of General Habitat Models. *Fish and Wildlife Research*, *8*. Retrieved from http://www.dtic.mil/dtic/tr/fulltext/u2/a322800.pdf.

Howes, J., Bakewell, D. & Yus Rusila, Noor. (2003). *Panduan Studi Burung Pantai.* Bogor: Wetlands International-Indonesia Programme.

Jensen, R. A., Wisz, M. S. & Madsen, J. (2008). Prioritizing refuge sites for migratory geese to alleviate conflicts with agriculture. *Biological Conservation*, *141*(7), 1806-1818.

Jenson, S. K. & Domingue, J. O. (1988). Extracting topographic structure from digital elevation data for geographic information system analysis. *Photogrammetric Engineering and Remote Sensing*, *54* (11), 1593-1600.

Jonzén, N., Hedenström, A., Hjort, C., Lindström, A., Lundberg, P. & Andersson, A. (2002). Climate patterns and the stochastic dynamics of migratory birds. *Oikos*, *97*, 329–336.

Kairo, J. G., Kivyatu, B. & Koedam, N. (1995). An integrated comparative approach to mangrove vegetation mapping using advanced remote sensing and GIS technologies: preliminary results. *Hydrobiologia*, *295*, 285-294.

Kathiresan, K. & Rajendran, N. (2005). Coastal mangrove forests mitigated tsunami. *Estuarine, Coastal and Shelf Science*, *65*(3), 601–606.

Keane, R. E., Burgan, R. & Wagtendonk, J. V. (2001). Mapping wildland fuels for fire management across multiple scales: integrating remote sensing, GIS, and biophysical modelling. *International Journal of Wildland Fire*, *10*, 301–319.

Kouki, J., Löfman, S., Martikainen, P., Rouvinen, S. & Uotila, A. (2001). Forest fragmentation in Fennoscandia: linking habitat requirements of wood-associated threatened species to landscape and habitat changes. *Scandinavian Journal of Forest Research*, *16*(S3), 27-37.

Larson, M. A., Thompson, F. R. III., Millspaugh, J. J., Dijak, W. D. & Shifley, S. R. (2004). Linking population viability, habitat suitability, and landscape simulation models for conservation planning. *Ecological Modelling*, *180*, 103-118.

Lennon, J. J., Greenwood, D. & Turner, J. R. G. (2000). Bird diversity and environmental gradients in Britain: a test of the species–energy hypothesis. *Journal of Animal Ecology*, *69*(4), 581–598.

Li, K. S., Guan, Y., Wang, J., Smith, G., Xu, K., Duan, L., et al. (2004). Genesis of a highly pathogenic and potentially pandemic H5N1 influenza virus in eastern Asia. *Nature*, *430*, 209–213 10.1038/nature02746.

Lum, K. (2005). *The Development and Use of GIS Database for the Study of Migratory Birds In Kuala Gula National Park*. University teknologi Malaysia.PSZ 19:16 (Pind. 1/97).

Marks, K. & Bates, P. (2000). Integration of high-resolution topographic data with floodplain flow models. *Hydrological Processes*, *14*(11-12), 2109-2122.

Martin, T. E. & Finch, D. M. (Eds.) (1995). *Ecology and Management of Neotropical Migratory Birds: A Synthesis and Review of Critical Issues*. New York: Oxford University Press.

Martin, M. E., Newman, S. D., Aber, J. D. & Congalton, R. G. (1998). Determining forest species composition using high spectral resolution remote sensing data. *Remote Sensing of the Environment*, *65*, 249–254.

McGuckin, C. P. & Brown, R. D. (1995). A landscape ecological model for wildlife enhancement of storm water management practice in urban greenways. *Landscape and Urban Planning*, *33*, 227-246.

McKee, K. L., Cahoon, D. R. & Feller, I. C. (2007). Caribbean mangroves adjust to rising sea level through biotic controls on change in soil elevation. *Global Ecology and Biogeography*, *16*(5), 545-556.

Medina, E., Cuevas, E., Popp, M. & Lugo, A. E. (1990). Soil salinity, sun exposure, and growth of Acrostichumaureum, the mangrove fern. *Botanical Gazette*, *151*(1), 41-49.

Milsom, T. P., Langton, S. D., Parkin, W. K., Peel, S., Bishop, J. D., Hart, J. D. & Moore, N. P. (2000). Habitat models of bird species' distribution: an aid to the management of coastal grazing marshes. *Journal of Applied Ecology*, *37*(5), 706–727.

Moore, F. R., Woodrey, M. S., Buler, J. J., Woltmann, S. & Simons, T. R. (2005). *Understanding the Stopover of Migratory Birds: A Scale Dependent Approach*. USDA Forest Service Gen. Tech. Rep. PSW-GTR-19.

Osborne, P. E., Alonso, J. C. & Bryant, R. G. (2001). Modelling landscape-scale habitat use using gis and remote sensing: a case study with great bustards. *Journal of Applied Ecology*, *38*, 458–471.

Oyana, T. J., Dai, D. & Scott, K. E. (2006). Spatiotemporal distributions of reported cases of the Avian Influenza H5N1 (Bird Flu) in southern China in early 2004. *Avian Diseases*, *50*(4), 508-515.

Pearson, D. L. & Carroll, S. S. (1998). Global patterns of species richness: Spatial models for conservation planning using bioindicator and precipitation data. *Conservation Biology*, *2*(4), 809-821.

Pearson, R. G. & Dawson, T. P. (2003). Predicting the impacts of climate change on the distribution of species: are bioclimate envelope models useful. *Global Ecology & Biogeography*, *12*, 361–371.

Pennington, D., Hansel, J. & Blair, R. (2008). The conservation value of urban riparian areas for landbirds during spring migration: Land cover, scale, and vegetation effects. *Biological Conservation*, *141*, 1235-1248. 10.1016/j.biocon.2008.02.021.

Peterson, A. T. (2001). Predicting species' geographic distributions based on ecological niche modelling. The *Condor*, *103*(3), 599-605.

Pfister, C., Harrington, B. A. & Lavine, M. (1992). The impact of human disturbance on shorebirds at a migration staging area. *Biological Conservation*, *60*(2), 115–126.

Phillips, R. A., Silk, J. R. D., Croxall, J. P., Afanasyev, V. & Bennett, V. J. (2005). Summer distribution and migration of nonbreeding albatrosses: individual consistencies and implications for conservation. *Ecology*, *86*(9), 2386–2396.

Phillips, S. J., Anderson, R. P. & Schapire, R. E. (2006). Maximum entropy modelling of species geographic distributions. *Ecological Modelling*, *190*(3, 4), 231–259.

Philpott, S. M., Arendt, W., Armbrecht, I., Bichier, P., Dietsch, T., Gordon, C., Greenberg, R., Perfecto, I., Soto-Pinto, L., Tejada-Cruz, C., Williams, G. & Valenzuela, J. (2008). Biodiversity loss in Latin American coffee landscapes: reviewing evidence on ants, birds, and trees. *Conservation Biology*, *22*, 1093–1105.

Piersma, T. (1998). Phenotypic flexibility during migration: optimization of organ size contingent on the risks and rewards of fuelling and flight? *Journal of Avian Biology*, *29* (4), 511-520.

Piersma, T. & Lindström, A. (2004). migrating shorebirds as integrative sentinels of global environmental change. *Ibis*, *146*(1), 61–69.

Pulido, F., Berthold, P., Mohr, G. & Querner, U. (2001). Heritability of the timing of autumn migration in a natural bird population. *Proceedings of the Royal Society B* *268*(1470), 953-959.

Putz, F. E., Blate, G. M., Redford, K. H., Fimbel, R. & Robinson, J. (2001). Tropical forest management and conservation of biodiversity: an overview. *Conservation Biology*, *15(*1), 7-20.

Robertson, A. I. (1992). *Tropical Mangrove Ecosystems*. Washington, DC: American Geophysical Union.

Rosenfeld, J., Porter, M. & Parkinson, E. (2000). Habitat factors affecting the abundance and distribution of juvenile cutthroat trout (Oncorhynchus clarki) and coho salmon (Oncorhynchus kisutch). *Canadian Journal of Fisheries and Aquatic Sciences*, *57*, 766–774.

Rotenberry, J. T. (1985). The role of habitat in avian community composition: physiognomy or floristics. *Oecologia*, *67*, 213-217.

Schaub, M. & Jenni, L. (2000). Fuel deposition of three passerine bird species along the migration route. *Oecologia*, *122*, 306–317.

Schaub, M., Pradel, R., Jenni, L. & Lebreton, J. D. (2001). Migrating birds stop over longer than usually thought: an improved capture-recapture analysis. *Ecology*, *82*(3), 852-859.

Shao, G. & Wu, J. (2008). On the accuracy of landscape pattern analysis using remote sensing data. *Landscape Ecology*, *23*(5), 505-511.

Siart, C., Bubenzer, O. & Eitel, B. (2009). Combining digital elevation data (SRTM/ASTER), high resolution satellite imagery (Quickbird) and GIS for geomorphological mapping: a multi-component case study on Mediterranean karst in Central Crete. *Geomorphology*, *112*, 106 -121.

Smith, B. & Sandwell, D. (2003). Accuracy and Resolution of Shuttle Radar Topography Mission Data. *Geophysical Research Letters*, *30*(9).

Smith, T. E. (2003). *Spatial Point Pattern Analysis. Notebook for Spatial Data Analysis.* ESE 502. Retrieved from https://www.seas.upenn.edu/~ese502/.

Stockwell, D. R. B. & Peterson, A. T. (2002). Controlling bias in biodiversity data. In Scott, J. M., Heglund, P. J., Morrison, M., Haufler, J. B., Raphael, M. G., Wall, W. A. & Samson, F. B. (Eds.), *Predicting Species Occurrences: Issues of Scale and Accuracy*, 537-546. Washington, DC: Island Press.

Store, R. & Jokimäki, J. (2003). A GIS-based multi-scale approach to habitat suitability modelling. *Ecological Modelling*, *169*(1), 1-15.

Sweitzer, J., Langaas, S. & Folke, C. (1996). Land use and population density in the Baltic Sea drainage basin: A GIS database. *Ambio*, *25*, 191–19.

Tankersley, R. Jr. & Orvis, K. (2003). Modelling the geography of migratory pathways and stopover habitats for neotropical migratory birds. *Ecology and Society*, *7*(1). Retrieved from https://www.ecologyandsociety.org/vol7/iss1/art7/main.html.

Tian, B., Zhou, Y., Zhang, L. & Yuan, L. (2008). Analyzing the habitat suitability for migratory birds at the Chongming Dongtan Nature Reserve in Shanghai, China. *Estuarine, Coastal and Shelf Science*, *80*, 296-302. 10.1016/j.ecss.2008.08.014.

Tuttle, B. T., Anderson, S. J., Sutton, P. C., Elvidge, C. D. & Baugh, K. (2013). It used to be dark here: Geolocation calibration of the Defense Meteorological Satellite Program Operational Linescan System. *Photogrammetric Engineering and Remote Sensing*, *79*, 287–297. doi: 10.14358/PERS.79.3.287.

Valiela, I., Bowen, J. L. & York, J. K. (2001). Mangrove forests: one of the world's threatened major tropical environments. *BioScience*, *51*(10), 807-815.

Wade, S. & Hickey, R. (2007). Mapping migratory wading bird feeding habitats using satellite imagery and field data, Eighty-Mile Beach, Western Australia. *Journal of Coastal Research*, 759-770.

Weng, K. C., Foley, D. G., Ganong, J. E., Perle, C., Shillinger, G. L. & Block, B. A. (2008). Migration of an upper trophic level predator, the salmon shark Lamna ditropis, between distant ecoregions. *Marine Ecology Progress Series*, *372*, 253–264.

Wiens, J. A. (1989). *The Ecology of Bird Communities.*, Vol. *2*, Processes and Variations. Cambridge: Cambridge University Press.

Wilson, R. R. & Twedt, D. J. (2003). Spring Bird Migration in Mississippi Alluvial Valley Forests. *American Midland Naturalist*, *149*(1), 163-175.

Yasue, M. (2006). Environmental factors and spatial scale influence shorebirds' responses to human disturbance. *Biological Conservation*, *128*, 47-54.

Yen, P., Ziegler, S., Huettmann, F., Onyeahialam, A. I. (2005). Change detection of forest and habitat resources from 1973 to 2001 in Bach Ma National Park, Vietnam, using remote sensing imagery. *International Forestry Review*, *7*(1), 1-8.

Zharikov, Y., Skilleter, G., Loneragan, N., Taranto, T. & Cameron, B. (2005). Mapping and characterising subtropical estuarine landscapes using aerial photography and GIS for potential application in wildlife conservation and management. *Biological Conservation.*, *125*, 87-100. 10.1016/j.biocon.2005.03.016.

In: Geomatics and Conservation Biology
Editor: Michael O'Neal Campbell
ISBN: 978-1-53614-468-0

Chapter 9

PROBLEMS OF VEGETATION IDENTIFICATION USING HISTORICAL AERIAL PHOTOGRAPHS IN WEST AFRICA

Michael O'Neal Campbell
Camosun College, Victoria, Canada

ABSTRACT

In the assessment of historic vegetation change, using time series images, older forms of data such as black and white, coarse-grained aerial photographs are frequently the only data available. Attempts at feature identification, and the production of maps illustrating landcover change must therefore use visual analysis of such photographs, in some cases with stereoscopic equipment. Useful parameters for such analysis include the visual elements of tone, shape, size, pattern, texture, shadow, and association. This is problematic for complex landcover, such as the tropical savanna, one of the dominant vegetation types in globally, especially in Africa. Savanna vegetation is problematic because there is no clear definition of what constitutes savanna vegetation; many scholars however agree that savanna stands between grassland and forest in vegetation size and density. Problems encountered in the assessment of vegetation in savanna include the difficulty in identifying which stands comprise permanent shrubland or stunted tree growths, natural grassland or fallow growths from recent cultivation, natural savanna configurations or derived savanna after human induced deforestation, or landcover recovering from bushfires. For historical studies, for example the assessment of landcover change over the past 50 years, aerial photographs, field data and social surveys may be the only information sources, as the more advanced techniques (fine resolution satellite images, RADAR, LIDAR, and GPS systems) were unavailable in the past. This chapter takes the example of the coastal savanna of Ghana, a savanna zone of disputed origin and classification (in some literature, it is classified as thicket rather than savanna) that has arguably undergone marked, multidirectional change, with overall deforestation over the past century. It examines the evidence for such change, which was derived from black and white aerial photographs at the scale of 1: 10,000 flown in September 1960 and August 1986 (towards the end of the second rainy season in the case study of the coastal savanna of Ghana). It is concluded that these older methods offer a solid information background, which may complement the modern techniques that provide more detailed assessments of problematic contexts.

INTRODUCTION

Tropical savanna is a biome which is defined as containing varying proportions of three types of vegetation; grasses, shrubs and trees. Tropical savannas exist in sub-Saharan Africa, northern South America and south and southeast Asia (Campbell, 2015, 2017a,b). Such savannas cover about half the African continent and one fifth of the land surface of the world (Scholes & Walker, 1993). Wetter savannas, usually nearer forest zones have more trees, while drier savannas, usually nearer deserts, have more grasses. In terms of climates, savannas are usually located in semiarid to semi-humid climate regions of subtropical and tropical latitudes. Some argue that the average annual rainfall is usually between 750 millimeters and 1,270 millimeters, with temperatures above 18°C (Campbell, 2015). Others argue for a wider range of annual rainfall totals, from 2000 to 250 millimeters, these being intermediate between moist forest and desert (Harris, 1980; Stott, 1994). The WWF (2018) defines the tropical and subtropical grassland, savanna, and shrubland biome as "characterized by rainfall levels between 90-150 centimeters per year." Under the classification of the World Wide Fund for Nature (WWF), there are 14 terrestrial biomes, where a biome is defined as a region of generalizable climatic conditions containing associated communities of animals and plants (Olson et al., 2001). The most relevant for savanna vegetation are Biome 7, Tropical and subtropical grasslands, savannas, and shrublands; Biome 2, Tropical and subtropical dry broadleaf forests (the biome one step wetter than the tropical savanna); and Biome 13, the deserts (the biome one step drier than the tropical savanna) (Campbell, 2015).

There is no definitive agreement about what constitutes savanna vegetation (Medina & Huber, 1994), because savannas are highly heterogeneous (Solbrig, 1993). Harris (1980, p. 3) points out that "the word savanna has had an ambiguous career since it entered the English language in the 16th century." As noted by Hopkins (1992, p. 21) this is "an immense subject". Nevertheless, Stott (1994, pp. 286-287) points out that further attempts to understand these contexts are worthwhile, because "the majority of people in the tropics, over one-fifth of the world's population, live in savannas" and such studies may contribute to "the long-term survival of so many people, who are already threatened with landscape degradation, drought, famine and disease." Scholes and Walker (1993, p. 3) argue that "the central concept - a tropical mixed tree-grass community- is widely accepted, but the delimitation of the boundaries has always been a problem". Therefore, they envisage the savanna as representing a "continuum of vegetation types which have trees and grasses as their main constituents", while "acknowledging that at the extremes, the distinction between savannas and woodlands is unavoidably arbitrary" (Scholes & Walker, 1993).

To attempt a concise definition, Scholes and Walker (1993, p. 4) define the savanna as a "tropical vegetation type in which ecological processes, such as primary production, hydrology and nutrient cycling, are strongly influenced by both woody plants and grasses, and only weakly influenced by plants of other growth forms", a definition based on that which emerged from the International Responses of Savanna to Stresses and Disturbance (RSSD) Programme (cited in Frost et al. 1986, in Scholes and Walker 1993, p. 4). Sankaran & Ratnam (2013, p. 58) define the savanna as "a mixed tree and grass system, with trees ranging from sparse to medium density, with an understory of grasses. Savannas are open, well-lit systems. Rainfall usually occurs in the warmer, summer months with a dry period of 2 - 8 months. Fires are typical across savannas during drier months and occur at intervals from 1 to 50 years." Solbrig and Young (1993, p. 3) add that savanna "designates tropical grassland with scattered trees" and is the "most common tropical landscape unit", a point supported by Werner et al. (1990, p. 343) who argue that savannas are "delineated both structurally (specifying a woody:grass composition) and climatically (seasonality of water availability)", and also by Menaut et al. (1990, p. 471) who note that "the structure and dynamics of savanna vegetation is generally expressed as the balance between trees and grasses, and more easily by tree density and distribution".

Because of these complexities, it has become an intriguing research topic to investigate long term change in savanna contexts. This, as noted by Werner (1990, p. 401), is of "intrinsic, heuristic value", but a balance must be struck between general models and contextual variations (Belsky, 1990). The key issue is to look at the long-term change in the savanna vegetation, to determine if it appears to be in a slow transition from a dry forest (possibly due to either climate change or human action or both) or it has always been savanna, at least for the last few decades. Over-generalizations, subjective definitions of scale and insufficiently detailed research are serious problems in the research on these topics (Campbell, 1998). Burrows (1990, p. 486), stressing the difficulties of field research, adds that "a lack of empirical data on dynamic processes in vegetation taken from actual censuses over time and long-term experiments in stands of vegetation has been a major hindrance to progress". To test such hypotheses there must be supportive field studies (Scoones et al., 1993).

A key problem concerns the fact that the complexity and heterogeneity of savanna vegetation has created several theoretical controversies, one being the tendency to see savanna vegetation as the product of forest degradation, although their complexity and spatial prevalence in variable climatic contexts merits an examination of their uniqueness from a broader and more objective perspective. Stott (1994, p. 290) notes that "the 'core' savannas are ecosystems in their own right..." not "simply secondary forms derived from forest communities", and there is need for research into their "peculiar ecologies and ethno-ecologies." Scholes and Walker (1993, p. 10) support this point by arguing that a "common misconception is that savannas are transitional between grasslands and forests. Even within grass and shrublands, Menaut (1983, p. 134) argues that it is difficult to distinguish "tree" savanna from "shrub" savanna, because the distinction between trees and shrubs in unclear. Trees may be defined as plants with a distinct bole or trunk exceeding seven to eight meters in height, while shrubs may be shorter, low branched with short, "contorted" stems, although such distinctions may be "unrealistic", as under varying conditions, tree/shrub definitions may overlap (ibid). An important point that Menaut (1983) does not mention is the role played by firewood cutting in the creation of low stemmed and branched trees that pass for shrubs: a good example being the 'stunted' neem trees (*Azadirachta indica*) that dominate certain 'thicket' contexts of the coastal savanna of Ghana (Campbell, 1998).

The case study in this chapter is based on the vegetation type classified by some as savanna, and by others as scrub, shrubland or coastal thicket (Campbell, 1998). Therefore, a relevant point raised by Scholes and Walker (1993, p. 9) is whether the vegetation type referred to as tropical thicket (such as that in the current research) is a type of savanna, as this is a "vegetation formation which does not fit comfortably into the savanna definition." Tropical thicket is described as vegetation dominated by multi-stemmed woody plants (in the current research most multi-stemmed woody plants owe their form to firewood cutting), and is some cases with no grasses, supporting the contention that they are "disqualified as savannas" (ibid). Scholes and Walker (1993, p. 9) however, acknowledge that due to the dynamism of compositional forms, which often sees savannas becoming thickets and vice versa, "in the absence of better understanding of the ecology of thickets, we treat them as a class of savannas rather than as miniature dry forests". This position is supported by Cole (1986, p. 16), who includes "thicket and shrub" in the vegetation types classified as savanna, and also by Menaut (1983, p. 128) who noted that thickets, which may be found in "a great diversity of environments, close to the rainforest as well as near the desert edge", are common aspects of coastal savannas.

The Coastal Savanna of Ghana

Although described by some as scrubland, shrubland or coastal thicket, in this chapter the coastal savanna will be described as savanna (Campbell, 2013). The location of this area is seen in Figure 1. Historical records record more forest in the coastal savanna than at present. For example, Joseph Dupuis (1824), described the coastal zone as "an entanglement of lofty bush, blended with tall trees." Thompson (1910, p. 33) described the coastal plain as composed of savanna forest, with more trees than the northern savanna. There were evidently local variations, as Macdonald (1898) described the plains behind Accra as "rather thinly wooded" and of a "bushy and scrubby nature" while the coast was "fringed for its entire length by enormous groves of the coconut palm, separated at intervals by clumps of low bushes and scrub." Irvine (1947, p. 24) noted that at a coastal village in eastern Accra, the fishermen reported oral traditions of the eighteenth century, when the forest was closer to the sea, so the people were able to make their own boats from trees. Later, forests retreated inland, so boat manufacture followed.

Taylor (1952, p. 11) writes that the coastal savanna was once a form of dry forest termed the Antiaris-Chlorophora association, the driest of the forest types of Ghana, hence part of the forest-savanna transition zone. Dickson and Benneh (1988, p. 33) note that "reports from the early European visitors to the coast indicate that the area used to carry a more luxuriant vegetation, probably a drier and more open variety of the moist deciduous forest." Sanford and Ischei (1986) cited in Lawson (1986, p. 127) concur, as do Hall and Swaine (1981) who classify the western portion of the coastal thicket (sometimes termed the Winneba plains) as Southern Marginal dry forest, possibly its original fire protected form. Forest stands within the savanna are sometimes termed relict forest, composed of trees still found in the inland forests, such as *Celtis zenkeri*, *Chlorophora excelsea*, *Cola gigantea glabresens*, *Entandrophragma angolense*, *Hildegardia barteri*, *Khaya grandifolia*, *Mansonia altissima*, *Nesogardonia paparentera* and *Triplochiton scleroxylon* (Nielson 1965, p. 70). Sacred groves, areas not farmed by traditional edict may be representative of the previous vegetation, thicker than nearby exploited areas. These issues were apparent in the selected study area of the Tuba-Aplaku plains in 1986. Figure 1 shows the vegetation distribution of 1986.

Traditionally protected sacred groves are held as evidence of past, denser vegetation. For example, Booth (1956, p. 124) argues that "forest of a very dry type which may formerly have been typical of the whole area is found in scattered, mainly juju (sacred) spots." Hall and Swaine (1981, p. 79) observed that "most surviving specimens of the heavily cultivated southern marginal forest lie in sacred groves." Becko (1992, p. 32) notes that the "sacred sacred groves or traditional conservation areas" occur throughout the ecological zones of Ghana, and the habitats thus conserved are in "their near pristine condition... the plant communities are primary climax communities which are still in their earliest condition." The open, savanna vegetation is believed to be due to "centuries of farming resulting in the impoverishment of the soil" (Lane (1962, p. 167), with the dry climate playing a secondary role. Taylor (1952, p. 7) writes "man has been the cause of much damage to the forest of the Antiaris Chlorophora Association." Furthermore, fire protection sometimes enables reforestation (Lane, 1962, p. 167). Lawson (1986, p. 54) also supports the forest fire hypothesis: "were it not for the fire factor, most of this area would probably be a dry forest and there are in fact occasional patches of relic forest."

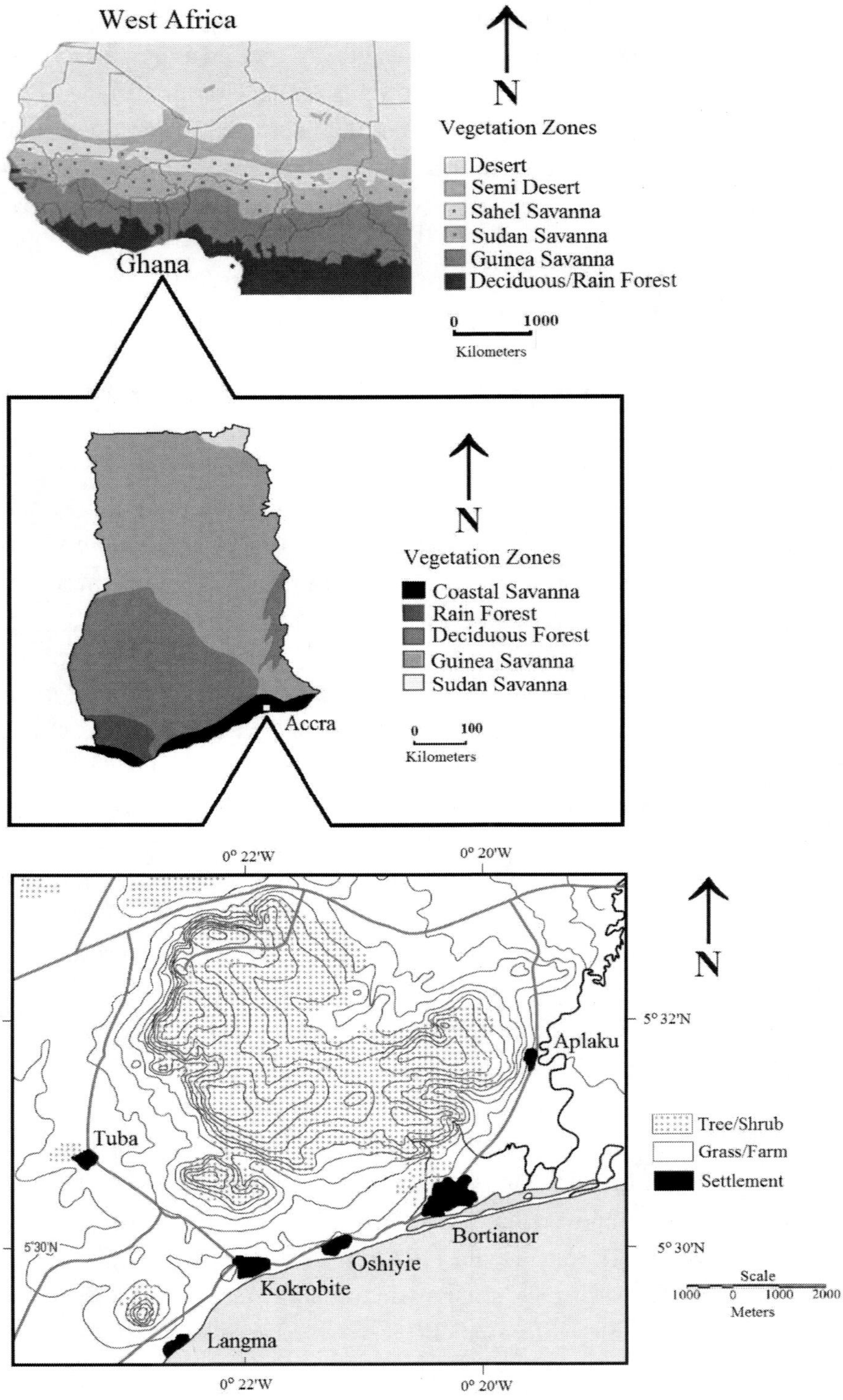

Figure 1. The Tuba-Aplaku plains, the study area of this chapter. Source: Campbell (1998, 2017b).

Rose Innes (1977, p. 6) argues that the vegetation form is due, not only to natural factors, but to "massive disturbance by man in the form of fire, overgrazing, cutting and cultivation, which is now occurring on an ever, increasing scale." Rose Innes (1977, p. 6) documents the typical vegetation succession after cultivation, which is the "commonest and most drastic form of disturbance." This passes through "several over lapping stages in its progress back to the fire pro-climax." The main stages are: (1) the "coloniser stage"; broad leaved, non- grassy weeds spread over the soil, lasting "a season or two"; (2) the "annual short grass stage," when short lived annuals and a few perennials predominate for one to two years; (3) the "mid grass stage," with mostly perennials, lasting for several years if "disturbed by trampling and grazing," typified by *Sporobolus pyramdois*, Heteropogon contortus and Rhynchelytum repens; (4) the "tall grass stage," when tall grasses such as *Andropogon* sp. and *Panicum* sp. are common (also be called the pre - thicket stage, especially as it tends to occur on the edges of thickets); (5) the "fire pro-climax stage," when thick scrub predominates; and finally (6) when in the absence of fire or other disturbance, there is eventually a "dry forest stage" (ibid.).

THE AERIAL PHOTOGRAPH METHODOLOGY

The main parameter for landcover assessment in the study area was vegetation type, which would first be classified through the field survey and ground-truthing, and then measured for area using aerial photographs and then GIS based analyses. The analysis was based on black and white aerial photographs at the scale of 1:10,000 flown in September 1960 and August 1986 (towards the end of the second rainy season). The aerial photographs were presented as a comparison between the two sets of photographs (1960 and 1986) to determine the extent of vegetation change over this period. The choice was based on the availability of photographs from the Survey Department in Accra, Ghana. These data were complemented by a field survey using quadrats. Figure 2 shows the coverage of the aerial photographs in the study area.

The hypothesis of the study was that

> Aerial photographs enable the identification of changes in visible representations of plant structures (crown height, position, perimeter canopy dynamics and open spots, etc. assessed by the parameters of tone, shape, texture, size and pattern of the features) over time giving powerful impressions of phytogeographical change (Elliott & Campbell, 2001, 223).

The vegetation was classified into four classes, trees, shrubs, grasses and crops, which were determined by both the field survey and the appearance of the features in the aerial photographs. The details of the classification of these vegetation types is shown in Tables 3a and 3b.

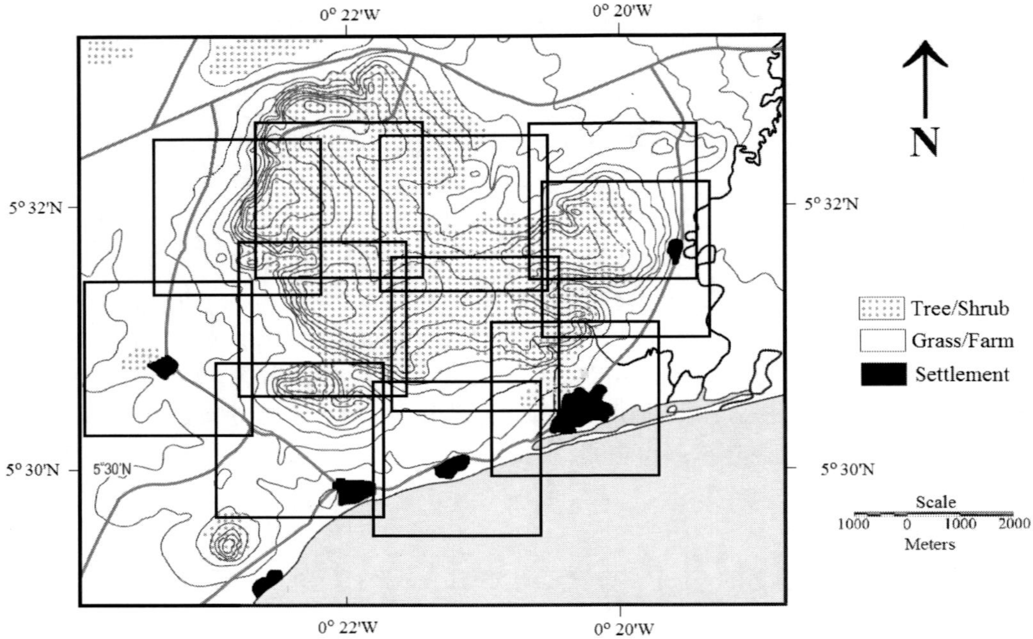

Source: Campbell (1998).

Figure 2. The positions of the aerial photographs in the study area of the Tuba-Aplaku plains.

The distinction between trees and shrubs is an important one, as the distinction between these two forms is difficult in landscapes were there are several intermediate forms of vegetation such as young, shrub-sized trees, trees stunted by branch-cutting for firewood, larger shrubs, and trees without a distinct bole (trunk), but low branches with the cutting of boles for various uses. Despite the common definition cited above, where a tree is defined as plants with a distinct bole or trunk exceeding seven to eight meters in height, and shrubs as shorter with "contorted" stems (Menaut (1983, p.134),, this author nevertheless argues that the distinction between "tree" savanna and "shrub" savanna can still be tedious. In the case of the current research (Campbell, 1998), firewood cutting contributed resulted in low stemmed and branched trees, usually neem (*Azadirachta indica*) that pass for shrubs and dominate certain 'thicket' contexts of the coastal savanna of Ghana.

The information derived from the aerial photographs, depicted in Table 3b, was complemented by the field research in 1995-1996 and the parameters were matched to derive a synthesis. This synthesis constructed four main landcover types (Campbell, 1998). Firstly, there were grass dominated areas. The main species were *Andropogon guyanus* L., *Sporobulus pyramidalis* L. and *Panicum maximum* L. Few shrubs occurred in these areas, usually *Azadirachta indica* A. Juss and *Fagara xanthoxyloides* L. Intensive farming is combined with intensive firewood cutting served to maintain this vegetation association. The few tree species, were mostly shrub-sized. Larger trees were tolerated for firewood or fruits (e.g., *Mangifera indica* L.). Secondly, there were shrub/grass mosaics, with shrub and shrub sized tree species in scattered clumps or individual specimens in grass dominated areas. The main tree and shrub species were *Azadirachta indica*, *Fagara xanthoxyloides*, *Baphia nitida* Lodd. and *Securinega virosa* (Roxb.ex Willd) Baillon. The main grasses were those mentioned above. This vegetation type existed mainly in the short fallow areas, where farming ceases either seasonally or for up to two years.

Table 3a. Vegetation identification in aerial photographs and in the field

Vegetation Classification	Field Identification	Aerial Photography Identification	Problems
Grasses	Long bladed leaves, 0.1 - 2m	Fine grained texture	Like ground texture/some crops
Shrubs	Branched stem, < 7m	Coarse grained texture, small canopies, < 3m diameter	Confusion between shrubs/stunted trees
Trees	Distinct trunk > 7 m	Larger canopies, distinct crown, > 3m diameter	Confusion from trees with missing branches
Crops	As grass/shrubs, arranged in rows	Like grass/shrubs, but planted in patterns e.g., rows	Confusion between crops/grass/shrubs

Table 3b. Details of photograph interpretation (Scale 1:10000)

Parameter	Grasses	Shrubs	Trees	Crops	Bare soil/stones
Location	Lowlands, slopes	Highlands, slopes, rivers	Highlands, slopes, rivers	Flatland, lowland	Flatland, steep slopes, rivers
Size	<0.5m	0.5-2m	2 - 10 m	< 0.5	< 0.1m
Shape	Irregular areas	Irregular patches	Large, irregular patches	Rows, squares	Large, irregular contiguous patches
Shadow	Very small	Small	Large	Linear	Slopes
Tone/color	Light grey	Medium grey	Medium grey, darker contours	Medium grey	Grey-white
Texture	Coarse, light grey	Coarse, round clumps	Large, rounded clumps	Coarse	Fine, grey-white
Pattern	Even	Clustered	Open canopy	Linear, patches	Patches
Height	< 0.5m	< 5m	> 5m	< 2m	Flat
Association	Flatland, rivers	Flatland, trees, grass	Highlands	Rivers, shrubs	Grass, rivers

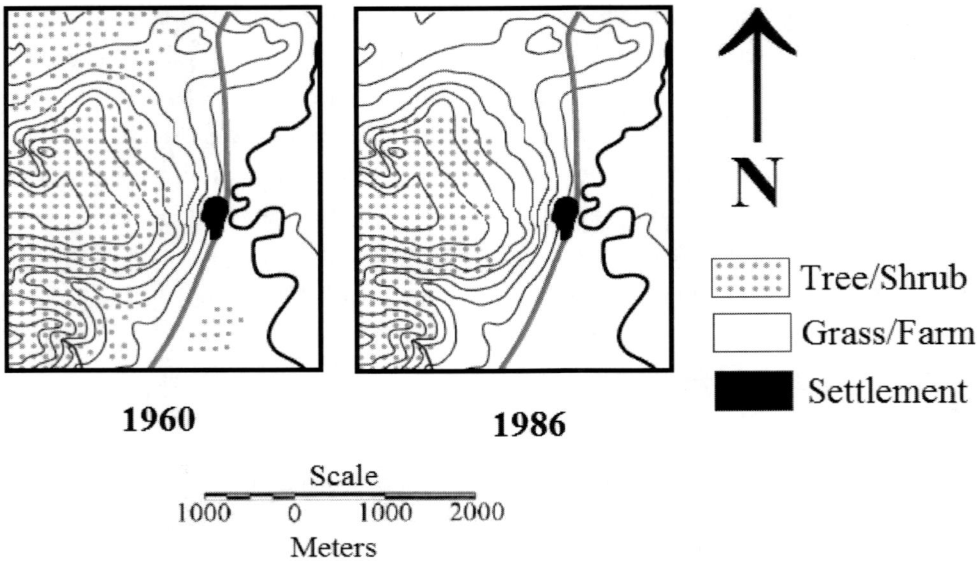

Figure 3. Landcover Change in the environs of Aplaku. There were major losses of trees and shrubs. This set of photographs showed an almost total conversion from dense shrubs and trees to shrub patches and grassland between 1960 and 1986. Some areas have tree losses, but in more local areas than the hinterland. In most situations dense tree stands have been replaced by farmland or fallow shrub/grass mosaics.

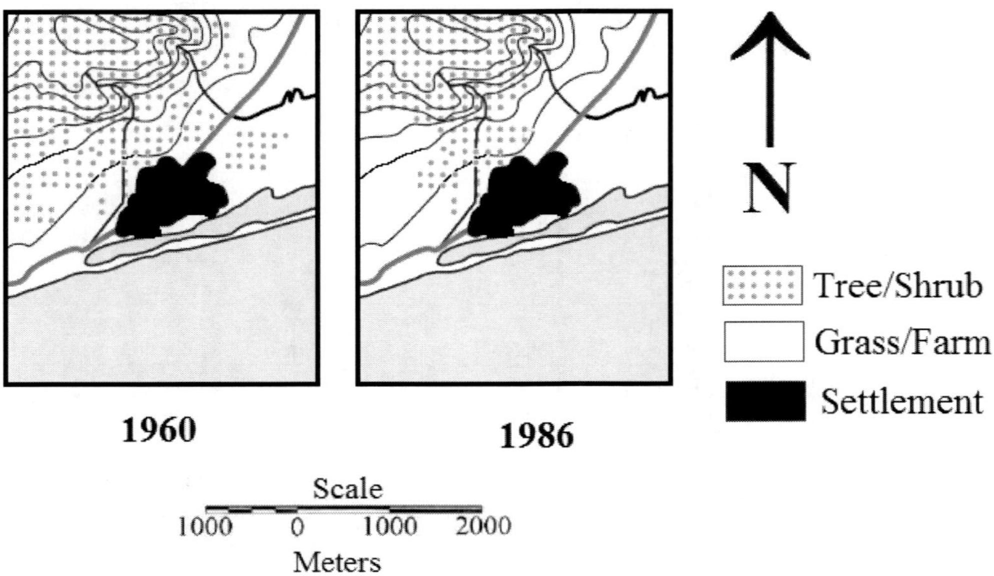

Figure 4. Landcover Change in the environs of Bortianor. There were major losses of trees and shrubs. This set of photographs showed the main change to be substantial losses of trees from the sacred groves, and tree losses in the area to the north of the village.

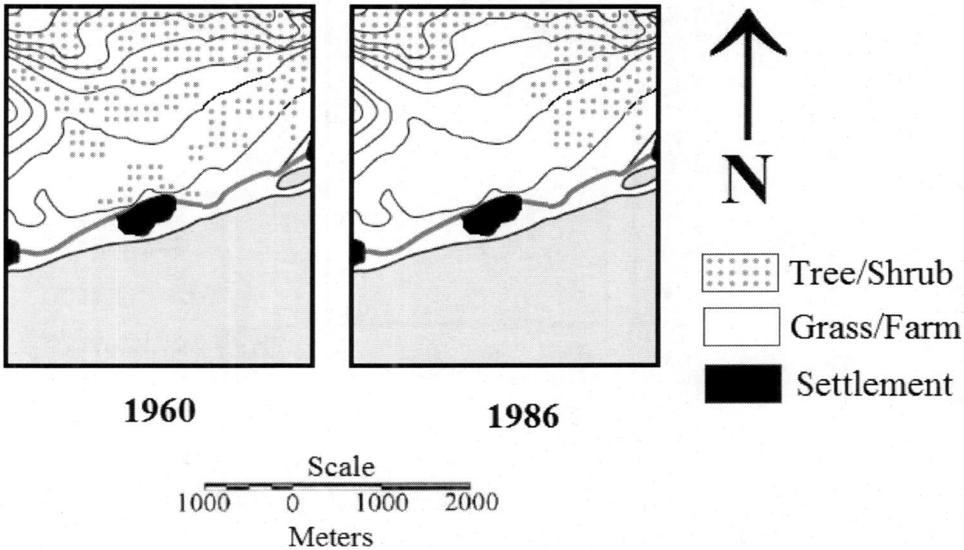

Figure 5. The environs of Oshiyie. This set of photographs showed some regrowth of long fallow areas, and a few new trees within farms and tree losses in the northern sections between 1960 and 1986. The main form of change was however still deforestation.

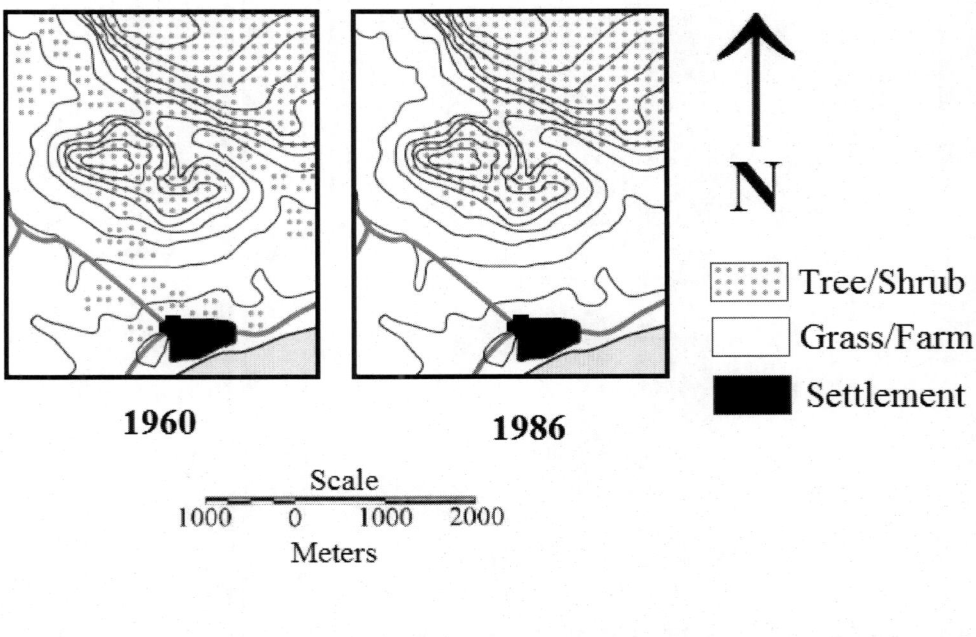

Figure 6. The environs of Kokrobite. This area shows a more dynamic change between 1960 and 1986, as farmland was replaced by long fallow vegetation, especially on the upland areas, there was tree loss in some areas and increased stands in others, and farms were more intensive on flat areas but less so on hills.

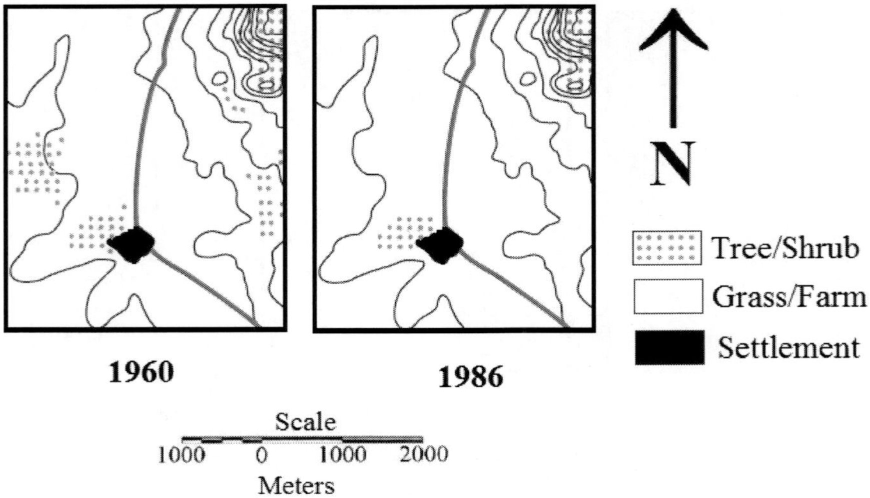

Figure 7. The environs of Tuba. There were minor tree-shrub losses.

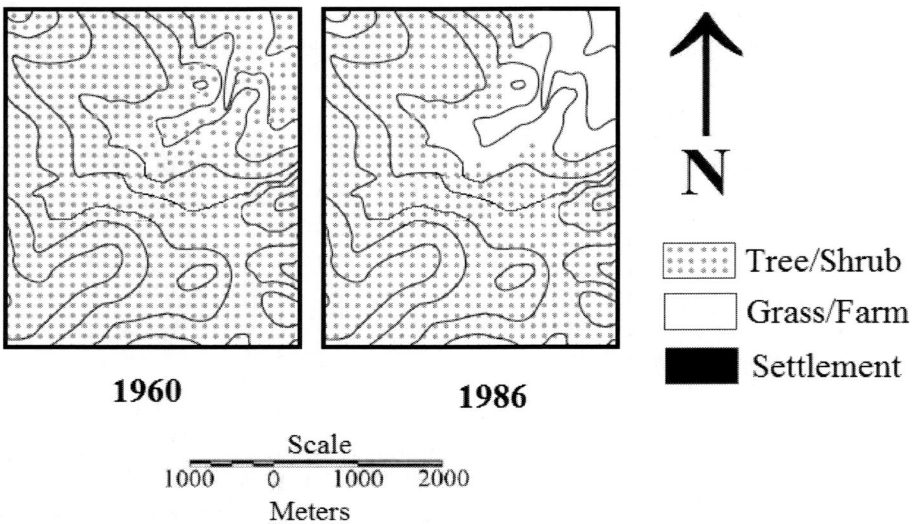

Figure 8. The Hinterland. This set of photographs showed substantial tree losses between 1960 and 1986, from a dense tree/shrub mosaic to a tree/shrub and short/long fallow mosaic.

Thirdly, there were thick shrub and shrub sized tree growths, occurring mostly in areas left fallow for more than two years. The main species were the tree and shrub species mentioned above. Tree species such as the neem (*Azadirachta indica*) were kept at shrub size (about two to three meters high) due to firewood cutting. Isolated patches of grass and occasional full-sized trees (usually *Mangifera indica*) occurred. Finally, there were forest clumps composed of full-sized trees, including sacred groves, graveyards and woodlots. Ttree species common in the inland forests occurred, including *Samarea saman* (Jacq.) Merr., *Entandrophragma angolense* C.E.C. Fish., *Cassia siamea* Lam., *Triplochiton sceleroxylon* K. Shum., *Antiaris africana* Lesch., *Sterculia tragacartha* L.,

Albizia zygia Durraz., *Ceiba pentandra* Mill., *Antiaris toxcana* Lesch., *Ficus thonningii* L. and *Diospyros mesopiliformis* L. Other species, common in local areaa outside the forest clumps (usually stunted to shrub size by firewood cutting) occurred in full size, including *Azadirahta indica*, *Mangifera indica*, *Milletia thonningi*, and *Fagara zanthoxloides*.

These classifications were examined in the field and then and used as parameters for landuse change by identifying similar landcover types in the two sets of aerial photographs. Trees were merged with shrubs for the maps produced, rather than shrubs with grass, as some 'shrubs' were stunted or young trees. Grass was merged with farm growths due to the fallow grass growths on farmland. Stereoscopic work was used determine the heights of vegetation, using heights of permanent features in the photographs and comparing these with the vegetation in the photographs, after field measurements of the permanent features during field surveys.

Other parameters included evidence of footpaths (necessary for farming and firewood cutting) and canopy size, measured by the photographic components of visible representations of plant structures (crown height, position, perimeter canopy dynamics and open spots, etc. assessed by the parameters of tone, shape, texture, size and pattern of the features). To measure these, the aerial photographs were enlarged from a scale of 1: 10,000 to 1: 5,000, 1: 4,000, 1: 2,500 and 1: 1,000. The areas covered by the aerial photographs were divided into six areas (Figure 2). These are described in Figures 3 to 8 below. In all areas there was some deforestation, with both tree and shrub losses and a few areas on reforestation.

The Details of Landcover Change

For the Aplaku area, the overall assessment was change from mostly dense shrub/tree growths to a grass dominated mosaic. In 1960 the vegetation was an almost continuous canopy of medium sized trees and shrubs, broken in places by grass/shrub mosaics to the west of the village. By 1986, there had been a vast expansion in the grassed area. Shrubland was reduced to isolated clumps, or as single plants in grass dominated areas. Farms were much more apparent, and the area was crisscrossed with footpaths. The greatest amount of grass colonization was within 1,500 meters of the western edge of Aplaku. There was little change in the shrubland on the hilly area in the south west of the village. Because of slope desiccation, this area was not heavily farmed, the result being thick shrub vegetation.

For the area surrounding Bortianor, the overall assessment was of a substantial decline in tree stands, but in most areas the shrub grass mosaic remained the dominant form of vegetation. The main tree vegetation was of sacred groves and single trees. There were two sacred groves in this area; a large forested area immediately north of Bortianor, and a smaller forest clump approximately 500 meters to the north-west. The large sacred grove was reduced in area during period under review. The main tree and shrub losses occurred on the extreme north-western, north-eastern and south-western edges of the grove. There was some expansion in the southwestern end of the grove, largely of oil palms. The smaller sacred grove was slightly reduced between 1960 and 1986. In 1960, farming and firewood cutting were already intensive. In 1986, there was an increase in farming intensity, the tree growths were replaced with farmland, short fallow grass and shrubland. The area in 1986 was crisscrossed by a dense network of footpaths. The grasslands to the south-west of the village hardly changed. In this area, farming and grass fallow contrasted sharply with the forested sacred groves. However, to the immediate north of this area, there was greater shrub density, which were more visibly part of a farm fallow mosaic in 1986 than in 1960, because there were clearer edges of farms, and linear distinctions between shrub and grass. Farming and building appeared to be the main factors for landcover change.

For the environs of Oshiyie, there was an increase in fallow shrubland, which was largely absent in 1960. Significantly, absolute numbers of shrubs increased along with the increased number of farms. There was an increase in tree growth in farmland just to the north of the village. In many farms in 1986 mostly new mango trees grew in the farmland and some trees were recent additions (after 1960). This form of dynamic shift, which resulted from the people's appraisal of trees (mango trees were more valued than other large trees) shows the importance of the social values and that shrubland can increase even with population and farm increases. There was increased shrub growth north and north-east of the village. These dense, closed canopy shrub clumps were replaced by open grassland with a few shrubs in 1986. There were also tree losses between 1960 and 1986. In 1960, Oshiyie was surrounded by dense stands of coconut trees on the seaward, western and eastern margins, and a line of large shrubs along the northeastern edge. In 1986, there was a drastic reduction in the number of coconut trees along the edges of the village, replaced by dense shrubland. The dominant vegetation for several hundred meters around the town was a shrub/grass mosaic with scattered trees, pointing to shifting cultivation with attendant long fallow areas as the main land use. There was no visible change in the area of the sacred grove. The sacred grove to the west of the village remained the same size as in 1960, which was the main remaining area of forested growth in the area.

In the environs of Kokrobite, the best description of the landscape change that it was patch-dynamic. Farming was reduced in certain areas, due to the constraining influence of increasing drought. This influenced the reduction of farmland, but there was little reforestation as the abandoned farmlands and long fallow areas were used by woodcutters. In the lowland areas, there was intensified farming, mechanization and substantial losses of shrubs, but also local expansion of shrubs. The general phytogeographical change, in terms of tree losses and grassland expansion, was not as marked as in the other areas surveyed, and like the Oshiyie area, there were areas of increased grassland and others of increased shrub density. There are two main topographical forms, and these are analogous to the two main forms of change. In the hilly upland area in the north center, where was a thickening of shrub stands and a decline of farmland. The central hilly area showed increased shrub density, because of an increase in the intensity of farming in this area. There was farm intensification and shrub loss in the lowland area to the south west of the village. Tree losses occurred in the area immediately north of Kokrobite. In this area, there was also evidence of increased tree and shrub growth to the north north-east of the village, the extreme east of the village, north of the road and in shrub land in the immediate south-west of Kokrobite. This occurred because some farmed areas were set aside for building in the 1980s, and as no farming or building took place, long fallow shrub growth developed.

For the hinterland, the vegetation in 1960 was composed entirely of dense tree stands within close canopied shrubland. The overall assessment for the hinterland area was that a closed canopy of small canopied trees and shrubs had been replaced with what may be termed a "patch dynamic" landscape. Although the general vegetation was tree stands, which in the scale of 1: 10,000 gave a canopy of about 5 to 10 meters, there were numerous smaller trees or shrubs, possibly shrubs recovering from shifting cultivation, the start of reforestation after firewood cutting, tree variation due to edaphic factors, and/or the results of intensive natural or human induced burning. There was also evidence of farming. Scattered farms were found nearer the settlements and shifting cultivation was apparent, because the farms showed substantial shrub growth within the perimeters of farms, pointing to long fallows. There was also evidence of firewood cutting, evidenced by long straight lines cut to allow firewood cutters access to the wooded areas, and to facilitate hunting.

It was difficult to classify these stands as secondary or primary forest, or as natural climax, either in form or species content. Comparisons of the canopies of the largest trees in the hinterland with those in the sacred groves showed that the former were substantially smaller than those in the latter area. Comparison with the photographs of 1986, revealed almost complete deforestation, as forested stands had been replaced by intensively cultivated landcover. The result for 1986 was a mosaic of long fallow, short fallow and farmland. The total absence of trees was apparent, even in the enlarged photographs. A network of footpaths showed the intensity of farming and firewood cutting. The western edge of the area, mostly upland areas, was visibly less farmed, due to difficultly of access.

A vegetation succession sequence after cultivation could be discerned in the 1986 photographs. The thicker vegetation represented long fallow shrub growths and thinner vegetation represented earlier stages of the succession sequence, or similar long fallow areas that have been subjected to intensive firewood cutting. The long fallow areas were generally areas that had not been farmed for more than two to three years. Because of the frequent occurrence of firewood cutting, the length of the fallow, and the reforestation possibilities were difficult to determine. In some cases, long fallow areas, which had not been farmed for up to eight years, had been subjected to such intensive cutting that they looked like short fallow areas (due largely grass invasions), and were only discernible as long fallow because in most cases roots were not dug out during wood cutting (unlike farmed areas) and the thick stumps were visible.

CONCLUSION

This chapter has demonstrated some of the methods and problems in the study of historic landcover change in complex environments such as the African savanna. Important issues were the lack of high resolution images covering periods before such technology became available; linking of features recorded in the old photographs with the objects identified in more recent field surveys; and the difficulty of vegetation classification, especially as people have modified the form of some vegetation types. Subjective classifications may be employed, for example merging trees with shrubs, rather than shrubs with grass, as some 'shrubs' may actually be stunted or young trees. Stereoscopic work may be needed to evaluate the heights of vegetation, which may be determined by comparing the heights of permanent features in the photographs with those of the vegetation, after field measurements of the features during field surveys. It may be concluded that such surveys using old images to examine historical changes offer useful starting points for studies of recent environmental change. Studies of such complex environments must however, examine details, such as human modifications of vegetation forms, which may affect assessments of landcover change.

REFERENCES

Becko, C. (1992). Studies of traditional conservation of biodiversity in Ghana. In Laing, E. (Ed.), *Proceedings of the Workshop on Biodiversity*, 32-33. Accra: Environmental Protection Council.

Belsky, A. J. (1990). Tree/grass ratios in east African savannas: A comparison of existing models. *Journal of Biogeography*, *17*, 483-489.

Booth, A. H. (1956). On the mammalian fauna of the Accra plain. *Journal of the West African Science Association*, *5*(1), 26-35.

Burrows, C. J. (1990). *Processes of Vegetation Change*. London: Unwin Hyman.

Campbell, M. (1998). *Interactions Between Biogeography and Rural Livelihoods In The Coastal Savanna Of Ghana.* PhD Thesis. University of London, Imperial College (London).

Campbell, M. (2013). *Biogeography, Environmental History and Sustainability in Coastal Ghana*. New York: Nova Science Publishers.

Campbell, M. (2015). *Vultures: Their Evolution, Ecology and Conservation.* New York: Taylor and Francis, CRC Press.

Campbell, M. (2017a). *Biological Conservation in the 21st Century: A Conservation Biology of Large Wildlife*. New York: Nova Science Publishers.

Campbell, M. (2017b). An applied biogeography of forest/savanna land cover change in Ghana. In Daniels, J. (Ed.), *Advances in Environmental Research, 61.* New York; Nova Science Publishers.

Cole, M. (1986). *The Savannas: Biogeography and Geobotany*. London: Academic Press.

Dickson, K. B. & Benneh, G. (1988). *A New Geography of Ghana*. London: Longman.

Dupuis, J. (1824). *Journal of a Residence in Ashantee*. London: H. Colburn.

Elliott, J. A. & Campbell, M. (2002). The environmental imprints and complexes of social dynamics in rural Africa: cases from Zimbabwe and Ghana. *Geoforum*, *33*(2), 221-237.

Frost, P. G. H., Menaut, J. C., Walker, B. H., Medina, E., Solbrig, O. & Swift, M. (1986). 'Responses of savannas to stress and disturbance', *Biology International Special*, Issue 10, 1 – 78.

Hall, J. B. & Swaine, M. D. (1981). *Distribution and Ecology of Vascular Plants in a Tropical Forest*. The Hague: DW Junk Publishers.

Harris, D. R. (1980). *Human Ecology in Savanna Environments*. London: Academic.

Hopkins, B. (1992) Ecological processes at the forest savanna boundary. In Furley, P. A., Proctor, J. & Ratter, J. A. (Eds.), *Nature and Dynamics of Forest-Savanna Boundaries*, 21-33. London: Chapman and Hall.

Irvine, F. R. (1947). *The Fish and Fisheries of the Gold Coast*. London: Longman.

Lane, D. A. (1962). The forest vegetation. In Wills, J. B. (Ed.), *Agriculture and Land Use in Ghana.*, 160 -169. London: Oxford University Press.

Lawson, G. (1986). *Plant Ecology on West Africa: Systems and Processes*. Chichester: John Wiley and Sons.

Macdonald, G. (1898). *The Gold Coast Past and Present: A Short Description of the Country and its People*. London: Longman, Green and Company.

Medina, E. & Huber, O. (1994). The role of biodiversity in the functioning of savanna ecosystems. In Solbrig, O. T., Van Emden, H. M. & Oordt, P. G. W. J. (Eds.) *Biodiversity and Global Change*, 141-160. Wallingford: CAB International.

Menaut, J. C. (1983). The vegetation of African savannas. In Bourliére F (Ed.) *Tropical Savannas. Ecosystems of the World*, *13*, 109–149. Amsterdam: Elsevier Scientific Publishing Company.

Menaut, J. C., Gignoux, J., Prado, C. & Clobert, J. (1990). "Tree Community Dynamics in a Humid Savanna of the Cote de Ivoire: Modelling the Effects of Fire and Competition with Grass and Neighbours. *Journal of Biogeography*, *17*, 471-481.

Nielsen, M. S. (1965). *An Introduction to the Flowering Plants of West Africa*. London: University of London Press.

Olson, D. M., Dinerstein, E., Wikramanayake, E. M., Burgess, N. D., Powell, G. V. N., Underwood, E. C., D'Amico, J. A., Itoua, I., Strand, H. E., Morrison, J. C., Loucks, C. J., Allnutt, T. F., Ricketts, T. H., Kura, Y., Lamoreux, J. F., Wettengel, W. W., Hedao, P. & Kassem, K. R. (2001). *Terrestrial Ecoregions of the World: A New Map of Life on Earth*, *BioScience*, *51*(11), 933–938.

Rose Innes, R. (1977). *A Manual of Ghana Grasses*. London: Ministry of Overseas Development.

Sanford, W. W. & Isichei, A. O. (1986). Savanna. In Lawson, G. (Ed.), *Plant Ecology in West Africa: Systems and Processes*, 119 -136. Chichester: John Wiley and Sons.

Sankaran, M. & Ratnam, J. (2013). *African and Asian Savannas. Encyclopedia of Biodiversity*, 58-74. 10.1016/B978-0-12-384719-5.00355-5.

Scholes, R. J. & Walker, B. H. (1993). *An African Savanna: Synthesis of the Nylsvley Study*. Cambridge University Press: Cambridge.

Scoones, I., Toulmin, C. & Lane, C. (1993). Land Tenure for Pastoral Communities. In Young, M. D. & Solbrig, O. T. (Eds.), *The Worlds Savannas: Economic Driving Forces, Ecological Constraints and Policy Options for Sustainable Land Use*, 49-66. Paris: Parthenon Publishing Group.

Solbrig, O. T. (1993). Ecological constraints to savanna land use. In Young, M. D. & Solbrig, O. T. (Eds.), *The World's Savannas: Economic Driving Forces, Ecological Constraints and Policy Options for Sustainable Land Use*, 21-47. Paris: The Pantheon Publishing Group.

Stott, P. A. (1994). Savanna landscapes and global environmental change. In Roberts, N., (Eds.), *The Changing Global Environment*, 287-303. Cambridge, Basil Blackwell Limited.

Taylor, C. J. (1952). *The Vegetation Zones of the Gold Coast*. Accra: Government Printing Department.

Thompson, H. N. (1910). *Gold Coast. Report on Forests*. London: British Parliamentary Papers No. 65, HMSO.

Werner, P. A. (1990). Ecological determinants of savannas: abiotic and biotic. *Journal of Biogeography*, *17*, 401-402.

World Wildlife Fund. (2018). *Tropical and Subtropical Grasslands, Savannas and Shrublands*. Retrieved from https://www.worldwildlife.org/biomes/tropical-and-subtropical-grasslands-savannas-and-shrublands.

In: Geomatics and Conservation Biology
Editor: Michael O'Neal Campbell
ISBN: 978-1-53614-468-0

Chapter 10

UNRAVELLING THE COMPLEXITY OF PROTECTED AREA MANAGEMENT IN TWO DEVELOPING COUNTRIES: ISSUES OF HUMAN DISPLACEMENT AND WILDLIFE CONFLICT

Kaushalendra Kumar Jha*[*] *and Pradeep Chaudhry
Tropical Forest Ecosystem and Environment Management Unit,
Technical Forestry Faculty Area,
Indian Institute of Forest Management, Bhopal, India

ABSTRACT

While the status of natural resources and protected areas (PAs) management now attracts widespread international attention, the practical implementation of related policies frequently falls short of expectations. This chapter emphasizes the intricate relationship between forest dwelling communities and biodiversity conservation in PAs on one hand, and the appropriate use of modern tools and technologies like geoinformatics in addressing land tenure related and human wildlife conflict issues on the other. The relocation of forest dependent communities from PAs and human wildlife conflicts in densely populated developing nations mostly keep PA managers on tenterhooks and need careful handling. India and Thailand, two similar but distinct developing Asian nations, are selected as case studies with similar problems in this vein. The PAs in these countries are in the midst of serious socio-economic and environmental issues, as the resources of the PAs may support a wide range of economic sectors. Assessment of such issues may require geomatics-based monitoring and evaluation, the better to record land use problems. Amicable and tactful handling of all such issues of PAs can be a win-win situation for park managers, local community members and other

[*] Corresponding Author Email: jhakk1959@gmail.com.

stakeholders. Improved governance and appropriate investments in these natural assets may ultimately lead directly to increased human wellbeing and indirectly maintain the ecological security, food security and sustainable development of such resources in the long run. This constitutes a contribution to conservation biology and management.

INTRODUCTION

The establishment and management of Protected Areas (PAs; mainly National parks and Wildlife sanctuaries) have received increased attention in developing countries worldwide. It is a common realization that PAs are playing a valuable role in maintaining biological diversity as well as providing immense economic benefits from tourism and other direct uses. They are the cornerstone of most conservation strategies. They protect biodiversity, safe-guard ecosystem health, and provide an array of ecosystem services, such as fresh drinking water, places for recreation and relaxation, storehouses of genetic material, and reservoirs of wild plants and animals that can contribute to species populations in surrounding areas. Hence, national governments are analysing various alternative ways of establishing and managing these areas (Dixon & Sherman, 1990). The global human population now exceeds 7.0 billion and is predicted to reach 9 billion by 2050. Mapping of the 'human footprint' on the earth has indicated that more than 80% of earth surface is directly influenced by human use (Sanderson et al., 2002). This is not commensurate with the Gandhian philosophy that the earth has everything to satisfy the genuine needs of all humans, rather than human greed. More than 200,000 designated PAs currently exist globally (Table 1), many of which are the last strongholds of the natural ecosystems of their local areas, and also have an important role in providing humankind with a range of valuable ecological services (Chape et al., 2008, UNEP-WCMC, 2014).

Table 1. Cumulative growth of the protected areas network in the world since 1962

Year	Number of Sites	Total area (km^2)
1962	9,214	2,400,000
1972	16,394	4,100,000
1982	27,794	8,800,000
1992	48,388	12,300,000
2003	102,102	18,800,000
2014	209,429	32,868,673

Source: UNEP WCMC 2014, with data from Chape et al., 2003).

However, the percentage of land covered by protected areas in Asia (mainly developing countries) is less than that of Europe, Africa, North America, South America and Central America (UNEP-WCMC, 2014). Among developing nations, domestic cattle often outnumber wild ungulates in PAs, densities reaching up to 1500 km.$^{-2}$ For example, there is an evidence that livestock graze in 73% of Wildlife Sanctuaries and 39% National Parks of India (Mishra, 1997). Livestock are therefore an important source of prey for wild predators. Situations like this are common in almost all the developing countries, making wildlife conservation difficult. In developing countries, poverty, high human density and human dependence on protected ecosystems are factors which may not exist in developed countries. By contrast, in the developed countries, for example, domestic cattle are not generally allowed to roam freely in PAs.

The continuous growth and extent of PAs, nearly 32 million km^2 globally in 2014 is a great achievement worldwide. However, the establishment of a PA is not enough to promote biodiversity conservation and achieve other objectives of PAs management (Pretty & Smith, 2004; Hayes, 2006; Ban et al., 2008). Larger social and environmental issues are also important. This chapter examines the main issues of displacement or relocation of people from PAs and human wildlife conflicts, which are causing hindrance for the various objectives (wildlife conservation, ecosystem services, elimination of conflicts etc.) of establishing PAs in two socio-politically different but developing nations, namely India and Thailand. The former has democratic governance while latter is a constitutional monarchy.

PROTECTED AREAS IN INDIA AND THAILAND

India and Thailand are located in the Indo-Malayan ecozone, but the biogeographic regions within the two countries are different due to varying physiographic and climatic conditions. India is a vast nation with an area of 3,287,469 km^2 and ranges in altitude from 0 m to 8586 m above mean sea level in Himalayas in the north (Kanchenjunga Peak). Biogeographically, the country is divided into ten different zones namely; the Trans-Himalayan Laddakh mountains, Indo-Gangetic plains, Himalayan region, the desert region, semi-arid region, Western Ghats, Deccan peninsula, north east India, Islands, and coastal region (Chauhan, 2008). India is also among one of the 17 megadiverse countries (Mittermeier & Mittermeier, 2005). Though occupying only 2.5% of world's geographical area, the country supports 16% of world's human population and 17% livestock population. The per capita availability of forest and productivity are among the lowest in the world, and the biotic pressure on country's forests is immense making biodiversity conservation a very challenging task. There are 771 PAs in India spread over an area of 162,099.47 km^2, or about 4.93% of the country's area. These PAs include various national parks, wildlife sanctuaries, conservation reserves and community reserves.

Thailand is located in the Indochina and Sundaic biogeographical subregions and adjoins several other biogeographical sub regions (TFSMP, 1992). The total land area of the country is 513,115 km^2 and ranges in altitude from 0 m to 2564 m above mean sea level in the northern Chiang Mai Province. The climate is influenced by the seasonal monsoon and local topography. Being under the influence of this monsoonal climate, the vegetation of the country consists broadly of highly diverse tropical evergreen and deciduous broadleaf forest types (Trisurat, 2007). Thailand's forest area diminished from 53.33% of the total land area in 1961 to 25.13% in 1998 (Charuppat, 1998; Lakanavichian, 2001), increasing up to 32.66% in 2003 (DNP, 2003). However, Thailand has made a significant public commitment by establishing and maintaining an extensive system of PAs. In 2004, Thai PAs consisted of 102 national parks (81 terrestrial parks and 21 marine parks), 55 wildlife sanctuaries, 55 non-hunting areas, and 69 forest parks, which covered ~18.2% of the country's land area (Trisurat, 2007). Nevertheless, Emphandhu & Chettamart (2003) recorded until the date of publication, there were 319 units in the PA system, covering 108,064 km^2 or over 21% of the country's land surface. The system comprised 145 national parks (both terrestrial and marine), 53 wildlife sanctuaries, 52 non-hunting areas and 69 forest parks.

Human population density, poverty, high rates of unemployment, dependence of people on agriculture and forests, and shifting cultivation in hills are some common features in India and Thailand, which make PA management challenging. Relocation and displacement of local communities in PAs remains one of the major issues in both countries besides human wildlife conflicts discussed in following sections.

Relocation and Displacement

Conservation of natural resources, reduction of human wild-animal conflict, provisioning of ecosystem services and other objectives warrant eviction of people from the PAs. Nonetheless, displacement of people from PAs has been a highly controversial subject, particularly in developing countries (Schmidt-Soltau & Brockington, 2007). Campaigners for wilderness preservation believe that the presence of people in a PA conflicts with their ideas of conservation (Nash, 1967; Oelschlaeger, 1991) and therefore human displacement is fundamental to wildlife conservation (Terborgh & Peres, 2002). Many ecologists maintain that hunting, habitat modification and/or over grazing by livestock reduces some wildlife like deer and other antelopes, which become more vulnerable to diseases and extinction (Bodmer et al., 1997). On the other hand, critics maintain that local residents are the best guardians of these PAs (Schwartzman et al., 2000a, b). Various non-governmental organisations, human rights activists and indigenous peoples'organizations accuse conservation organizations of evicting people from parks either forcibly or by showing false rosy picture of alternate sites (Chapin, 2004; Colchester, 2003; Dowie, 2005).

Studies record both satisfaction and dissatisfaction among people who are relocated from PAs, and serious implications of such policies. In one such resettlement process, 419 families were relocated from Bhadra Wildlife Sanctuary and Tiger Reserve of Karnataka state, India. As a part of this study, a survey of 60% households was conducted in 2002. Later in 2006, after the relocation of 11 villages, one more survey was conducted in which 55% households were covered. Prior to resettlement, people were facing challenges, like crop loss and livestock predation. Substantial human-wildlife conflict occurred, and people faced many losses. Moreover, the people faced difficulties related to the lack of basic amenities like drinking water, electricity supply and medical facilities.

The resettlement scenario largely began in 1970s, when Bhadra people residing inside or near the forest requested the government to help them by relocating and resettling them. Then, in 1974 a resettlement project was proposed by the Karnataka government. Funds were provided by Project Tiger Steering Committee between 1992 to 2002. In October 2001, 419 families moved when deeds were received by the people. Many positive and negative feedbacks were given after the resettlement, but overall the Bhadra resettlement process resulted in positive outcomes. Households received access to electricity, drinking water, phones, solar lights etc. at alternate sites. A health care centre and a nursery were also established. Earning sources for the people were also expanded (Karanth, 2007).

Another example from another Indian national park (Kanha National Park, Madhya Pradesh) showed that displaced people were not happy with the type of agricultural lands given to them at alternate sites (*Pers. Comm.* with the Baiga tribe, a local community in the locality). Ganguly (2004) described the difficulties faced by displaced people, particularly women, around Gir National Park, the last home of Asiatic lion in Gujarat state of India. Volunteers and the Indian government mediated recent relocations of several villages outside other National Parks in Madhya Pradesh, India (Satpura National Park, Kuno Wildlife Sanctuary etc.) with positive results. After successful relocations, attempts were made towards habitat restoration in evacuated areas by raising palatable grasslands (Raman, 2010; Singh, 2015; Figure 1). With time, there has been a remarkable increase in herbivore populations in such PAs (*Pers. Comm.* Shrivastava, B. and Singh, RP; Divisional Forest Officer of Kuno Wildlife Division and Field director of Satpura National Park, respectively).

In Thailand, there have been conflicts between the Karen community in northern hilly region and Government agencies, including the Royal Forest Department (RFD). This dispute centred on the RFD's desire to maintain forested areas as PA, which deprived the Karen community of agricultural land (Sato, 2002; Chatty & Colchester, 2002; Roth, 2004, 2008; Hares, 2009). The RFD considered park inhabitants to be "illegal occupants" but this claim was contested by the Karen and Hmong (local indigenous peoples) as they had been living in Doi Inthanam National Park (DINP) for over 100 years. They challenged the assertion of the RFD that the local communities had in fact migrated more recently into the park. Arguably, local communities should have been included in planning activities in and around DINP. The park management should have engaged in a serious discussion over the local people's property claims at least on the land classified as protected, which belonged to them before the park's designation (Beltrán, 2000).

Marine parks are also important PAs in both Thailand and India, with similar problems concerning local residents and land use. Thailand has an extensive system of Marine Protected Areas (MPAs), which require significant improvements in management to achieve their conservation potential. Methods to improve management have included increased attention to offshore settlers and local development. Enhanced National Marine Park (NMP) governance and management processes have been suggested to build trust and ameliorate relationships with local communities, with possible improvements in conservation outcomes. These issues were investigated in a study undertaken by Bennett and Dearden (2014) when they covered 17 NMPs around Andaman coast of Thailand.

India has 131 MPAs along its long coastline. The Andaman & Nicobar and Lakshadeep islands account for 106 National Parks and Wildlife Sanctuaries (Saravanan et al., 2011), whereas Thailand has 28 MPAs, mostly along Andaman sea coast (Manopawitr, 2012). Sedimentation, nutrient pollution from development on land, and over fishing (particularly by large trawlers) contribute to reef degradation in both the countries. The conversion of mangrove forest to shrimp farms, excessive tourism activities, and improper management of pollution (sewage and industrial) have further contributed to the degradation of coastal habitats and resources (Wilkinson, 1998; Cheung et al., 2002). Traditional uses of coastal water, such as fishing and usage of landing sites by local fishermen also create problems for the PA management (Singh, 2003).

Figure 1. Grassland areas developed in PAs of India: (a) Kuno Wildlife Sanctuary, Madhya Pradesh on left; rainy season and (b) Melghat Tiger Reserve, Maharashtra on right; winter season.

Legal Framework

Dislocation and resettlement of local communities from PAs in India is legally supported by the Wildlife (Protection) Act 1972 and the Forest Rights Act 2006. Under these Acts, wildlife authorities are to demarcate critical 'tiger' habitats and critical 'wildlife' habitats in PAs and settle traditional forest dwellers and scheduled tribes outside such demarcated areas to create territory for tigers and other wildlife. Most such actions have been completed, with some local communities already resettled outside critical tiger habitats (core area of Tiger reserve), with some compensation while other communities are still in the process of relocation. There are two main objectives: wildlife conservation and local community welfare/societal development. However, the concept of the national park as an inviolate haven of untouched nature controlled by an all-powerful central government agency, may be modified by the realities of local contexts, which may require more attention to careful manipulation to achieve both conservation and local human development objectives (Hough, 2009).

Thailand has similar forest and wildlife related legislative Acts. The Forest Act 1941, Wildlife Preservation and Protection Act 1960, National Parks Act 1961, National Forest Reserve Act 1964, The Reforestation Act 1992 etc. are some such examples. However, the Joint Management in Protected Area (JoMPA) project is a landmark participatory natural resource management project for PA management. Under this project, the main objective was the development of positive engagements between PA authorities and villagers, countering past conflicts between these groups. JoMPA activities enabled interactive learning between these groups and their tasks, responsibilities and problems. Relationships between PA authorities and villagers improved tremendously, as community owned committees were formed, capacitated and strengthened. Land use demarcation involved villagers, communities, government and stakeholders, resulting in the creation of land use maps and community rules which were agreed by PA authority to govern land use within the boundary. Training was provided on the use of GPS equipment and GIS technology in boundary surveys and land use demarcation, for PAs staff and villagers. The model of the JoMPA project represents a good start for the solution of problems of natural resource management in PAs. Keeping in view the merit and success of JoMPA, the Government may allocate funds and implement joint management projects in other PAs of Thailand (RECOFTC, 2017).

Human Wildlife Conflict

Human–wildlife conflict is the competition between humans and wildlife in relation to habitat, livelihood assets (e.g., crops and livestock) and physical safety. Human-wildlife conflicts are reported to be increasing in both frequency and intensity and are expected to continue to do so (Madden, 2004; Bond, 2015). Human conflict with wildlife in South-East Asia involves numerous species, including elephants *Elephas maximus*, tigers *Panthera tigris* (Nyhus & Tilson, 2004), Sun bears *Helarctos malayanus* (Fredriksson, 2005) and several primate species; particularly *Macaca* spp. (Linkie et al., 2007; Riley, 2007; Gumert, 2011). Increasing numbers or activities of humans in a PA enhances the likelihood of encounters with large carnivores, with negative effects on the survival of sensitive carnivorous species through several mechanisms like local culture, government policy and international trade (Woodroffe & Ginsberg, 1998; Woodroffe, 2000). The importance of these factors may mean that extinction risks for carnivores will continue to increase, even though human population growth is projected to decelerate during the new millennium (Woodroffe, 2000). However, following cases of human wildlife conflicts increase the understanding of management complexity with respect to crop raiding, livestock damage and even human losses.

For conservation practitioners, one of the most challenging issues is to address the intricacies of human-wildlife conflicts. There can be many ecological and social factors responsible for these conflicts, nevertheless, there is a requirement to develop preventive strategies to avoid these conflicts. A study was conducted in five important reserves of Karnataka's Western Ghats of India, to examine the patterns of conflict losses and compensation. It was found that the crop raiding incidents were experienced by villagers throughout the year, but the frequency of such incidents was slightly higher from October to December. Some species, especially wild pig, elephant and spotted deer were responsible for the high crop losses in that region. The local people evolved their own mitigation measures to cope up with the problems, but these were ineffective. Individual measures like night watching, guarding animals, etc. were adopted. Like the crop raiding pattern, it was found that longer distances to forest reserve were associated with lower livestock loss. Livestock loss was positively associated with grazing animals that were inside the reserve. Another disabling factor for conservation was the compensation provided by the government in lieu of the loss. It was found that the compensation process was too long. Compensation reached the affected people after more than one year (Karanth et al., 2013).

The Indian government organized compensation schemes to compensate people for their losses suffered, but the implementation of the programmes and proper compensation varied in different areas. Many studies carried out in different national parks across the country brought out different results. By and large, despite heavy losses suffered by the people living in and around PAs, few households applied for compensation, due to inadequate remuneration, processing delays, corruption and excessive red tape (Ogra & Badola, 2008). One such example is from Western Ghats, India where only 31% households received compensation in five PAs (Karanth et al., 2013). In another study in Bhadra Tiger Reserve of South India, Madhusudan (2003) estimated that each household around the reserve lost 12% of their livestock annually to large felines and approximately

11% of annual grain production to elephants. Compensation awarded covered about 5% of livestock loss and 14% of crop losses. People were again unhappy with procedural delays in claim settlement. Further, Maikhuri et al., (2000) in a survey near Nanda Devi Biosphere Reserve, Indian Himalaya found that 6 annual crops and 4 horticultural crops on private farms were damaged by wildlife and some livestock were killed, but the reserve management provided compensation only for livestock damage and not for crop loss. The compensation was also minimal, amounting to only 4–10% of the total monetary value of killed livestock.

In north east region of India, there are common conflicts between people and large wild animals like tigers, rhinoceros and elephants. Of these animals, elephants have become the focal point of conflict and conservation issues. According to Choudhury (2004), more than 1150 humans and 370 elephants were killed between 1980 and 2003 due to conflicts. Burgeoning human populations, their increasing needs for housing, agriculture etc. are the main causal factors for these problems. Due to reduced habitat, elephants usually come down to paddy fields and destroy the crop. In the year 2001, angered residents in northeast India (Assam) selectively targeted their paddy fields with poison for crop-raiding elephants; a mutilated elephant carcass was subsequently discovered in the field with the words, "Paddy Thief Bin Laden" scrawled upon its body (Sethi, 2003; Ogra, 2008). Such a comparison with of an infamous name like Bin Laden explains the anguish and gravity of the situation.

Crop raiding by wild animals, especially elephants can result in the displacement of poor families from local areas. Loss of livestock due to carnivore predation can destroy a family's wealth and way of life. Livestock forms a substantial proportion of the socioeconomic capital of communities in many areas of both India and Thailand. Ultimately, people have to bear the consequences. As carnivores mostly attack cattle at night, people must guard their livestock at this time, and this activity can create exhaustion and stress (Barua et al., 2013).

The Gir National Park of Gujarat, India is the main conservation area for Asiatic lions, and people and lions exist together in the areas near the park. Here, Saberwal et al., (1994) found 14.8 attacks on human by lions on an average and 2.2 the lion related deaths annually between 1978-1991. Most of these attacks (82%) occurred outside PAs, especially during the dry season. However, a major factor for such attacks was hostile attitudes of humnas toward lions, based on the threat of personal injury and economic hardship (mainly livestock losses). The government's compensation system for lion-depredated livestock was highly dissatisfactory in this area. Nonetheless, the researchers advocated reducing the lion population in the park by relocating or culling them.

Many conservationists advocate plans to reduce the economic impacts of elephant-human conflicts, while conserving viable populations of elephants. Recommendations include the use of maps by park and revenue authorities and non-governmental organisations (NGOs) to target preventive actions in the most vulnerable conflict zones.

Science based mitigation measures should be investigated. Compensation must be in accordance with the scales of losses and observed/monitored locally as well (Karanth et al., 2013). A number of actions are required in this regard. A systematic assessment of the extent and scale of hidden impacts (like diminished psychosocial well-being, disruption of livelihoods and food insecurity) is needed based on different conflict scenarios. It should also be determined that how such conflict impacts the nutrition, physical and psychological well-being of the people. The risk of replicating other conservation conflicts must be avoided. Additionally, a strong link between conservation, health and social sciences is required (Barua et al., 2013) to ascertain the goal.

Human-wildlife conflict is one of the most significant conservation issues in many Asian countries, exacerbated by increasing human populations, loss and fragmentation of natural habitats, and in some cases, increased wildlife populations. For example, although, the leopard (*Panthera pardus*) is considered to be a generalist large carnivore and usually insensitive to disturbance (Sunquist & Sunquist, 2002), a study conducted by Ngoprasert et. al. (2007) clearly demonstrated adverse effects on leopards from human presence and activity in Kaeng Krachan National Park, Thailand. Conflicts between people and wild elephants is not a new phenomenon in Thailand. Most significant events, like crop raiding by wild elephants, occurred during the second half of the 19th and 20th century, when people encroached into the prime elephant habitats in lowland forests and in the uplands to grow rice and other economic plants. Elephants also affected rail transport, this being another type conflict at the turn of the 20th century. The known causes of human-elephant conflict reported in Thailand are: (i) elephant movements into agricultural land adjacent to their park habitats; (ii) elephant movement between habitats; (iii) and when young elephants, mostly bulls stray out of the forests. Thai inhabitants use direct and indirect measures to reduce the damage and control wild elephants by setting fire around rice paddies, capturing, chasing or relocating elephants, elephant captures, chasing, relocation and poaching for tusks etc. (Srikrachang, 2008).

DISCUSSION

Sometimes, the PA management objectives translate into the practices that ultimately lead to the displacement of local people. Thus, conservation and displacement are closely related to each other. Displacement may involve use of force, and may lead to impoverishment, and the political and social isolation and disempowerment of people. The PA may become inaccessible to local people and, consequently, the basic amenities may not be available to them. Livestock too, may be badly affected as grazing is banned in the PA. Since the main aim of displacement is conservation, it is very important to know the extent to which the conservation targets were met after displacement. Though it is quite difficult to fully assess the effects of displacement on conservation, various studies suggest that displacement affects biodiversity positively. The balance between human costs and conservation benefits must be assessed carefully (Agrawal & Redford, 2009).

Deforestation is also important in addressing the issue under discussion. In northern Thailand, deforestation has led the government to implement conservation policies aimed at maintaining forest cover of 50% or more and promoting agribusiness, forestry, and tourism development in the region (Trisurat, 2010). The Indian government is also working to achieve to maintain forest cover over 33% of the country and to increase the percent of the national areas in PA from the current level of less than 5%. Some researchers have suggested various conservation measures and habitat management activities to minimize the impacts of deforestation, and to restore original habitat; such measures include the enrichment and assisted regeneration of shrubs and grasses, and soil and moisture conservation structures (Hossain et al., 2009; Wijitkosum, 2012; Sarma, et al., 2008, Reddy et al., 2007; Imam et al., 2009). Also suggested are more detailed studies of the impacts of land use change in areas near PAs (including marine PAs) using modern tools of Information Technology (IT) like Remote sensing (RS) and Geographic Information System (GIS) in both the countries.

The compliance of local people with PA regulations is also important. For example, Andrade & Rhodes (2012) conducted a meta-analysis of 55 published case studies from developing countries and found that local community participation in the PA management and decision-making process was the only variable that was significantly related to the level of compliance with PA polices. In general, the higher the level of participation, the higher was the level of compliance. This had important implications for PA management and suggested that greater inclusion of local communities in management should be a key strategy for ensuring the integrity of PAs. Nepal (2002) also found similar results from his studies conducted on various PAs located in developing countries. His findings indicated that indigenous people were frequently in conflict with park authorities where park management was centralised and non-participatory. Therefore, he concluded that the livelihood issues of these people need to be addressed properly for achieving success in wildlife conservation and park establishment objectives. Most findings from other studies concur, as they indicate that local communities are more likely to comply and to commit themselves to long-term conservation strategies when their knowledge and opinions are incorporated into PA decision-making processes (Mascia, 2003; Fu et al., 2004; Pretty & Smith 2004; Gelcich et al., 2005).

Establishing an ideal PA system requires long-term political and financial commitments that go far beyond simply declaring new parks (Hockings et al., 2000). These commitments remain frequently unmet as thousands of PAs in the developing world currently suffer an extreme funding deficit (James et al., 1999 and 2001; Wilkie et al., 2001). Insufficient funding means that many PA systems have inadequate staff, equipment, and other management necessities (Bruner et al., 2004). The world famous first Tiger reserve of India, Jim Corbett National Park along with Rajaji National Park, both located in the North Indian state of Uttarakhand, face shortage of frontline staff. According to latest update, at least 40% posts of Rangers and Forest guards are lying vacant in these two parks (The Tribune, 2018). Another bottleneck is insufficient and untimely budget allocation for PAs which mainly comes from Central Governments in India and Thailand. Nevertheless, the principal fund appropriated each year for financing protected areas, including marine national parks, is from the respective governments in both these countries. The budget per unit area is normally allocated on the basis of area coverage and management needs. However, revenue earned from gate entry fees is more wisely utilized in Thailand, as annual income from each national park is used in four ways. The first 5 per cent of total revenue is transferred to the Sub-district (Tambon) Administration Organisation (TAO) through the Department of Local Administration, Ministry of Interior. Ten per cent of the remaining revenue returns to the park for improving its visitor services, for example, through the hiring of temporary staff as guest house cleaners or nature guides. A further 50 per cent of funds also returns to the park for protection activities and staff capacity building. The remaining thirty-five per cent goes to the National Park Revenue Fund and is distributed to each park accordingly to project proposals submitted and approved by the Department's Revenue Committee (ICEM, 2003). It is suggested that the nearby private hotels and lodges in both the countries must be made to pay a "conservation tax" since they profit from the publicly funded PA. In most PAs the entry fee is quite low, not even inflation adjusted over time, therefore, a slight increase is unlikely to affect the ability of ordinary citizens to visit their national parks (Banerjee, 2010). Some studies advocate the assessment of average 'consumer surplus' enjoyed by visiting tourists to determine the optimum entry fee and other related expenses in PAs incurred by tourists (Seenprachawong, 2003; Bharali & Mazumdar, 2012; Chaudhry & Tewari, 2016).

Conservative estimates are that between 56% to 86% of PAs in developing countries have people living inside them (Brockington & Igoe, 2006; Berkes, 2009). Both, national parks and wildlife sanctuaries legally prohibit agricultural use and the extraction of forest products or mineral resources. However, these regulations and definitions of resources significantly conflict with reality on the ground. Protected Areas overlap with and enclose large areas of agricultural land in Thailand. According to the International Center for Environmental Management (ICEM) 2003 report, more than 500,000 people live inside national parks and wildlife sanctuaries of Thailand. This incongruity is the result of rapid protected area designation, initial weak enforcement encouraging in-migration, and population growth of communities inside protected areas (Dearden et al., 1996, 1998; Vandergeest, 1996). Such conflicts between legal definitions and *de-facto* use are not unique to Thailand but is a common problem for PAs in many developing countries (Naughton-Treves et al., 2005).

Protected Areas in developing countries are generally associated with high levels of poverty. But there are certain success stories where local communities have benefitted economically from nearby PAs. Ferraro et al. (2011) concluded that after accounting for the selective placement of PAs in locations with lower development potential, the PAs had contributed to economic development and reduced poverty in Thailand and Costa Rica. The most likely explanation lies in economic benefits from increased tourism, large enough to offset the costs of restrictions on land use. Andam et. al. (2010) also got similar findings in their research study around PA systems of Thailand. Their results suggested that protecting biodiversity could contribute to both environmental sustainability and poverty alleviation, two of the United Nations Millennium Development Goals (Sachs et al., 2009; United Nations, 2008). Some other studies in other developing countries, including India, suggest a positive correlation between PA establishment and poverty reduction among nearby communities (Clements et al., 2014 in Cambodia; Gurney et al., 2014 in Indonesia; Karanth & Nepal, 2012 in India and Nepal). However, without replications of such studies in the larger areas of PA networks in most of the other developing countries, it would be rather unfair to draw similar conclusions as the enabling conditions vary in different countries. Wildlife tourism (particularly tiger tourism) is on the rise in India and Thailand, but the recent experience of Jim Corbett Tiger Reserve, north India has shown that such tourism is negatively impacting PA ecology and monetary benefits are received by middle men, rather than local communities (Rastogi et al., 2015). Nevertheless, in order to achieve PA success in long run, managers of parks should focus on building social capital and strengthening local institutions instead of increasing tourist pressure unwisely. The larger challenge is to allow human society to meet its potential and share the fruits of economic growth while sustaining a biosphere that not only sustains full ecological functions but retains its living diversity (Adams et al., 2004).

Although there is widespread agreement on the importance of maintaining natural areas and their associated flora and fauna, the total amount of area given protected status is frequently insufficient in developing countries, and the level of funding provided for management is almost always inadequate. Establishing and maintaining PAs require both political and financial commitment in the long term (Bruner et al., 2004). A major reason for this is that the benefits to society from PAs are often grossly underestimated, and the immediate costs of protection appear large in comparison. Some researchers have tried to estimate such intangible benefits to society from PAs. For example, economic valuation of tiger reserves, invariably National Parks in India, may provide the economic rationale for such resources and support biodiversity conservation and human wellbeing. Some however argue the contrary, that there are social costs incurred by the dislocated people due to formation of PAs (Menon & Rai, 2017). One study examined six tiger reserves of India in different landscapes and found that the flow benefits from these reserves ranged from INR 50,000/ha/year (US $ 769/ha/year) to INR 1,90,000/ha/year (US $

2923/ha/year). The lower value was of a reserve in the tropical dry deciduous forest region, where Ranthambore Tiger Reserve, Rajasthan is located while the higher value was of a reserve in the tropical moist evergreen forest region where Periyar Tiger Reserve, Kerala, is located (Verma et al., 2017). Therefore, nearly 5% of India's geographical area covered by PAs is responsible for providing ecosystem services or benefits flow worth millions annually. This shows the huge significance of these areas for the ecological and social security of society and other living systems of the country.

GEOINFORMATICS IN PAS

Geoinformatics or geomatics is considered the science and technology dealing with the structure and character of spatial information. This field uses remote sensing data and geographical information system to portray and record features of the Earth's surface on paper or a computer screen. This technology has been applied to the study of PAs of many countries, for example: the selection of priority areas for preservation in the Ahaggar National Park, Algeria (Hamadouche et al., 2014); the mapping the past and the future of indigenous territories in the Peruvian Amazon through geomatics (Smith et al., 2003); the recording of land use and land cover mapping, change detection and conflict analysis of the Nagzira-Navegaon corridor, India using geospatial technology (Yadav et al., 2012); and conflict resolution using participatory management and remote sensing and GIS (Sandström et al., 2003) etc.

Certain issues concerning the PAs in India and Thailand are particularly amenable to analysis using geomatics technology. For example, cases of land encroachments where accurate recording of landcover, borders and land features may be needed to settle related disputes. In many cases, there is insufficient evidence to settle such disputes, pointing to the need for maps and diagrams of sharp resolution. Geomatics techniques have been abundantly used in several states of India (Gujarat, Kerala, Andhra Pradesh, Madhya Pradesh, Maharashtra etc.) as a successful e-governance model in dealing with conflicting claims of forest dwellers over reserve forest and PAs (Goswami, 2014). The Promulgation of "The Scheduled Tribes and Other Traditional Forest Dwellers (Recognition of Forest Rights) Act 2006", in India induced numerous conflicting land tenure claims by the Forest Dwelling Schedule Tribes and Other Tribal Forest Dwellers. These were very effectively studied and analysed with the help of the geomatics technology, especially for the verification of certain criteria concerning the authenticity of claims. Claim eligibility criteria was fixed as per the provisions of the Act. Figures 2 to 5 illustrate the evidence for and against the claims in the state of Maharashtra, India using geoinformatics.

Figure 2. The claimed area, white polygon, in two CARTOSAT pictures bearing different signatures: (a) Forested area in contiguity outside it in November, 2005 (left); and (b) tree cleared area with agriculture practices in December, 2008. The claim was ineligible based on qualifying criteria. Photo courtesy: Dr AK Jha IFS (Retd).

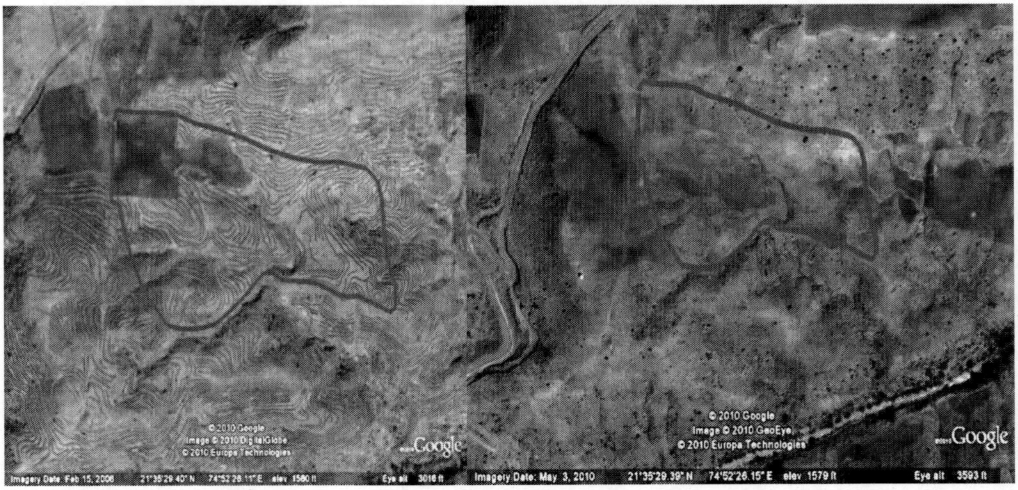

Figure 3. Another claimed area, red polygon, in two google earth images bearing varying signatures: (a) Continuous Contour Trenches of afforestation activities on February 15, 2006 (left); and (b) Plantation cleared area with agriculture activities on May 3, 2010. A case of untenable claim. Photo courtesy: Dr AK Jha IFS (Retd).

Figure 4. Three CARTOSAT images are from 2005 (top), 2007 (bottom left) and 2012 (bottom right). The yellow polygons are the claimed area which have different signatures. In 2005 they were not under cultivation, in 2007 there was not much change but in 2012 cultivation signature is evident. These claims were found ineligible. Photo courtesy: Dr AK Jha IFS (Retd).

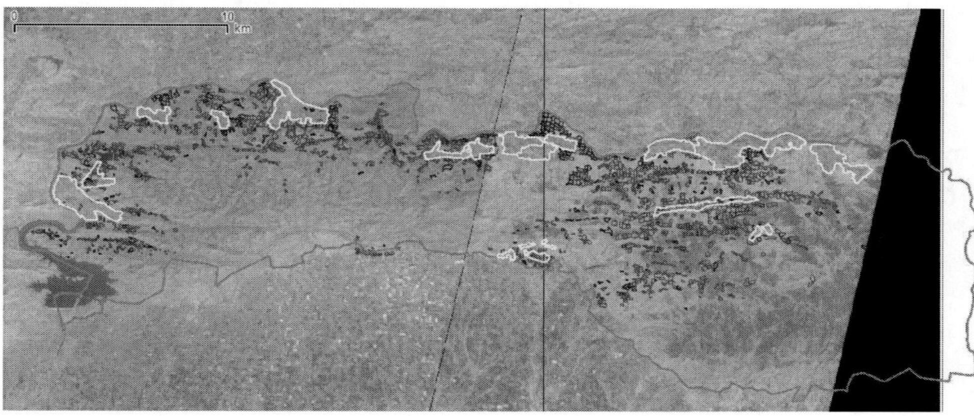

Figure 5. Post FRA promulgation status of Yawal Wildlife Sanctuary, Maharashtra, India during the claim settlement process. The green polygon (Sanctuary) encompassing yellow polygons (the villages) and red polygons (ineligible claimed area). Very few claims were found eligible also but not clear in this small picture. (Image of the sanctuary for the dark area on the right is not available). Photo courtesy: Dr AK Jha IFS (Retd).

Management documents for forests in India have a history of over 150 years. Such documents of reserve forests (RF) are known as working plans and those of PAs as management plans. In recent years PA management has become more specific and based on flagship species, for example, tigers, elephants, rhinos etc. In Tiger Reserves, tiger conservation plans (TCP) are prepared aimed at the conservation of this flagship species and its umbrella effects on sympatric species. Generally, this divides the studied area into core and buffer zones. For the TCP, the main focus is core management exclusively for the wildlife and buffer management for various purposes, mostly similar to that of the RF. Major activities include habitat management, ecodevelopment, ecotourism, conflict control, community forestry, resource protection etc. Therefore, to achieve the management objectives various maps are required. These include forest cover maps, change detection (degradation as well as improvement) maps, forest fire detection and burnt area maps, biodiversity maps, afforestation planning maps, wildlife habitat suitability maps, boundary demarcation maps, land encroachment maps etc. For many of these applications, there are no practical alternatives to geomatics-based methods (MoEF, 2014). Thailand also has park management plans in place. The current evidence suggests that PAs were under encroachment and with variable conflicts (Vandergeest, 1996).

CONCLUSION

Forest dwelling communities and biodiversity conservation of PAs have an intricate relationship (Figure 6). Separating (relocation of the communities) them may lead to better conservation of resources but vagaries to the communities if not addressed well. Therefore, including local community in decision-making processes can potentially create a sense of stewardship, where local residents collaborate with PA managers and act together to conserve biodiversity in PAs and local livelihoods (Horowitz, 1998) and also reducing the wild-animal conflict. Such decision making must be supplemented with modern tools and technologies to maintain transparency and achieve accuracy. However, putting such concepts into practice is not an easy task. There is no simple rule for conservation objectives with local community needs. What has worked in one PA may not have worked in another. Understanding the unique characteristics of each PA and the people who live in and around them is vital for the success of each PA's conservation program (Andrade & Rhodes, 2012). We, further, infer that the PAs of India and Thailand support a wide range of economic sectors and investment in this natural capital would lead to maintaining ecological security and food security, thereby leading to overall sustainable development. These investments can be cost effective responses to the climate change crisis, for creating jobs, for supporting local economies and maintaining ecosystem benefits on a long term basis. Amicable and tactful handling of all contentious issues of protected areas can be a win-win situation to park managers, local communities and other stakeholders.

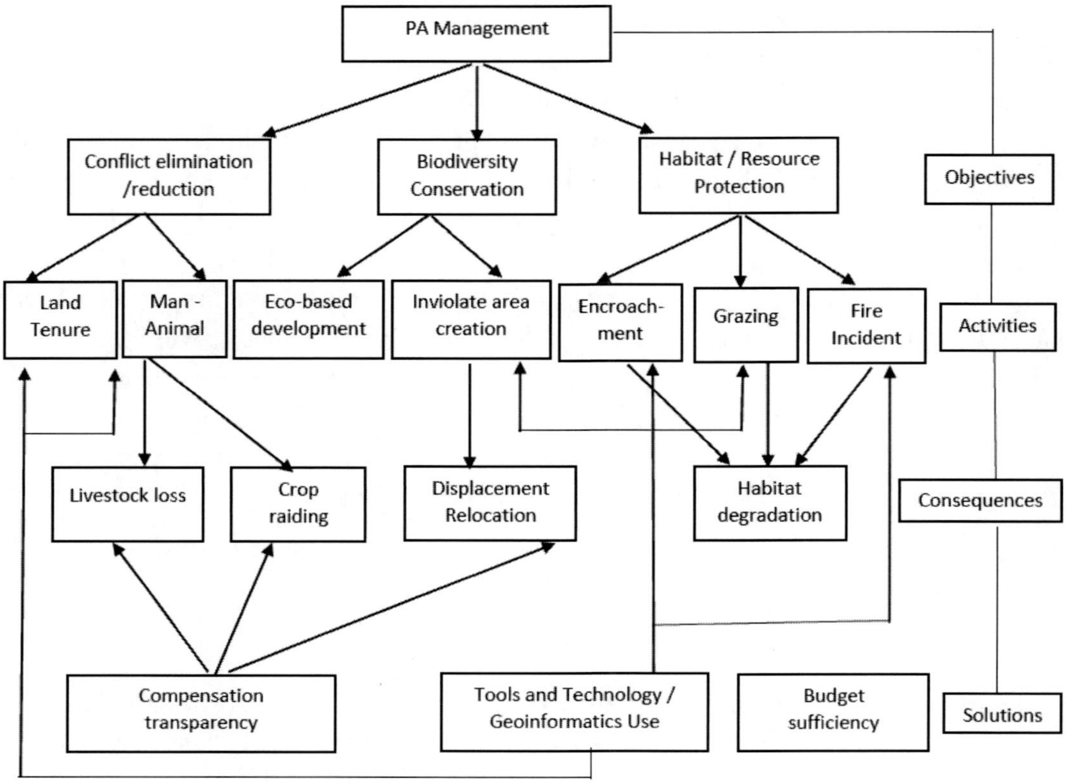

Figure 6. Shows the complexity of PA management and suggestion under the scope of the text. Boxes in a row are at same levels of the categories mentioned at extreme right. Arrows are source to destination link.

ACKNOWLEDGMENTS

The senior author is thankful to Dr A.K. Jha, retired PCCF, Maharashtra for providing photographs and making him understand the geoinformatics practiced in his state for solving the problems of intended encroachment in the forest and PAs. He is also thankful to Dr Rejani and Shri Bob of RECOFTC, Bangkok for helping to understand PA management in Thailand during field visits. The contribution of Shri Sridhar in preparation of this document is also acknowledged. Both the authors appreciate editing effort of Dr Michael O'Neal Campbell.

REFERENCES

Adams, W. A., Aveling, R., Brockington, D., Dickson, B., Elliott, J., Hutton, J., Roe, D., Vira, B. & Wolme, W. (2004). Biodiversity Conservation and the Eradication of Poverty. *Science 306*: 1146-1148.

Agrawal, A & Redford, K. (2009). Conservation and displacement: an overview. *Conservation and Society. 7*(1): 1-10.

Andam, K. S., Ferraro, P. J., Sims, K. R. E., Healy, A. & Holland, M. B. (2010). Protected areas reduced poverty in Costa Rica and Thailand. *Proceedings of the National Academy of Sciences of the United States of America 107*: 9996–10001.

Andrade, G. S. M., & Rhodes, J. R. (2012). Protected areas and local communities: an inevitable partnership toward successful conservation strategies? *Ecology and Society, 17*(4): 14.

Ban, N., Picard, C & Vincent, A. (2008). Moving toward spatial solutions in marine conservation with indigenous communities. *Ecology and Society 13*(1): 32.

Banerjee, A. (2010). Tourism in Protected Areas: Worsening Prospects for Tigers? *Economic and Political Weekly, 65*(10): 27-29.

Barua, M., Bhagwat, S. A & Jadhav, S. (2013). The hidden dimensions of human-wildife conflict: Health impacts, opportunity and transaction costs. *Biological Conservation 157*: 309-316.

Beltrán, J. (Ed.) (2000). *Indigenous and Traditional Peoples and Protected Areas: Principles, Guidelines and Case Studies*. Gland, Cambridge: IUCN.

Bennett, N. J. & Dearden, P. (2014). Why local people do not support conservation: Community perceptions of marine protected area livelihood impacts, governance and management in Thailand. *Marine Policy 44*: 107-116.

Berkes, F. (2009). Community conserved areas: Policy issues in historic and contemporary context. *Conservation Letters 2*: 19–24.

Bharali, R. & Mazumdar, R. (2012). Application of travel cost method to assess the pricing policy of public parks: the case of Kaziranga National Park. *Journal of Regional Development and Planning 1*(1): 41–50.

Bodmer, R. E., Eisenberg, J. F., & Redford, K. H. (1997). Hunting and the likelihood of extinction of amazonian mammals. *Conservation Biology 11*: 460-466.

Bond, J. (2015). Making sense of human–elephant conflict in Laikipia County, Kenya. *Society & Natural Resources: An International Journal,* DOI: 10.1080/08941920.2014.948238.

Brockington, D., & Igoe, J. (2006). Eviction for conservation: A global overview. *Conservation and Society 4*:424–470.

Bruner, A., R. Gullison, & A. Balmford. (2004). Financial cost and shortfalls managing and expanding protected-area systems in developing countries. *Bio Science 54*(12): 1119– 1126.

Chape, S., Blyth, S., Fish, L., Fox, P & Spalding, M. (2003). *United Nations list of Protected areas*.

Chape, S., Spalding, M. & Jenkins, M. (2008). *The World's Protected areas. Prepared by UNEP World Conservation Monitoring Centre*, University of California Press, Berkeley, US.

Chapin, M. (2004). A Challenge to Conservationists. *World Watch Magazine, (11/12)*, 17–31.

Chauhan, B. S. (2008). *Environmental Studies*. Book by University Science Press, 113, Golden House, Daryaganj, N Delhi.

Chaudhry, P & Tewari, V. P. (2016). Estimating recreational value of Mahatama Gandhi Marine National Park, Andaman and Nicobar Islands. *Interdisciplinary Environmental Review. 17*(1): 47-59.

Choudhury, A. (2004). Human elephant conflicts in Northeast India. *Human Dimensions of Wildlife 9*:261-270.

Clements, T., Suon, S., Wilkie, D. S. & Milner-Gulland, E. J. (2014). Impacts of protected areas on local livelihoods in Cambodia. *World Development 64*: 125-134.

Colchester, M. (2003). *Salvaging nature. Indigenous peoples, protected areas and biodiversity conservation.* World Rainforest Movement. International Secretariat, Maldonado 1858, Montevideo, Uruguay.

Charuppat, T. (1998). *Forest Situation in the Past 37 Years* (1961−1998). Bangkok, RFD. 116 pp. (in Thai).

Chatty, D. & Colchester, M. (2002). *Conservation and Mobile indigenous people: Displacement, forced settlement and sustainable development*. Book edited by Chatty and Colchester. Publisher Berghahn Books, New York.

Cheung, C. P. S., Aliño, P. M., Uychiaoco A. J. & Arceo H. O. (2002). *Marine Protected Areas in Southeast Asia.* ASEAN Regional Centre for Biodiversity Conservation, Department of Environment and Natural Resources, Los Baños, Philippines.

Dearden, P., Chettamart, S. & Emphandu, D. (1998). Comment: protected areas and property rights in Thailand. *Environmental Conservation 23*(3): 195-197.

Dearden, P., Chettamart, S., Empanada, D. & Tanakanjana, N. (1996). National parks and hill tribes in northern Thailand: A case study of Doi Inthanon. *Society and Natural Resources 9*(2):125-141.

Dixon, J. A. & Sherman, P. B. (1990). *Economics of Protected Areas: A New Look at Benefits and Costs*. Washington: Island Press.

Dowie, M. (2005). Conservation refugees. When protecting nature means kicking people out. *Orion 24*: 16–24.

DNP (2003). *Forest Type Map year 2000.* Bangkok: Department of National Park, Wildlife and Plant Conservation.

Emphandhu, D. & Chettamart, S. (2003). *Thailand experience in protected area management.* Paper presented at Vth IUCN World Park Congress, Durban, South Africa, 8-17 September 2003.

Ferraro, P. J., Hanauer, M. M., & Sims, K. R. E. (2011). Conditions associated with protected area success in conservation and poverty reduction. *PNAS. 108*(34): 13913–13918.

Fredriksson, G. (2005). Human–sun bear conflicts in East Kalimantan, Indonesian Borneo. *Ursus* 16: 130–137.

Fu, B., Wang, K., Lu, Y., Liu, S., Ma, K., Chen, L. & Liu. G. (2004). Entangling the complexity of protected area management: the case of Wolong Biosphere Reserve, south western China. *Environmental Management 33*: 788–798.

Gelcich, S., Edward-Jones, G. & Kaiser, M. (2005). Importance of attitudinal differences among artisanal fishers toward co-management and conservation of marine resources. *Conservation Biology 19*(3): 865–875.

Ganguly, V. (2004). *Conservation, displacement and deprivation. Maldharis of Gir forest of Gujarat.* Indian Social Institute, New Delhi.

Goswami, V. (2014). Geo-informatics for Forest Rights: Harnessing technology for better livelihood support and access to forest rights. *Case Studies on e-Governance in India – 2013 – 2014. Ministry of Communications and Information Technology*, New Delhi.

Gumert, M. (2011). The common monkey of Southeast Asia: Longtailed macaque populations, ethnophoresy, and their occurrence in human environments. In Gumert, M., Fuentes, A. & L. Jones-Engel (Ed.), *Monkeys on the Edge: Ecology and Management of Long-tailed Macaques and their Interface with Humans,* 3–44. Cambridge: Cambridge University Press.

Gurney, G. G., Cinner, J., Ban, N. C., Pressey, R. L., Pollnac, R., Campbell, J., Tasidjawa, S. & Setiawan F. (2014). Poverty and protected areas: An evaluation of a marine integrated conservation and development project in Indonesia. *Global Environmental Change. 26*: 98-107.

Hamadouche, M. A., Mederbal K., Kouri, L., Reagan, Z., Fekir, Y. & Anteur, D. (2014). GIS-based multicriteria analysis: an approach to select priority areas for preservation in the Ahaggar National Park, Algeria. *Arabian Journal of Geosciences 7* (2): 419–434.

Hares, M. (2009). Forest conflict in Thailand: Northern minorities in focus. *Environmental Management*, *43*(3): 381-395.

Hayes, T. M. (2006). Parks, people, and forest protection: an institutional assessment of the effectiveness of protected areas. *World Development 34*(12): 2026–2075.

Hockings, M., Stolton, S. & Dudley, N. (2000). *Evaluating Effectiveness: A Framework for Assessing Management of Protected Areas.* Gland: IUCN–The World Conservation Union.

Hough, J. L. (2009). Obstacles to effective management of conflicts between national parks and surrounding human communities in developing countries. *Environmental Conservation. 15*(2): 129-136.

Horowitz, L. S. (1998). Integrating indigenous resource management with wildlife conservation: a case study of Batang Ai National Park, Sarawak, Malaysia. *Human Ecology 26*: 371–403.

Hossain, M. Z., Tripathi, N. K, & Gallardo, W. G. (2009). Land use dynamics in a marine protected area system in Lower Andaman Coast of Thailand, 1990–2005. *Journal of Coastal Research 25*(5): 1082 – 1095.

ICEM. (2003). *Thailand National Report on Protected Areas and Development. Review of Protected Areas and Development in the Lower Mekong River Region.* Indooroopilly, Queensland, Australia.

Imam, E., Kushwaha, S. P. S., & Singh, A. (2009). Evaluation of suitable tiger habitat in Chandoli National Park, India, using spatial modelling of environmental variables. *Ecological Modelling. 220*(24): 3621-3629.

James, A., Gaston, K., & Balmford, A. (1999). Balancing the earth's accounts. *Nature 401:* 323–324.

James, A., Gaston, K. & Balmford, A. (2001). Can we afford to conserve biodiversity? *BioScience 51*: 43–52.

Karanth, K. K. (2007). Making resettlement work: The case of India's Bhadra wildlife sanctuary. *Biological Conservation. 139*: 315-324.

Karanth, K. K. & Nepal, S. K. (2012). Local residents' perception of benefits and losses from protected areas in India and Nepal. *Environmental Management. 49*(2): 372-386.

Karanth, K. K., Gopalaswamy, A. M., Prasad, P. K. & Dasgupta, S. (2013). Patterns of human wildlife conflicts and compensation: Insights from Western Ghats protected areas. *Biological Conservation166*: 175-185.

Lakanavichian, S. (2001). Impacts and effectiveness of logging bans in natural forests: Thailand. In FAO, *Forests Out of Bounds: Impacts and Effectiveness of Logging Bans in Natural Forests in Asia-Pacific*. Bangkok: FAO Regional Office for Asia and the Pacific.

Linkie, M., Dinata, Y., Nofrianto, A. & Leadrer-Williams N. (2007). Patterns and perceptions of wildlife crop raiding in and around Kerinci Seblat National Park, Sumatra. *Animal Conservation, 10:* 127–135.

Locke, H. & Dearden, P. (2005). Rethinking protected area categories and the new paradigm. *Environmental Conservation. 32*(1): 1–10.

Madden, F. (2004). Creating coexistence between humans and wildlife: Global perspectives on local efforts to address human–wildlife conflict. *Human Dimensions Wildlife 9*(4): 247–257.

Madhusudan, M. D. (2003). Living amidst large wildlife: Livestock and crop depredation by large mammals in the interior villages of Bhadra Tiger Reserve, South India. *Environmental Management. 31*(4): 466-475.

Maikhuri, R. K., Nautiyal, S., Rao, K. S., Chandrashekhar, R. G. & Saxena, K. G. (2000). Analysis and resolution of protected area - people conflicts in Nanda Devi Biosphere Reserve, India. *Environmental Conservation. 27* (1):43-53.

Manopawitr, P. (2012). *Thailand's marine and coastal protected area management effective evaluation.* ICRI meeting, Cairns, Australia July 2012.

Mascia, M. B. (2003). The human dimension of coral reef marine protected areas: recent social science research and its policy implication. *Conservation Biology 17*(2): 630–632.

Menon, A. & Rai, N. D. (2017). Putting a price on Tiger reserves. Creating conservation value or green grabbing*? Economic and Political Weekly 52*(52): 23-26.

Mishra, C. (1997). Livestock depredation by large carnivores in the Indian trans-Himalaya: conflict perceptions and conservation prospects. *Environmental Conservation 24*(4): 338-343.

Mittermeier, R. A. & Mittermeier C. G. (2005). *Megadiversity: Earth's Biologically Wealthiest Nations.* Cemex.

MoEF (Ministry of Environment and Forests), (2014). *National Working Plan Code – 2014 for sustainable management of forests and biodiversity in India.* Ministry of Environment and Forests, Government of India, New Delhi.

Narain, S., Singh, S., Panwar, H. S. & Gadgil, M. (2005). *Joining the Dots: The Report of the Tiger Task Force* (New Delhi: Ministry of Environment and Forests, Government of India.

Nash, R. F. (1967). *Wilderness and the American Mind.* New Haven: Yale University Press.

Naughton-Treves, L., Holland, M. B. & Brandon, K. (2005). The role of protected areas in conserving biodiversity and sustaining local livelihoods. *Annual Review of Environment and Resources 30*, 219-252

Nepal, S. K. (2002). Involving indigenous peoples in PA management: Comparative perspectives in Nepal, Thailand and China. *Environmental Management 30*(6): 748-763.

Ngoprasert, D. & Gale, G. (2007). Human disturbance affects habitat use and behaviour of Asiatic leopard Panthera pardus in Kaeng Krachan National Park, Thailand. *Oryx 41*: 343 - 351. 10.1017/S0030 605 307001102.

Nyhus, P. & Tilson, R. (2004). Agroforestry, elephants, and tigers: balancing conservation theory and practice in human-dominated landscapes of Southeast Asia. *Agriculture, Ecosystems & Environment*, *104,* 87–97.

Oelschlaeger, M. (1991). *The Idea of Wilderness*. New Haven: Yale University Press.

Ogra, M. V. (2008). Human-wildlife conflict and gender in protected area borderlands: A case study of costs, perceptions, and vulnerabilities from Uttrakhand, India. *Geoforum 39*(3): 1408-1422.

Ogra, M. & Badola, R. (2008). Compensating human-wildlife conflict in protected area communities: Ground level perspectives from Uttrakhand, India. *Human Ecology. 36*: 717-724.

Pretty, J. & Smith, D. (2004). Social capital in biodiversity conservation and management. *Conservation Biology 18* (3):631–638.

Raman, K. (2010). *Management Plan of Kuno Wildlife Division, 2010-11 to 2019-20.* Forest Department, Government of Madhya Pradesh, India.

Rastogi, A., Gordon, M. H., Anand, A., Badola, R. & Hussain, S. A (2015). Wildlife-tourism, local communities and tiger conservation: A village-level study in Corbett Tiger Reserve, India. *Forest Policy and Economics 61*, 11-19.

Reddy, C. S., Pattanaik, C. & Murthy, M. S. R. (2007). Assessment and monitoring of mangroves of Bhitarkanika Wildlife Sanctuary, Orissa, India using remote sensing and GIS. *Current Science. 92(10)*, 1409-1415.

RECOFTC (2017). *Participatory management of natural resource, livelihood development and climate change mitigation and adaptation in Thailand.* RECOFTC-the Centre for People and Forests. Bangkok.

Riley, E. (2007). The human–macaque interface: conservation implications of current and future overlap and conflict in Lore Lindu National Park, Sulawesi, Indonesia. *American Anthropologist 109*: 473–484.

Roth, R. J. (2004). On the colonial margins and in the global hotspot: Park–people conflicts in highland Thailand. *Asia Pacific Viewpoint. 45*(1): 13-32.

Roth, R. J. (2008), Fixing the forest: The spatiality of conservation conflict in Thailand. *Nature and Society. 98*(2): 373-391.

Saberwal, V. K., Gibb, S. J. P., Chellam, R. & Johnsingh, A. J. T. (1994). Lion human conflict in the Gir forest, India. *Conservation Biology 8*(2): 501-507.

Sachs, J. D., Baillie, J. E. M., Sutherland, W. J., Armsworth, P. R., Ash, N., Beddington, J., Blackburn, T. M., Collen, B., Gardiner, B., Gaston, K. J., Godfray, H. C. J., Green, R. E., Harvey, P. H., House, B., Knapp, S., Kümpel, N. F., Macdonald, D. W., Mace, G. M., Mallet, J., Matthews, A., May, R. M., Petchey, O., Purvis, A., Roe, D., Safi, K., Turner, K., Walpole, M., Watson, R. & Jones, K. E. (2009). Biodiversity conservation and the Millennium Development Goals. *Science 325:* 1502-1503.

Sanderson, E. W., Jaiteh, M., Levy, M. A. & Redford, K. H. (2002). The human footprint and the last of the wild: The human foot print is a global map of human influence on the land surface which suggests that human beings are stewards of nature, whether we like it or not. *Bioscience. 52*(10): 891-904.

Sandström, P., Pahlén, T. G., Edenius, L., Tømmervik, H., Hagner, O., Hemberg, L., Olsson, H., Baer, K., Stenlund, T., Brandt, L. G., & Egberth, M. (2003). Conflict resolution by participatory management: remote sensing and GIS as tools for communicating land-use needs for reindeer herding in Northern Sweden. *AMBIO: A Journal of the Human Environment, 32*(8): 557-567.

Sarma, P. R., Lahkar, B. P., Ghosh, S., Rabha, A., Das, J. P., Nath, N. K., Dey, S. & Brahma, N. (2008). Land-use and land-cover change and future implication analysis in Manas National Park, India using multi-temporal satellite data. *Current Science 95*(2): 223-227.

Saravanan, K. R., Sivakumar, K. & Choudhury, B. C. (2011). *Status of Marine and Coastal environments and developing a Marine Protected Area Network in India.* Wildlife Institute of India. Dehradun. pp 334.

Sato, J. (2002). People in between: Conversion and conservation of forest lands in Thailand. *Development and Change. 31*(1): 155-177.

Schmidt-Soltau, K. & Brckington, D. (2007). Protected areas and resettlement: What scope for voluntary relocation? *World Development. 35*(12): 2182-2202.

Saout, S. L., Hoffmann, M., Shi, Y., Hughes, A., Bernard, C., Brooks, T.M., Berzky, B., Butchart, S. H. M., Stuart, S. N., Badman, T. & Rodrigues, A. S. L. (2013). Protected Areas and Effective Biodiversity Conservation. *Science 342*(6160): 803-805.

Schwartzman, S., Moreira, A., & Nepstad, D. (2000a). Arguing tropical forest conservation: People versus parks. *Conservation Biology 14*: 1370–1374.

Schwartzman, S., Moreira, A., & Nepstad, D. (2000b). Rethinking tropical forest conservation: Perils in parks. *Conservation Biology 14*: 1351–1357.

Sethi, N. (2003). Battle zones: afterwards, an eerie silence. *Down to Earth,* March issue.

Seenprachawong, U. (2003). Economic valuation of coral reefs at Phi Phi islands, Thailand. *International Journal of Global Environmental Issues 3*(1):104–114.

Sims, K. R. E. (2009). Conservation and Development: Evidence from Thai Protected areas. *Working paper, Department of Economics*, Amherst College, Amherst.

Singh, H. S. (2003). Marine Protected Areas of India. *Indian Journal of Marine Sciences 32*(3): 226-233.

Singh, R. P. (2015). *Tiger conservation Plan, 2015-16 to 2024-25, Volume-I,* Saptura Tiger Reserve, Madhya Pradesh, India.

Smith, R. C., Benavides, M., Pariona, M. & Tuesta, E. (2003). Mapping the past and the future: Geomatics and Indigenous Territories in the Peruvian Amazon. *Human Organization: Winter 62 (*4): 357-368.

Srikrachang, M. (2008). Management of human–elephant conflict in Thailand: Governmental perspectives. *Proceedings, The 15th Congress of FAVA-OIE Joint Symposium on Emerging Disea*ses 27-30 October Bangkok, Thailand.

Sunquist, M. & Sunquist, F. (2002). *Wild Cats of the World.* Chicago and London: University of Chicago Press.

Terborgh, J., & Peres, C. A. (2002). The problem of people in parks. In C. van Schaik, L.J. Terborgh, & Rao, M. (Eds.), *Making Parks Work: Strategies for Preserving Tropical Nature.* Washington: Island.

TFSMP. (1992). *Thai Forestry Sector Master Plan Volume 5: Subsectoral Plan for People and Forestry Environment.* Bangkok: Thai Forestry Sector Master Plan Project, Royal Forest Department.

The Tribune (2018): Corbett, Rajaji reserves face staff shortage. *The Tribune*, Aug 6, 2018, Chandigarh edition.

Trisurat, Y. (2007). Applying gap analysis and a comparison index to evaluate protected areas in Thailand. *Environmental Management 39*: 235–245.

Trisurat, Y., Alkemade, R. & Verburg, P. H. (2010). Projecting land-use change and its consequences for biodiversity in Northern Thailand. *Environmental Management 45*(3):626-639.

UNEP-WCMC (2014). *Global statistics from the World Database on Protected Areas (WDPA), August 2014.* Cambridge, UK: UNEP- WCMC.

United Nations. (2008). *Millennium Development Goals.* Retrieved from http://www.un.org/millenniumgoals/2008highlevel/pdf/newsroom/mdg%20reports/MDG_Report_2008_ENGLISH.

Vandergeest, P. (1996). Property rights in protected areas: Obstacles to community involvement as a solution in Thailand. *Environmental Conservation 23*(3): 259-268.

Verma, M., Negandhi, D., Khanna, C., Edgaonkar, A., David, A., Kadekodi, G., Costanza, R., Gopal, R., Bonal, B. S., Yadav, S. P. & Kumar, S. (2017). Making the hidden visible: Economic valuation of tiger reserves in India. *Ecosystem Services. 26*: 236-244.

Wilkie, D. S., Carpenter, J. F. & Zhang, Q. (2001). The under-financing of protected areas in the Congo Basin: So many parks and so little willingness to pay. *Biodiversity and Conservation 10*: 691–709.

Wilkinson, C.W. (1998). *Status of Coral Reefs of the World: 1998. Global Coral Reef Monitoring Network,* Cape Ferguson: Australian Institute of Marine Science.

Wijitkosum, S. (2012). Impacts of land use changes on soil erosion in Pa Deng Sub-district, adjacent area of Kaeng Krachan National Park, Thailand. *Soil and Water Research. 7*(1): 10-17.

Woodroffe, R. (2000) Predators and people: using human densities to interpret declines of large carnivores. *Animal Conservation 3*: 165–173.

Woodroffe, R. & Ginsberg, J. R. (1998) Edge effects and the extinction of populations inside protected areas. *Science 280*: 2126–2128.

Yadav, P. K., Kapoor, M. & Sarma, K. (2012). Land use land cover mapping, change detection and conflict analysis of Nagzira-Navegaon Corridor, Central India Using Geospatial Technology. *International Journal of Remote Sensing and GIS, 1* (2): 90-98.

About the Editor

Michael O'Neal Campbell

Lecturer

Camosun College

Simon Fraser University

Victoria, British Columbia, Canada

ecol55@hotmail.com

INDEX

A

B

C

D

E

F

G

H

Q

R

S

T

U

V

W

Z

Biological Conservation in the 21st Century: A Conservation Biology of Large Wildlife

Editor: Michael O'Neal Campbell (Camosun College, Victoria, Canada)

Series: Wildlife Protection, Destruction and Extinction

Book Description: This edited book, composed of chapters written by scholars of the environmental and biological sciences, examines selected topics from the vast field of conservation biology, with a focus on some of the issues that dominate the current discourses and practices on the conservation biology of large wildlife.

Hardcover ISBN: 978-1-53612-073-8
Retail Price: $195

The Political Ecology of Soil Erosion in West Africa: An Historical and Contemporary Perspective

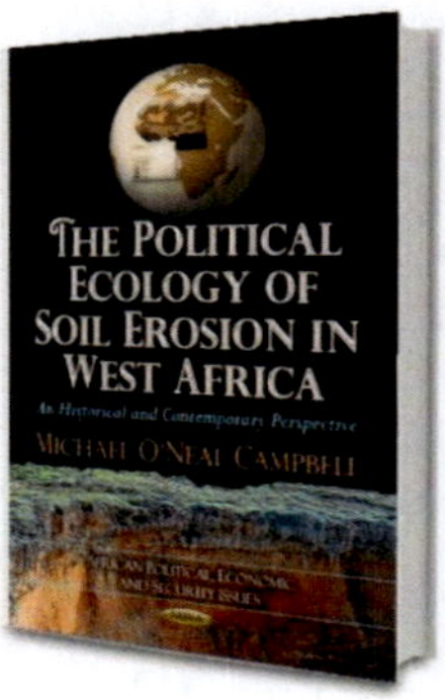

Author: Michael O'Neal Campbell (Thompson Rivers University, Department of Geography, Kamloops, BC, Canada)

Series: African Political, Economic, and Security Issues

Book Description: This book looks at theories of nature-society relations and development, applied to soil erosion and environmental change in West Africa, and Ghana. Acknowledging the importance of power relations for environmental management, the book begins with an examination of the field of political ecology, alternately the "political economy of the environment."

Hardcover ISBN: 978-1-62417-258-8
Retail Price: $110

The Political Ecology of Agricultural History in Ghana

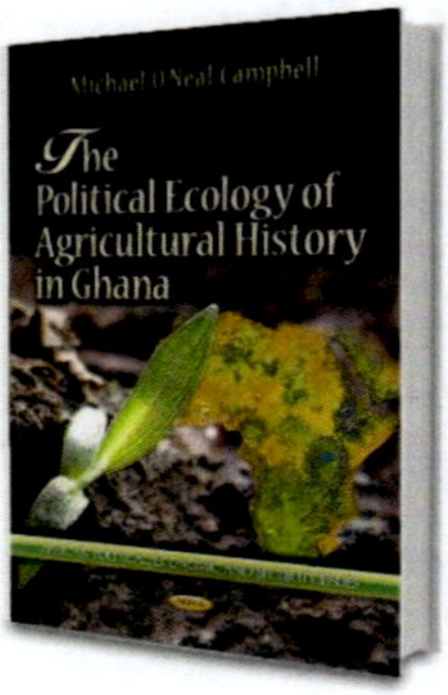

Author: Michael O'Neal Campbell (Thompson Rivers University, Kamloops, BC, Canada)

Series: African Political, Economic, and Security Issues

Book Description: This book takes an incisive, political ecology perspective of the development of agriculture in the West African nation of Ghana (formerly the British colony of the Gold Coast), focusing on agricultural, socio-cultural, economic, political and environmental issues during the pre-colonial (pre-16th century), colonial and post colonial periods (after 1957, up until 1992, the start of the last Republic).

Hardcover ISBN: 978-1-62417-276-2
Retail Price: $209

Biogeography, Environmental History and Sustainability in Coastal Ghana

Author: Michael O'Neal Campbell (Kamloops, BC, Canada)

Series: African Political, Economic, and Security Issues

Book Description: This book examines the socio-environmental relations and environmental changes in the African savanna. Socio-environmental relations in sub-Saharan Africa: the externalities of deforestation, floods, droughts, fires, socio-political upheavals, diseases and tribally, regionally relevant wars; still capture the headlines of the global media.

Hardcover ISBN: 978-1-62257-953-2
Retail Price: $110